Substrate-Integrated Millimeter-Wave Antennas for Next-Generation Communication and Radar Systems

IEEE Press
445 Hoes Lane
Piscataway, NJ 08854

Substrate-Integrated Millimeter-Wave Antennas for Next-Generation Communication and Radar Systems

Edited by

Zhi Ning Chen
Xianming Qing

The IEEE Press Series on Electromagnetic Wave Theory

Published by John Wiley & Sons, Inc., Hoboken, New Jersey.
Published simultaneously in Canada.

Library of Congress Cataloging-in-Publication Data:

Names: Chen, Zhi Ning, editor. | Qing, Xianming, editor. | John Wiley & Sons, publisher.
Title: Substrate-integrated millimeter-wave antennas for next-generation communication and radar systems / edited by Zhi Ning Chen, Xianming Qing.
Other titles: IEEE Press series on electromagnetic wave theory
Description: Hoboken, New Jersey : Wiley, [2021] | Series: IEEE Press series on electromagnetic wave theory | Series information from CIP data view. | Includes bibliographical references and index.
Identifiers: LCCN 2021003274 (print) | LCCN 2021003275 (ebook) | ISBN 9781119611110 (hardback) | ISBN 9781119611127 (adobe pdf) | ISBN 9781119611158 (epub)
Subjects: LCSH: Antennas (Electronics). | Millimeter wave communication systems.
Classification: LCC TK7871.6 .S83 2021 (print) | LCC TK7871.6 (ebook) | DDC 621.382/4–dc23
LC record available at https://lccn.loc.gov/2021003274
LC ebook record available at https://lccn.loc.gov/2021003275

Cover Design: Wiley

Set in 9.5/12.5pt STIXTwoText by SPi Global, Chennai, India

Contents

Editor Biographies

Zhi Ning CHEN (Fellow, IEEE and Fellow, SAEng) received his B.Eng., M.Eng., and Ph.D. degrees in electrical engineering from the Institute of Communications Engineering (ICE), China, in 1985, 1998, and 1993 and a second Ph.D. degree from the University of Tsukuba, Tsukuba, Japan, in 2003.

From 1988 to 1995, Dr. Chen was a lecturer and later a professor with ICE and a post-doctoral fellow and later an associate professor with Southeast University, Nanjing, China. From 1995 to 1997, he was a research assistant and later a research fellow with the City University of Hong Kong, Hong Kong. In 2001 and 2004, he visited the University of Tsukuba twice under the JSPS Fellowship Program (senior fellow). In 2004, he joined the IBM Thomas J. Watson Research Center, Ossining, NY, USA, as an academic visitor. In 2013, he joined the "Laboratoire des SignauxetSystèmes," UMR8506 CNRS—Supelec—University Paris Sud, Gif-sur-Yvette, France, as a Senior DIGITEO Guest Scientist. In 2015, he joined the Center for Northeast Asian Studies, Tohoku University, Sendai, Japan, as a senior visiting professor. From 1999 to 2016, he was a principal scientist, the head/manager of the RF and Optical Department, and a technical advisor with the Institute for Infocomm Research (I^2R), Singapore. In 2012, he joined the Department of Electrical and Computer Engineering, National University of Singapore, Singapore, as a tenured full professor, the program director (Industry), and the founder and deputy director of the Advanced Research and Technology Innovation Center. He is holding/held the concurrent guest professorships at Southeast University (Changjiang Chair Professor), Nanjing University, Nanjing, China, Tsinghua University, Beijing, China, Shanghai Jiaotong University, Shanghai, China, Tongji University, Shanghai, University of Science and Technology of China, Hefei, China, Fudan University, Shanghai, Dalian Maritime University, Dalian , China, Chiba University, Chiba, Japan, National Taiwan University of Science and Technology, Taipei, Taiwan, South China University of Technology, Guangzhou, China, Shanghai University, Shanghai, Beijing University of Posts and Telecommunications, Beijing, Yunnan University, Kunming, China, Beijing Institute of Technology, Beijing, and City University of Hong Kong. He is currently the member of the State Key Laboratory of Millimeter Waves, Southeast University and the State Key Laboratory of Terahertz and Millimeter Waves, City University of Hong Kong. He has been invited to deliver 100+ keynote/plenary/invited speeches at international academic and industrial events. He has authored 660+ academic articles and five books entitled *BroadbandPlanar Antennas* (Wiley, 2005), *UWB Wireless Communication* (Wiley, 2006), *Antennas for Portable Devices* (Wiley, 2007), *Antennas for Base Stations in Wireless Communications* (McGraw-Hill, 2009), and *Handbook of Antenna Technologies* with 76 chapters (Springer References, 2016, as an editor-in-chief). He has also contributed the chapters to the books entitled *Developments in Antenna Analysis and Design* (IET, 2018), *UWB Antennas and Propagation: For Communications, Radar and Imaging* (Wiley, 2006), *Antenna Engineering Handbook* (McGraw-Hill, 2007), *Microstrip and Printed Antennas*(Wiley, 2010), and *Electromagnetics of Body Area Networks* (Wiley, 2016). He is holding 36 granted/filed patents and completed 40+ technology licensed deals with industry. He is pioneering in developing small and wideband/ultrawideband antennas, wearable/implanted medical antennas, package antennas, near-field antennas/coils, 3-D integrated LTCC arrays, microwave lens antennas, microwave metamaterial-metasurface (MTS)-metaline-based antennas for communications, sensing, and imaging systems. He is currently more interested in the translational research of MTSs into antenna engineering.

Dr. Chen was elevated a Fellow of the Academy of Engineering, Singapore in 2019 for *the contribution to research, development, and commercialization of wireless technology* and a Fellow of the IEEE for *the contribution to small and broadband antennas for wireless applications* in 2007. He was a recipient of the International Symposium on Antennas and Propagation Best Paper Award in 2010, the CST University Publication Awards in 2008 and 2015, the ASEAN Outstanding Engineering Achievement Award in 2013, the Institution of Engineers Singapore Prestigious Engineering Achievement Awards in 2006, 2013 (two awards), and 2014, the I^2R Quarterly Best Paper Award in 2004, the IEEE iWAT Best Poster Award in 2005, several technology achievement awards from China during 1990–1997 as well as more than 23 academic awards by his students under his supervision. In 1997, he was awarded the Japan Society for the Promotion of Science (JSPS) Fellowship to conduct his research at the University of Tsukuba. He has provided nine local and overseas telecommunication and IT MNCs and SMEs with technical consultancy service as a technical advisor, a guest professor, and a chief scientist. He is the Founding General Chair of the International Workshop on Antenna Technology (iWAT in 2005), the International Symposium on InfoComm & Mechatronics Technology in Biomedical & Healthcare Application (IS 3Tin3A in 2010), the International Microwave Forum (IMWF in 2010), and the Asia-Pacific Conference on Antennas and Propagation (APCAP in 2012).

Dr. Chen has also been involved in many international events as general chairs, chairs, and members for technical program committees and international advisory committees. He served as a vice president for the IEEE Council on RFID (2015–2020), deputy editor in chief for *IEEE Journal of RFID* (2018–2020), and a distinguished lecturer since 2015. He served as an associate editor for the *IEEE TRANSACTION ON ANTENNAS AND PROPAGATION* (2010–2016) and as a distinguished lecturer for the IEEE Antennas and Propagation Society (2009–2012). He also served as the chair for the IEEE MTT/AP Singapore Chapter (2008) and as the founding chair for the IEEE RFID Singapore Chapter (2015 and 2019–2020). Dr. Chen is currently servicing the member for IEEE MTT-29 on aerospace Committee as well as the general chair of 2021 IEEE International Symposium on Antennas and Propagation and USNC-URSI Radio Science Meeting in Singapore.

Xianming QING (Fellow, IEEE) received his B. Eng. degree in electromagnetic field engineering from University of Electronic Science and Technology of China (UESTC), China, in 1985, and Ph.D. degree from Chiba University, Japan, in 2010.

From 1985 to 1996, Dr. Qing was with UESTC as a teaching assistant first, then a lecturer, and later an associate professor for teaching and research. He joined National University of Singapore (NUS) in 1997 as a research scientist for studying high temperature superconductor antennas. Since 1998, he has been with the Institute for Infocomm Research (I^2R, formerly known as Center for Wireless Communication, CWC, and Institute for Communication Research, ICR), Agency for Science, Technology and Research (A*STAR), Singapore. He is currently holding the position of senior scientist and the leader of the RF Group in Signal processing, RF, and Optical Department. His main research interests are antenna design and characterization for wireless applications. In particular, his current R&D focuses on beam-steering antennas, new material/metamaterial-based antennas, near-field antennas, medical antennas, and millimeter-wave/sub-millimeter-wave antennas.

Dr. Qing has authored/co-authored 270+ technical papers published in international journals or presented at international conferences. He is the editor of *Handbook of Antenna Technologies* (Springer, 2016). He has also contributed 10 chapters to the books entitled *Antennas for Small Mobile Terminals* (Artech House, 2018), *Developments in Antenna Analysis and Synthesis* (IET, 2018), *Electromagnetics of Body-Area Networks: Antennas, Propagation, and RF Systems* (Wiley-IEEE press, 2016), *Antenna Technologies Handbook* (Springer, 2015), *Microstrip and Printed Antennas: New Trends, Techniques and Applications* (Wiley, 2010), *Antennas for Fixed Base-Stations in Wireless Communications* (McGraw-Hill, 2009), *Ultra-Wideband Short-Pulse Electromagnetics 7* (Springer, 2007), *Antenna for Portable devices* (Wiley, London, England, April 2007). He is holding 21 granted and filed patents. He was a recipient of the IEEE International Symposium on Antennas and Propagation Best Paper Award in 2010, the CST University Publication Award in 2015, and the IEEE Asia Pacific Conference on Antenna and Propagation Best Student Paper Award in 2016. He was also a recipient of Electronic Industry Advancement Award (1st Class, Sichuan Province, P. R. China) in 1993 and 1995, Electronic Industry Advancement Award (3rd Class, Sichuan Province, P. R. China) in 1995 and 1997, Science and Technology Advancement Award (2nd Class, Ministry of Electronic Industry, P. R. China) in 1994, Science and Technology Advancement Award (3rd Class, Ministry of Electronic Industry, P. R. China) in 1997, IES Prestigious Engineering Achievement Award (Singapore) in 2006, 2013, and 2014, and Singapore Manufacturing Federation Award in 2014.

Dr. Qing is a fellow of IEEE. He is serving as the associate editor of *IEEE Open Journal of Vehicular Technology* and the *International Journal of Microwave and Wireless Technologies* (Cambridge University Press/EuMA), the editorial board member of *Chinese Journal of Electronics* (Chinese Institute of Electronics). He also served as the general co-chair of the IEEE Asia-Pacific Conference on Antennas and Propagation 2015/2018, the International Symposium on Info-Comm & Mechatronics Technology in Biomedical & Healthcare Application 2016 (IS 3Tin3A in 2016). He is currently servicing as the member for MTT-26 RFID, Wireless Sensor and IoT Committee as well as the general co-chair of 2021 IEEE International Symposium on Antennas and Propagation and USNC-URSI Radio Science Meeting.

Contributors

Zhi Ning Chen
National University of Singapore
4 Engineering Drive 3, Singapore 117583
Republic of Singapore
E-mail: eleczn@nus.edu.sg; znchen@ieee.org

Ke Gong
Xinyang Normal University
237#, Nanhu Road, Xinyang, 464000, People's Republic of China
Email: gongkexynu@163.com

Teng Li
Southeast University
Sipailou 2, Nanjing, 210096, People's Republic of China
Email: liteng@seu.edu.cn

Yue Li
Tsinghua University
Rohm Building 8-302, Tsinghua University
Beijing, 100084, People's Republic of China
Email: lyee@tsinghua.edu.cn

Wei Liu
National University of Singapore
4 Engineering Drive 3, Singapore 117583
Republic of Singapore
E-mail: wei.liu@nus.edu.sg;
jocieu.ustc@gmail.com

Xianming Qing
Institute for Infocomm Research (I^2R)
1 Fusionopolis Way, #21-01 Connexis (South Tower), Singapore 138632, Republic of Singapore
Email: qingxm@i2r.a-star.edu.sg

Lei Wang
Heriot-Watt University
Room 2.02, Earl Mountbatten Building
Edinburgh, EH14 4AS, United Kingdom
Email: wanglei@ieee.org

Xiaoxing Yin
Southeast University
Room South 608, Liwenzheng Building
Sipailou 2, Nanjing 210096, People's Republic of China
Email: xxyin@seu.edu.cn

Yan Zhang
Southeast University
Room A3415, Bldg A3, Wireless Valley (Wu Xian Gu), Mo Zhou Dong Lu 9
Nanjing, 211111, People's Republic of China
Email: yanzhang_ise@seu.edu.cn

Foreword

There are now massive research articles and presentations on the design and analysis of millimeter wave antennas, providing evidence that the topic has reached an age of maturity. Professor Zhi Ning Chen and his colleagues and former graduate students have made significant contributions in this area for years. This book is very timely as many upcoming wireless systems will be operated at millimeter wave frequencies, including 5G mobile communications, collision avoidance systems for cars, autonomous control systems for unmanned aerial vehicles, satellite radar and communication systems, and so forth.

The readers will find the book coverage both wide and deep. After describing the details on the features of millimeter wave technology and unique challenges of antenna design as well as millimeter wave measurement techniques and experimental setups, various techniques for improving the performance of classical antennas for the operation at millimeter waves, realizing by LTCC technologies, are reviewed and disclosed. Those antenna designs include low-profile substrate-integrated waveguide slot antennas, broadband antenna arrays on metamaterials, substrate-integrated cavity antennas, cavity-backed substrate-integrated waveguide slot antennas with large apertures, circularly polarized substrate integrated waveguide slot antennas, microstrip antennas with suppressed surface wave losses, substrate integrated antennas for automotive radars, and substrate edge radiating antennas. Pattern synthesizing techniques for achieving low sidelobe are also reviewed.

The book will have widespread appeal to practicing engineers, research scholars, and postgraduate students. I would like to congratulate Professor Zhi Ning Chen and his co-authors on the production of this important text, which will be of great benefit to the antenna community globally.

Professor Kwai-Man Luk *FREng, FIEEE*
City University of Hong Kong

Preface

Millimeter-wave technologies have a long history. In early 2000, the topic became hot again due to the unlicensed 60-GHz wireless communications for short-range links (later IEEE 802.11ad). As antenna researchers we thought how we can contribute to the research and development of the new wave of millimeter-wave systems. After the comprehensive study of the unique challenges of antenna design at millimeter-wave bands, we decided to focus on three issues: loss control, integration, and testing setups. Since 2008 we have conducted the loss analysis of all designs and proposed technologies to control the losses caused by materials, surface waves, and feeding structures. We have explored almost all ways to integrate the antennas and arrays into a variety of substrate from printed circuit board (PCB), low-temperature co-fired ceramic (LTCC), to integrated circuit package (IC package). We also configured and built up three measurement systems to test the impedance, radiation pattern, and gain of the antennas from 60 GHz to 325 GHz. Our works have been widely recognized with tens of papers published in prestigious journals, filed patents, and completed industry projects.

With the deployment of millimeter-wave technology in 5G, the research and development of antenna technologies at millimeter-wave bands are fast advancing to industry applications. The technologies we developed for alleviating fundamental challenges should have more opportunities to be further developed and applied.

The major contents of the book stem from the works of millimeter-wave antennas in the past decade when the editors as well as the authors worked in Institute of Infocomm Research (I2R), Singapore. The relevant research and development were fully supported by Agency for Science, Technology and Research (A*STAR).

The team has worked hard to complete this project in a short duration. All authors would like to appreciate their colleagues as well as their family members for their generous support when they were preparing the manuscripts, in particular, during the COVID-19 period.

Zhi Ning Chen, National University of Singapore
Xianming Qing, Institute for Infocomm Research, Singapore

1

Introduction to Millimeter Wave Antennas

Zhi Ning Chen

Department of Electrical and Computer Engineering, National University of Singapore, Singapore 117583, Republic of Singapore.

1.1 Millimeter Waves

Millimeter waves are regulated by the International Telecommunication Union (ITU) as the electromagnetic waves at the wavelength range of millimeter order, namely, 1–10 mm; the corresponding frequency range is from 30 to 300 GHz, or extremely high frequency (EHF), as listed in Table 1.1. However, the systems operating at the frequencies lower than 30 GHz, such as 24 GHz, are also categorized as millimeter wave (mmW) systems simply because the behaviors of the electromagnetic waves at such frequencies are very similar to those at the defined mmW frequencies. Furthermore, the waves at the wavelength of sub-millimeter order, namely 0.1–1 mm, or the frequency range from 300 to 3000 GHz, are regulated as "terahertz (THz) wave," and the waves at the wavelength of 1 mm–1 m, or the frequency range from 300 MHz to 300 GHz, are regulated as "microwave" by ITU [1]. Therefore, the mmW band is located at the upper edge of the microwave band. Accordingly, the wavelengths at the mmW bands are shorter than those at lower microwave bands but longer than those at infrared bands.

The majority of existing wireless communication and radar systems have been long operating at the lower microwave bands. This book will focus on the waves over the mmW bands at the frequency range from 24 to 300 GHz for wireless applications.

1.2 Propagation of Millimeter Waves

The high frequencies or short wavelengths of the mmWs make their propagation characteristics very unique. The propagation characteristics directly determine the behaviors of waves propagating to desired destinations through a certain path and media. In a long-distance wireless communication, radar, or imaging/sensing application, the propagation properties of the wave fully determine the system design requirements, in particular the selection of the adequate operating frequency and bandwidth [2].

As shown in Table 1.2, the dominant propagation modes of the waves vary against operating frequencies. Furthermore, the types of propagation modes determine the distance of wave propagation. It can be found that:

1. the wave mainly propagates in ionospheric modes like a *skywave* when the frequencies are lower, for instance, at very high frequency (VHF) and below;

Substrate-Integrated Millimeter-Wave Antennas for Next-Generation Communication and Radar Systems, First Edition.
Edited by Zhi Ning Chen and Xianming Qing.

Table 1.1 Allocation of the radio frequency bands by ITU.

ITU band number	Designated band	Frequency	Wavelength in air
1	Extremely low frequency (ELF)	3–30 Hz	9993.1–99 930.8 km
2	Super low frequency (SLF)	30–300 Hz	999.3–9993.1 km
3	Ultra low frequency (ULF)	300–3000 Hz	99.9–999.3 km
4	Very low frequency (VLF)	3–30 kHz	10.0–99.9 km
5	Low frequency (LF)	30–300 kHz	1.0–10.0 km
6	Medium frequency (MF)	300–3000 kHz	0.1–1.0 km
7	High frequency (HF)	3–30 MHz	10.0–100.0 m
8	Very high frequency (VHF)	30–300 MHz	1.0–10.0 m
9	Ultra high frequency (UHF)	300–3000 MHz	0.1–1.0 m
10	Super high frequency (SHF)	3–30 GHz	10.0–100.0 mm
11	Extremely high frequency (EHF)	30–300 GHz	1.0–10.0 mm
12	Tremendously high frequency (THF or THz)	300–3000 GHz	0.1–1.0 mm

Note:
1. Hz: hertz
2. k: kilo (10^3), M: mega (10^6), G: giga (10^9), T: tera (10^{12}).

2. the wave can propagate in surface modes like a *groundwave* when the frequencies are at low frequency (LF) to high frequency (HF) bands; and
3. at higher frequencies, typically VHF and above, the wave just travels in direct modes, that is, the *line-of-sight (LOS),* where the propagation is limited by the visual horizon up to about 64 km on the surface of the earth.

The LOS refers to the waves directly propagating in a line from one transmitting antenna to the receiving antenna. However, it is not necessary for the wave to travel in a clear sight path. Usually, the wave is able to go through buildings, foliage, and other obstacles with diffraction or reflection, in particular at lower frequencies such as VHF and below.

On the other hand, like a light wave, also an electromagnetic wave, the mmWs with shorter wavelengths in millimeter orders, in particular, at EHF and above, always propagate in LOS modes. Their propagation is significantly affected by the typical phenomena of reflection, refraction, diffraction, absorption, and scattering so that a clear path without any lossy or/and wavelength comparable obstacles in the traveling path is required. Such a propagation feature of waves will be reflected in the design considerations of antennas in mmW systems.

Besides the blocking of obstacles in the traveling path, the propagation of the mmWs are also affected by the interaction between the waves and the medium, for example, the atmosphere on the earth. Figure 1.1 shows the average atmospheric absorption of the waves at sea level (i.e., a standard atmospheric pressure of 1013.24 millibar), a temperature of 20 °C, and a typical water vapor density of 7.5 g m^{-3} [3]. The absorption is frequency dependent and ignorable when the frequency is lower than, for instance, 20 GHz with an attenuation less than 0.1 dB km^{-1} or 50 GHz with an attenuation less than 1.0 dB km^{-1}. This is one of the key reasons that almost all existing long-distance wireless systems operate at lower frequencies, for instance, sub-6 GHz bands.

Table 1.2 Dominant propagation modes and typical applications of electromagnetic waves at various frequencies.

Frequency	Wavelength in air	Dominate propagation modes	Typical applications
Extremely low frequency (ELF): 3–30 Hz	9993.1–99 930.8 km	Guided between the Earth and the ionosphere	Very long-distance wireless communication (under water/ground)
Super low frequency (SLF): 30–300 Hz	999.3–9993.1 km	Guided between the Earth and the ionosphere	Very long-distance wireless communication (under water/ground)
Ultra low frequency (ULF): 300–3000 Hz	99.9–999.3 km	Guided between the Earth and the ionosphere	Very long-distance wireless communication (under water/ground)
Very low frequency (VLF): 3–30 kHz	10.0–99.9 km	Guided between the Earth and the ionosphere	Very long-distance wireless communication (under water/ground)
Low frequency (LF): 30–300 kHz	1.0–10.0 km	Guided between the Earth and the ionosphere; ground guided	Very long-distance wireless communication and broadcasts
Medium frequency (MF): 300–3000 kHz	0.1–1.0 km	Ground guided; refracted wave in ionospheric layers	Very long-distance wireless communication and broadcasts
High frequency (HF): 3–30 MHz	10.0–100.0 m	Ground guided; refracted wave in ionospheric layers	Very long-distance wireless communication and broadcasts
Very high frequency (VHF): 30–300 MHz	1.0–10.0 m	Line-of-sight refracted in ionospheric	Wireless communication, radio, and television broadcasts
Ultra high frequency (UHF): 300–3000 MHz	0.1–1.0 m	Line-of-sight	Wireless communication, television broadcasts, heating, positioning, remote controlling
Super high frequency (SHF): 3–30 GHz	10.0–100.0 mm	Line-of-sight	Wireless communication, direct satellite broadcasts, radio astronomy, radar
Extremely high frequency (EHF): 30–300 GHz	1.0–10.0 mm	Line-of-sight	Wireless communication, radio astronomy, radar, remote sensing, energy weapon, scanner
Tremendously high frequency (THF): 300–3000 GHz	0.1–1.0 mm	Line-of-sight	Radio astronomy, remote sensing, imaging, spectroscopy, wireless communications

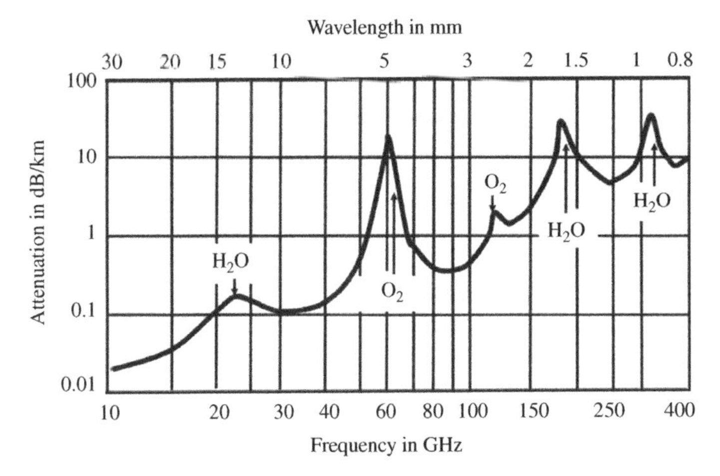

Figure 1.1 The average atmospheric absorption of waves at a sea level at the temperature of 20 °C, standard atmospheric pressure of 1013.24 millibar, and a typical water vapor density 7.5 g m^{-3} [3].

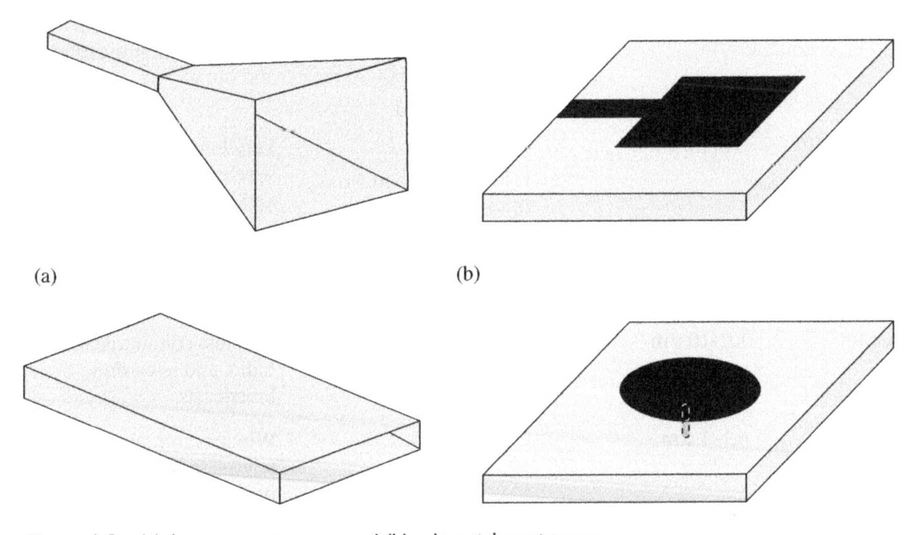

Figure 1.2 (a) Aperture antennas and (b) microstrip antennas.

The wave attenuation is caused by the absorption of water (H_2O) and/or oxygen (O_2) in the atmosphere. There are several absorption peaks across the frequency band up to 400 GHz. The lowest two peaks appear around the 25 and 60 GHz bands, respectively. In particular, the attenuation at the 60 GHz band is 10 times that of the 30 GHz band. In addition, the temperature, pressure, and water vapor density also significantly affect the absorption. It suggests that the wave attenuation at the mmW bands may increase greatly when it is raining, snowing, or foggy. Such an observation must be considered in the calculation of link budget of mmW systems. As a result, the selection and design of antennas should meet the requirements of mmW systems with particular attention to uniqueness of wave propagation.

1.3 Millimeter Wave Technology

1.3.1 Important Features

mmW technology has long been developed for various wireless systems in the past decades because of the apparent advantages over the systems operating at the lower frequency bands, that is, their shorter operating wavelength and wider operating bandwidth with the same fractional bandwidth. The shorter operating wavelength is, for instance, good for an imaging system with higher spatial resolution. Physically the resolution limitations in an imaging system restrict the ability of imaging instruments to distinguish between two objects separated by a lateral distance less than approximately half an operating wavelength of waves used to image the objects.

With the shorter operating wavelengths, mmW systems also enjoy an advantage over the systems operating at lower frequencies, namely, a tiny component size. In particular, the overall volume of the mmW devices can be greatly reduced because the performance of some key radio frequency (RF) components are determined by the electrical size of the design, for instance, antennas and filters. The smaller size of the RF components definitely benefits the device design significantly, especially for applications requiring tiny devices such as handsets, wearables, and implants. For example, it is very challenging to install more antennas, typically more than two antennas operating at the bands of 690–960 MHz in existing handsets with limited overall space. However, it is easy to install multiple antennas and even arrays operating at, for example, 28 or 39 GHz bands for the mobile phones in future fifth generation (5G) networks.

Another attractive advantage is the wide operating bandwidth of the mmW systems. The 10% operating bandwidth at a 60 GHz band offers a bandwidth of 6 GHz, 10 times the 10% bandwidth at 6 GHz, namely 600 MHz. The wider absolute operating bandwidth is able to support the transmission at much higher data-rates according to the Shannon–Hartley theorem. The fundamental information theory tells us that the maximum transmission rate or capacity over a communication channel in the presence of noise is directly proportional to a specified bandwidth [4]. Therefore, the mmW wireless communications can easily achieve the data-rate of a few gigabits per second (Gbps).

1.3.2 Major Modern Applications

The mmW technology has a long history in wireless applications since 1890, Hertz's days [5]. Selected key milestones of mmW research and technology development are briefed in Table 1.3.

With the rapid development of materials, processing, fabrication, and measurement at mmW bands, the mmW technology has been fast applied in modern wireless communications, radar, imaging scanning, and imaging systems. The following sections provide recent examples of new applications of mmW technology.

1.3.2.1 Next-Generation Wireless Communications

Wireless communications are rapidly progressing toward high data-rate and ultra-low latency for the Internet of Things (IoT). Due to the requirement to support higher data-rate, mmW technology is promising for 5G networks over the frequency range from 24 to 86 GHz. Investment on the research and technology development of mmW cellular mobile networks and WLAN/WiFi infrastructures is increasing exponentially. For instance, mmW technology is the main candidate for the

Table 1.3 Selected key milestones of research and development of mmW technology by 1980s.

Period	Important activities	Typical applications	Selected references
1890–1945	• Hertz and Lebedew's experiments in centimeter / millimeter wavelength • Nichols, Tear, and Glagolewa-Arkadiewa developed instruments extended to 0.22 and 0.082 mm using spark-gap generator • Cleeton and William developed vacuum tube sources • Boot and Randall developed cavity magnetron for radar	• Confirmed Maxwell's prediction • Radiometer • mmW sources • 10 and 24-GHz radar	[5–10]
1947–1965	• Atmospheric attenuation measurement by Beringer, Van Vleck's and Gordy • All circular-electric mode transmission with all RF components by Bell Labs • Geodesic lens antenna by Georgia Technology • 58-GHz broad-brand helix traveling-wave amplifiers by Bell Labs • 150-GHz backward-wave oscillators by Thomson-CSF, France • Imaging line, its associated components and surface-wave propagation on Goubau line or Sommerfeld wave on uncoated metal wire by Wiltse • First IRE Millimeter and Sub-millimeter Conference held in Orlando, FL, USA in 1963 • First special issue of The IEEE Proceedings published in April 1966	• Point-to-point transmission • The first 14-km long transmission system • Spectroscopy • 70-GHz radar • mmW sources • High-power mmW sources	[11–17]
1965–1984	• Development of components at 35, 94, 140, and 220-GHz by US Army Ballistic Research Laboratory as well as Royal Radar Establishment, UK • The first special issue of the IEEE Transaction on Antennas and Propagation about millimeter wave antennas and propagation. • Solid state source	• Radiometers • Radars • Missile guidance • Communications	More details can be found at [18–20]

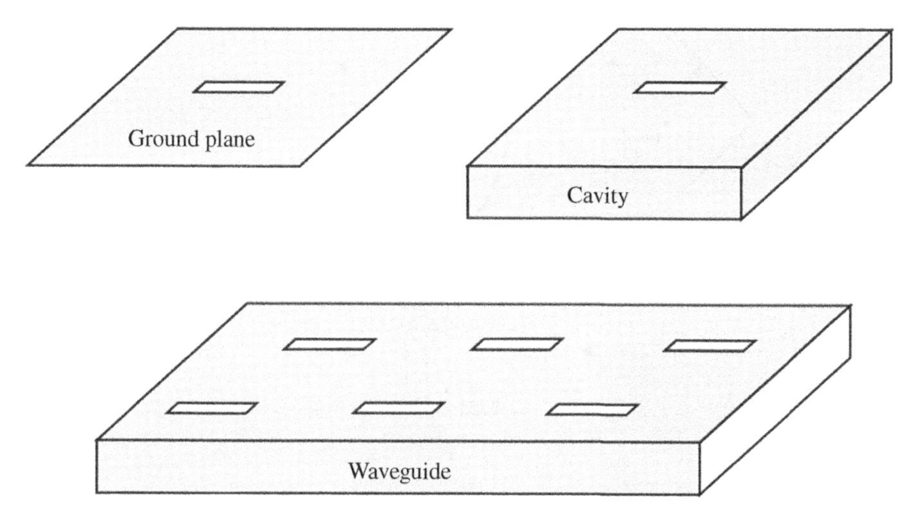

Figure 1.3 Slot antennas.

concept of small cells for future cellular network implementation. The mmW technology can be used not only for mobile terminals but also for point-to-point backhauls, to some degree to replace traditional fiber optic transmission lines by connecting mobile base stations (BSs).

1.3.2.2 High-Definition Video and Virtual Reality Headsets

The transmission of 1080p high-definition (HD) video needs the data-rate up to gigabits per second. None of the existing wireless microwave links can support such high speed at any sub-6 GHz bands. The 60-GHz mmW technology operating, for instance, with unlicensed bandwidths up to 5–7 GHz (for example, US: 57.05–64 GHz and Europe: 57.0–66.0 GHz) can be used to transmit HD video from digital set top boxes, laptops, digital video disc (DVD) players, HD game stations, and other HD video sources to HD television (TV) wirelessly. Furthermore, small transmission devices can be integrated into TV sets invisibly.

Similar to the HD video applications scenarios, virtual reality (VR) applications need ultra-high data-rate wireless links in a short range for future multimedia applications. The wireless links will support the high-speed transmission of video and audio data from mobile devices such as headsets to controlling computers or other VR devices. The mmW wireless communications are the only solution to meet such requirements so far.

1.3.2.3 Automotive Communications and Radars

The mmW radar operating at 24 GHz may be the first mmW system in the history of mmW technology, as shown in Table 1.3. Recently, autonomous driving is being developed very fast. Such applications require the ultra-high-resolution detection of pedestrians and other obstructions as well as communication with other vehicles through the network in real time and low latency. Ultra-high-resolution radars have been developed with the mmW radar systems operating at a range from 77 to 81 GHz. The detection range varies from 0.15 to 200 m. The communications can be built up to achieve the gigabits per second links through 5G networks.

1.3.2.4 Body Scanners and Imaging

Due to the short wavelength for possible high-resolution images and high frequency for fast imaging, the mmW technology has been extensively applied in human body scanners currently in the market. The mmW body scanners have achieved high-precision scanning with much less harm

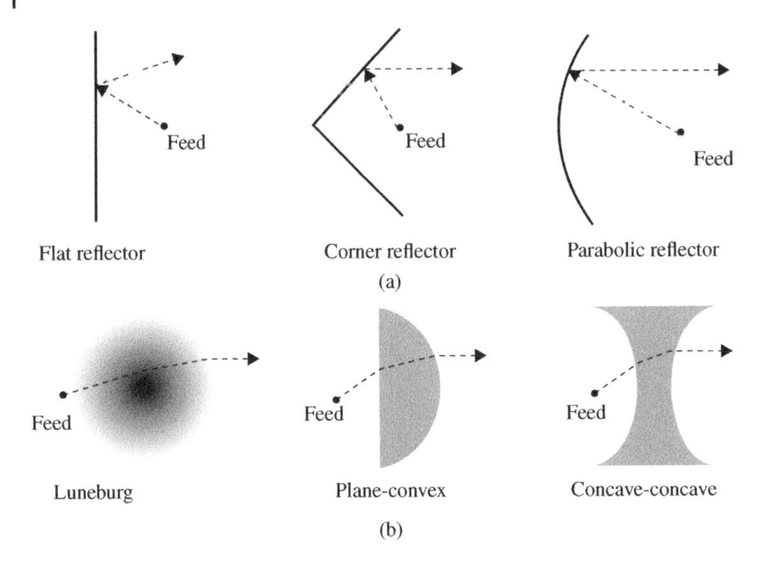

Figure 1.4 (a) Reflector antennas and (b) lens antennas.

to the human body. In particular, such mmW full-body scanners have been very popular for airport security. They use the transmitted power of less than 1 mW and operate at a frequency range between 70 and 80 GHz.

In conclusion, with its unique features, the mmW technology has a very promising future in the applications of high-speed wired/wireless communications as well as radar detection and imaging.

1.4 Unique Challenges of Millimeter Wave Antennas

An antenna is the only means to transfer the electric power from the circuits of a wireless system to a medium and vice versa. There have been many design challenges for conventional antennas such as wide operation bandwidth, high gain and radiation efficiency, desired radiation performance, small/compact size and conformal shapes, and low-cost material utilization and fabrication. However, the design of antennas for mmW systems faces unique challenges because of its higher frequencies, or shorter wavelengths.

From the design point of view:

1. *Wide Bandwidth:* At mmW bands, usually we enjoy wide available spectra for wireless communications and radar applications. The typical operating bandwidth is more than 10% for both required impedance matching and desired radiation performance such as radiation patterns, radiation efficiency, and polarization. It is indeed challenging for the antennas to meet all design requirements simultaneously.
2. *High Gain:* Due to high operating frequencies, the path-loss in propagation at mmW bands, in particular in rainy, foggy, or snowy weather, becomes much more critical than that at microwave bands. To compensate for the high path-loss, we must design antennas with much higher gain. For example, the gain of the antennas should reach up to 15 dBi for 10 m LOS wireless links at 60 GHz bands for some systems under the IEEE 802.11ad standard.
3. *High Radiation Efficiency:* Due to high operating frequencies, the ohmic losses caused by materials and connections in the design become severe. Usually, the loss of the dielectric and metals increase against frequencies. For instance, FR4 has loss tangents of 0.016 at 1 MHz and

> 0.1 at 16 GHz, respectively. The other loss can be caused by surface waves due to electrically thick dielectric used in antenna design. For instance, the dielectric substrate of a patch antenna increases its electrical thickness from a typical 0.01 operating wavelength at a frequency of 2.4 GHz to 0.1 wavelength at 24 GHz if the same dielectric substrate is used for the patch antenna design. The thicker substrate inevitably causes severer surface waves. The increased surface waves guide part of the radiated power to other directions rather than the desired direction, for instance, boresight. Such loss lowers the radiation efficiency or gain of the antenna.
4. *Beam-Scanning Functionality:* Due to the need of high gain, the beamwidth of the radiation pattern is narrow so that it is difficult to achieve a wide coverage, but it is easy for the radiation to be blocked by unwanted obstacles. To keep propagation links unbroken, the functionality of beam-scanning of antennas is one solution. Therefore, much more expensive and complicated antenna designs are necessary.

From the fabrication and measurement point of view:

1. *High-End Material and Tight Tolerance Fabrication:* To keep low the ohmic loss caused in materials, high-end materials, usually at higher prices, must be used. For instance, FR4 is replaced by more expensive but higher performance Rogers or ceramic or liquid crystal polymer (LCP). Meanwhile, due to the shorter wavelengths at higher frequencies, higher-precision fabrication processes are needed to achieve results within the acceptable tolerance. For example, conventional printed circuit board (PCB) process can guarantee a tolerance of 0.2 mm, which is 0.2% wavelength at 3 GHz but 4% wavelength at 60 GHz!
2. *Testing Setup:* There have been many antenna testing setups up to 40 GHz in the market and laboratories. It is necessary to upgrade the setups and system for the antenna test for the frequencies higher than 40 or 67 GHz. Expensive frequency up-converter heads are necessary for the frequency extension of such mmW systems. For the measurement of impedance or S-parameters, high-quality but expensive connectors and cables must be used to keep the insertion loss acceptable. In addition, the on-wafer mmW antenna measurement must be carried out using precise and expensive probes where the measurement is usually conducted on expensive specific probe stations. For the measurement of radiation performance, due to the lack of link budget and the concern of testing environments, it is necessary to design the setup by customizing the mechanical structures with high-precision location and orientation.

In short, the design of mmW antennas faces much more lossy, complicated, and costly issues than those at lower microwave frequencies. The research and development activities of mmW antenna technology should be done by taking all such unique challenges into account.

1.5 Briefing of State-of-the-Art Millimeter Wave Antennas

From an antenna operation point of view, mmW antennas can be any type of radiators, such as wire, aperture, slot, microstrip, reflector, and Lens as shown in Figures 1.2, 1.3, and 1.4. Also the radiators can be arranged as arrays to enhance their radiating performance and achieve more functionalities. However, due to the unique challenges caused by their physically short wavelengths, namely 1 mm at 300 GHz to 10 mm at 30 GHz in free space as mentioned above, some types of radiators are more suitable for mmW antenna designs, as shown in the general discussion listed in Table 1.4. The discussion is based on the basic versions of all types of antenna designs. A variety of variations of the antenna have long been proposed for performance enhancement as presented in the References [21, 22].

Table 1.4 mmW antennas.

Type of radiator	Fabrication/Testing	Performance/Applications	Sketches
Aperture: • Horn • Open-end waveguide • Waveguide horn	• Easy fabrication and test • Bulky three-dimensional geometry for metal structures • Difficult to be integrated with circuits • Low loss for air-filled design • Possibly fabricated using PCB[a] and LTCC[b] processes	• Moderate bandwidth • High gain • Point to point link • Standard antennas in measurement systems	Figure 1.2
Microstrip patch	• Easy fabrication and test • Flat and low-profile geometry • Lossy at high frequencies • Wide feeding strips • Conformal configuration • Easy to form arrays • Easy to be integrated with circuits • Easy to be fabricated using PCB and LTCC processes	• Narrow bandwidth • Low gain • Broadside radiation • Wide applications	
Slot: • On ground • On cavity • On waveguide	• Easy fabrication and test • Flat geometry • Easy to form arrays • Conformal geometry • Difficult to be integrated with circuits	• Narrow bandwidth • Low gain • Broadside radiation	Figure 1.3
Reflector: • Corner • Flat plane • Curved plane	• Not easy fabrication and test • Bulky and three-dimensional geometry with feed • Not easy to form arrays • Difficult to be integrated with circuits	• Wide bandwidth • Ultra-high gain • Directional radiation	Figure 1.4
Lens: • Luneburg • Convex-plane • Concave-plane • Concave-concave • Convex-concave • Convex-concave • Convex-convex	• Not easy fabrication but easy test • Bulky, heavy and three-dimensional geometry with feed • Not easy to form arrays • Difficult to be integrated with circuits	• Wide bandwidth • Ultra-high gain • Directional radiation	
Wires and their variations on substrate	• Easy fabrication and test • Flat and low-profile geometry • Lossy at high frequencies • Wide feeding strips • Conformal configuration • Easy to form arrays • Easy to be integrated with circuits • Easy to be fabricated using PCB and LTCC processes	• Moderate bandwidth • Low gain • Broadside/endfire radiation	

a) PCB: printed circuit board.
b) LTCC: low-temperature co-fired ceramic.

Also the antennas can be categorized into two classes based on physical geometry: flat/planar and three-dimensional structures. For high-gain applications, which are always required for mmW systems as mentioned previously, three-dimensional designs such as reflector and lens antennas are perfect options if there is no space and installation constraints; the large-scale arrays of planar elements such as microstrip antenna arrays and slot antenna arrays usually suffer from the difficulty to form the large-scale feeding network in a limited physical space and high loss caused in the feeding networks.

As an alternative, a technique of laminated waveguides on PCB substrate was invented [23, 24], which is to some degree considered the extension of the work based on post-rod to form air-waveguides [25]. Later the structure was comprehensively studied and named as substrate integrated waveguide (SIW) and widely applied in mmW antenna designs [26, 27], where an electromagnetic waveguiding structure is constructed by the two walls formed by two arrays of metalized vias. The spacing between the adjacent vias must meet the criteria to stop the leak of wave propagating in the structure. Such a substrate-integration technology provides much flexibility for waveguide designs and relevant antenna designs, in particular, at mmW bands.

1.6 Implementation Considerations of Millimeter Wave Antennas

At mmW bands, the integration of antennas and substrate by using a multilayer substrate process are desired for planar or flat design of system boards. The substrate integrated antenna (SIA) can be fabricated exactly as a conventional circuit as for printed circuits on layered boards, where the antennas become part of circuit boards or package of integrated circuits. Such integration greatly reduces the loss caused by the connection between the circuits and the antennas, miniaturizes the size of the system, lowers the fabrication cost, and increases the robustness of the system without additional installation of antennas. The integration of the antennas on the substrate is critically determined by fabrication including the selection of substrate materials and the applicable fabrication process.

1.6.1 Fabrication Processes and Materials of the Antennas

With the shorter operating wavelengths in the order of a millimeter, the fabrication of mmW antennas needs a tolerance usually tighter than microwave to achieve the desired performance. For example, for a straight thin half-wave dipole antenna operating at 60 GHz, the overall length of the antenna is about 2.5 mm. The acceptable fabrication tolerance is typically 0.2% wavelength, namely 0.05 mm, which nearly reaches the limit of the conventional commercial PCB process. The tighter tolerance of fabrication is needed if the antennas and feeding network are printed on the PCBs with the relative dielectric constant larger than unit. Therefore, the selection of fabrication for mmW antennas with feeding networks is more critical than that at lower microwave bands, not only because of the tolerance but also costs including processing, materials, and assembling.

At the mmW bands, multilayered substrates such as polytetrafluoroethylene (PTFE), a synthetic fluoropolymer of tetrafluoroethylene, and PTFE composite filled with random glass or ceramic such as RT/duroid® are commonly used for laminating circuits and antennas. PTFE-based substrates usually feature a low and stable loss tangent typically of 0.0018 at 10 GHz and even higher and high resistance to chemical processing and are waterproof and thermally stable. PTFE-based substrates, however, suffer from a higher cost compared to FR4 glass epoxy, are softer materials, and have a higher thermal expansion coefficient. FR4 glass epoxy is most commonly used in frequencies lower than 3 GHz because of its increasing loss tangent against frequency.

LCPs are a class of aromatic polymers. The unique feature of the LCP substrate is its softness although it has similar properties to PTFE-based substrates such as extreme chemical resistance, high mechanical strength at high temperatures, and inertness. Its poor thermal conductivity and surface roughness should be taken into account in electrical applications, in particular, at mmW bands.

To meet the requirements of fabrication tolerance, electrical, and other mechanical properties, low temperature co-fired ceramic (LTCC) has long been used as a cost-effective substrate technology in electrical and electronic engineering, especially at higher frequencies. LTCC is a multilayered glass ceramic substrate. It is co-fired with low-resistance metal conductors, such as Au or Ag, at low firing temperatures, usually ~850–900 °C, compared with high temperature multilayered ceramic sintered at ~1600–1800 °C. There have been many ceramic materials developed by commercial companies. More detailed information can be found in the book [28], which studies a variety of electrical materials for mmW applications. In particular, the information about the ceramic materials used in LTCC is comprehensive.

For example, Ferro A6M has been widely used in applications at mmW bands. It has a relative dielectric constant of 5.9–6.5 and loss tangent of ~0.001–0.005 at 3 GHz. In particular, the electrical properties are stable against frequency. The relative dielectric constant and loss tangent of DuPont 951 ceramic are ~7.85 and 0.0063 at 3 GHz, respectively. It should be noted that the ceramic used in LTCC usually has the relative dielectric constants of ~6–10, sometimes ~18 [29]. High relative dielectric constants are usually not desired for antenna design at mmW bands because they will shrink the dimensions of antennas so that the fabrication needs much higher accuracy [30].

With an LTCC process, the LTCC ceramic substrate can host almost an infinite number of layers. The thin layers are stacked one on the top of another. The conducting paths of gold or silver thick film pastes are printed on each surface layer by layer using the silk-screen printing method. When the multilayer setup has been stacked and printed, it is then fired in the process oven where the low sintering temperature allows the use of gold and silver as conducting traces. The simplified description of process includes:

Step I: via punch, via conductor fill, and trace printing;
Step II: layer stack and lamination; and
Step III: layer co-fire.

The PCB and LTCC processes are concisely compared in Figure 1.5. From a waveguide feeding network and antenna design point of view, the most important difference between the PCB and LTCC processes is that the LTCC process is able to implement the blind via and embedded cavity while the PCB process is unable to do it.

Furthermore, the LTCC used for SIA designs also increases the advantages such as low loss tangents, low permittivity tolerance, good thermal conductivity, multilayered substrate, cavities/embedded cavities, low material costs for silver or gold conductor paths, easy integration with other circuits, and low production costs for medium and large quantities.

In our experience, the PCB process is preferred for SIA designs when an operating frequency is lower than 60 GHz, while LTCC is a good candidate antennas operating at frequencies higher than 60 GHz and up to 300 GHz. At frequencies higher than 300 GHz, the LTCC fabrication becomes quite challenging because of its process limit such as via-hole pitch.

1.6.2 Commonly Used Transmission Line Systems for Antennas

Like any antenna systems, their feeding structures will be a critical issue in the implementation of the antenna design. In particular, the losses caused in the feeding networks greatly degrade the

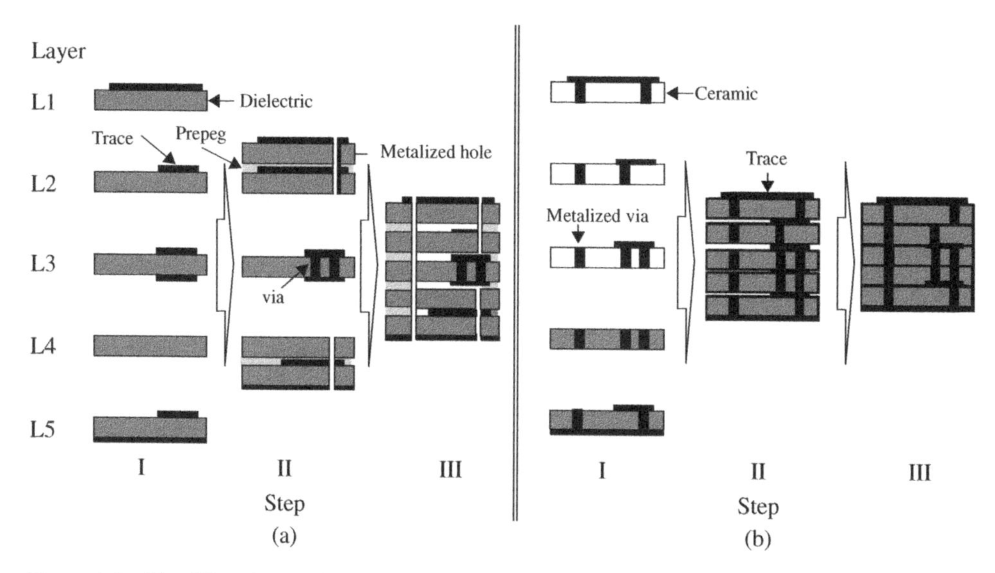

Figure 1.5 Simplified descriptions of PCB and LTCC processes. (a) PCB and (b) LTCC.

performance of the antenna arrays at mmW bands. Unlike the antennas at microwave bands, the losses can be caused by not only the dielectric substrate but also metals used as conductors for transmission and radiation. Like any dielectric at microwave bands, the loss of a dielectric substrate is measured by its loss tangent. Different from the designs at microwave bands, the metal loss that may be caused by the conductivity and surface roughness of the conductors can't be ignored.

Besides microstrip transmission lines, the waveguide-type transmission line systems are popularly used because they may enjoy the lower losses caused by dielectric and metals at mmW bands [23–27, 31]. Accordingly, for instance, the loss analyses have been conducted for microstrip lines, solid-metal-wall waveguides, and post-wall or laminated waveguide or SIW [32]. The study shows that in general the solid-metal-wall waveguides without dielectric loss enjoy less metal loss while microstrip lines suffer from several dielectric losses. The post-wall waveguides or SIWs feature acceptable total losses caused by both dielectric and metal losses at mmW bands. However, it should be noted that the causes of losses of transmission line systems can be complicated because they will be determined by the materials such as dielectric and metals as well as the types or configurations of transmission lines.

The transmission line systems can be in the form of microstrip lines and coaxial lines. Compared with conventional cylindrical versions, the substrate integrated coaxial line (SICL) is a type of planar rectangular coaxial lines. The lines comprise a strip sandwiched between two grounded dielectric layers and laterally shielded by the arrays of metallized vias [33]. Similar to the conventional coaxial line, the propagation of SICL is still in the dominant mode of transverse electromagnetic (TEM).

The SICLs can be realized using a traditional multilayer PCB or LTCC process. Therefore, SICLs feature the combined advantages of the coaxial lines and the planar transmission lines, including the wideband unimodal operation, low cost, non-dispersive performance, good electromagnetic compatibility, and easy integration with other planar circuits. It has been used for high-speed data transmission [34] and various other applications such as antennas, couplers, baluns, and filters at mmW bands [35–41].

Moreover, the substrate integrated gap waveguide (SIGW) or printed ridge gap waveguide (PRGW) is proposed for the transmission line systems at mmW bands. The SIGW or PRGW is the combination of the microstrip-line and gap-waveguide technology based on the PCB or LTCC process [42–44]. The inverted printed strip line is arranged on or above the periodic mushroom structures where the unwanted surface waves are suppressed and only the quasi-TEM mode is permitted over the operating band. Unlike SIW or SICL, the top and bottom grounds of a SIGW are unconnected. Therefore, the processing complexity is greatly reduced. The SIGW/PRGW technology has been widely used in the antennas and arrays at mmW bands [44–51].

It should be noted that the selection of the materials and transmission line systems significantly affects the antenna efficiency. The loss analyses of antennas including their feeding structures are strongly suggested to understand the main causes of the losses in order to control the overall loss by properly selecting the materials and the types of transmission systems, as well as optimizing the design configurations [52].

1.7 Note on Losses in Microstrip-Lines and Substrate Integrated Waveguides

As previously mentioned, to compensate for the path-loss at higher frequencies, usually very large-scale antenna arrays are required in mmW systems. In such large-scale antenna arrays the feeding network inevitably becomes complicated with a labyrinth of feeding network. The long current or power paths in the network are the critical causes for transmission losses. The additional unignorable transmission losses may be the stopper to limit the achievement of high gain of larger-scale antenna arrays when the insertion loss cancels the increase in the gain by increasing the number of the elements of arrays. For example, if the insertion loss caused by the increase of the power path of the feeding network reaches nearly 3 dB, the antenna array with doubled number of elements will achieve very little gain enhancement. Therefore, it is important to check the transmission line systems in terms of insertion loss before the design of the arrays at mmW bands.

Next, the insertion losses in microstrip-lines (MSLs) and SIWs in LTCC at 60 GHz are compared as an example. The LTCC is Ferro A6-M with relative dielectric constant $\varepsilon_r = 5.9 \pm 0.20$, loss tangent $\tan\delta = 0.002$ at 100 GHz. The conductor used for metallization and vias is Au, whose conductivity is $4.56 \times 10^7\ \mathrm{S \cdot m^{-1}}$.

Figure 1.6 shows a 10-mm long bent MS transmission line on an LTCC board. The 50-Ω MSL is with two ports in the simulation. Figure 1.7 compares the insertion losses for varying thickness of the LTCC board over a frequency range of 0–70 GHz. It is seen that when the thickness increases, the insertion at higher frequency edge quickly increases. For instance, the loss per centimeter reaches up to 13 dB when the thickness is larger than 0.7 mm.

Figure 1.8 clearly shows the causes of the insertion losses at higher frequencies or mmW bands. The losses caused by the dielectric substrate and conductor in the system are just a small percentage of the total losses. It is believed that at 60 GHz, the higher-order modes excited by the discontinuity of the MS cause large surface wave (SW)/leaky losses as previously discussed. This issue is even severer for the thicker substrate. So the SW of MS at mmW is definitely a big problem for practical antenna design.

Figure 1.9 shows a 10-mm long bent SIW in an LTCC board. Figure 1.10 shows the main losses at 60 GHz of a bent SIW in an LTCC board with varying thickness. It is clear that on the contrary, the SIW system does not suffer from such a dilemma, with the highest loss less than 1 dB per centimeter at smaller thicknesses and total losses lower than 0.6 dB for a thickness larger than 0.3 mm. The low-loss feature is quite stable for all the thicknesses. But actually for the very thin thickness of 0.1 mm, the conductor loss is high for the SIW. Fortunately, the thickness of 0.5 mm is usually selected for SIW at 60 GHz. In particular, the majority of losses are caused by both the dielectric and conductors, which is different from the MS lines.

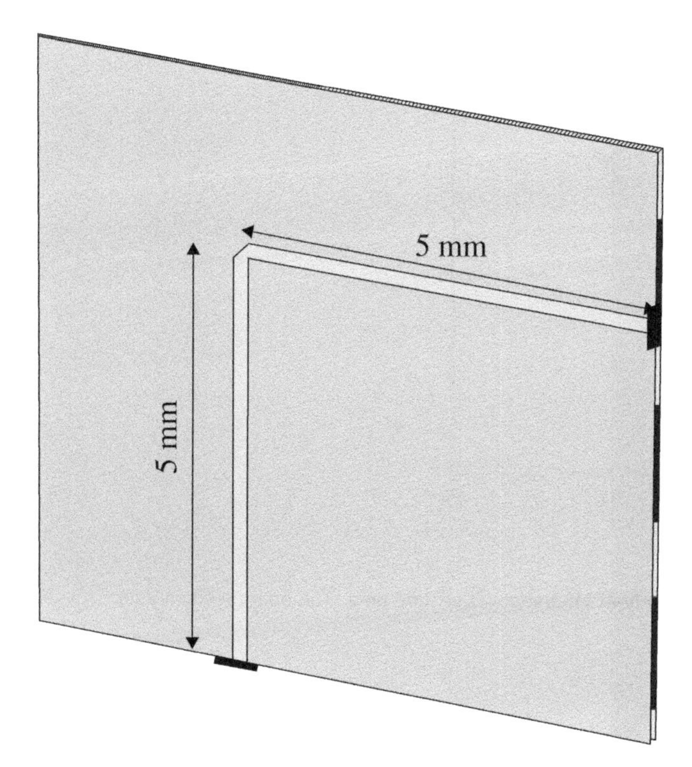

Figure 1.6 A bent MS transmission-line on a LTCC board.

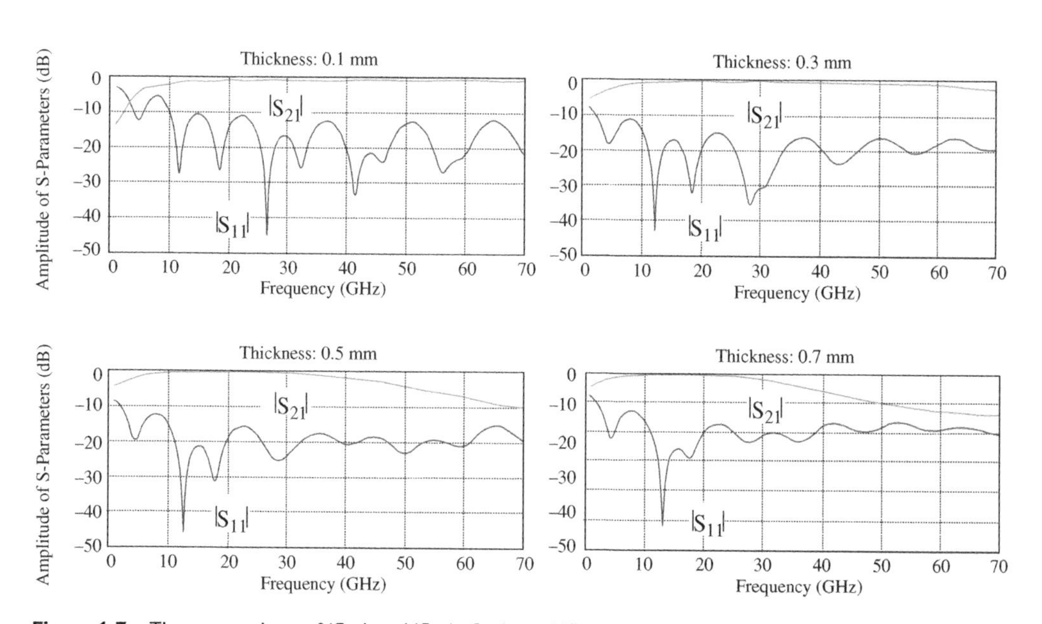

Figure 1.7 The comparison of $|S_{11}|$ and $|S_{21}|$ of a bent MS transmission-line on a LTCC board.

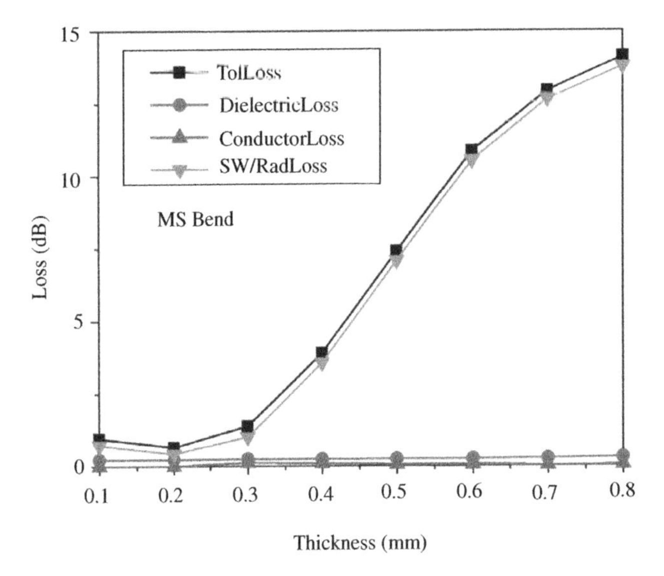

Figure 1.8 The main losses at 60 GHz of a bent MS transmission-line on a LTCC board with varying thickness.

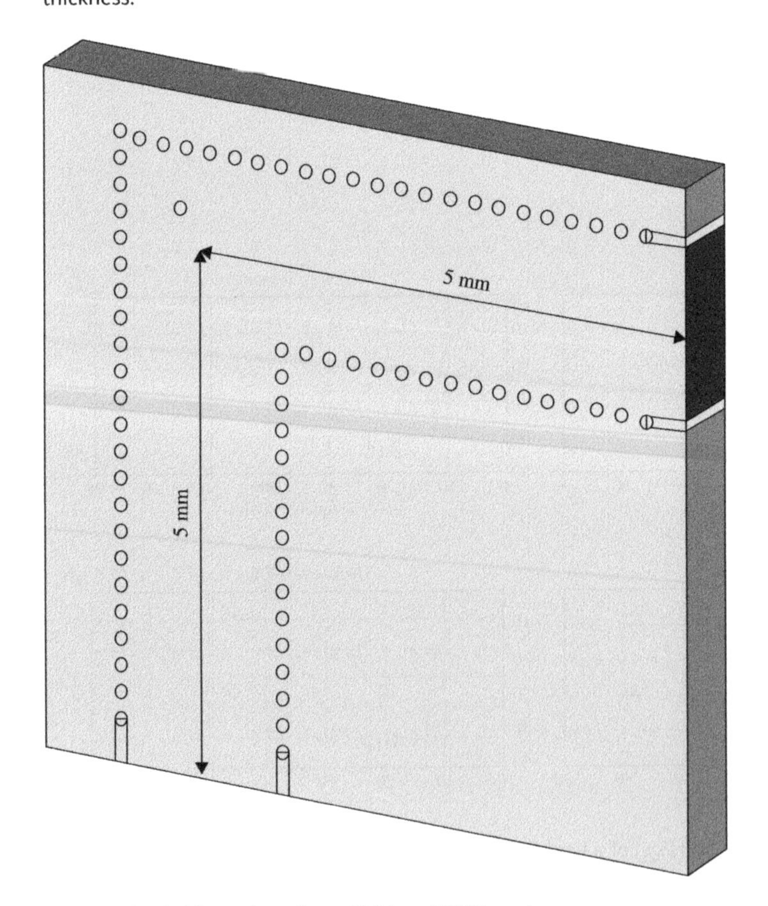

Figure 1.9 A 10-mm long bent SIW in a LTCC board.

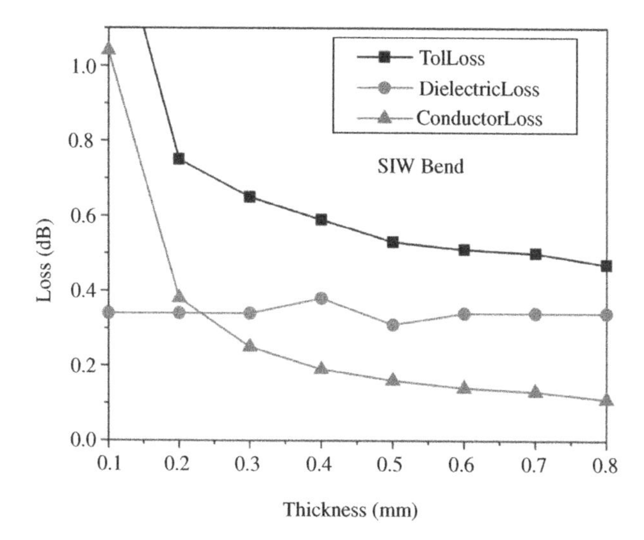

Figure 1.10 The main losses at 60 GHz of a bent SIW in a LTCC board with varying thickness.

In short, the SIW transmission line systems feature lower insertion loss at 60 GHz and above compared with the MS-lines suffering from several surface wave losses. However, it should be mentioned that the SIW has a more complicated fabrication and higher fabrication cost compared with the MS lines. The evaluation of performance and implementation cost is necessary to make a trade-off in the design.

1.8 Update of Millimeter Wave Technology in 5G NR and Beyond

Recently mmW has attracted much attention of the industry although it is not a new technology. In particular, the mobile communication networks are significantly increasing their capacity by increasing the usage efficiency of existing spectra such as massive multiple-input-multiple-output (massive MIMO), the number of small cell base stations, and the operating bandwidth by extending operating frequencies from sub-6 GHz bands to mmW bands to enhance the peak data rates of mobile broadband use case up to 20 Gbps in the downlink (DL) and 10 Gbps in the uplink (UL). With such high data rates, many new applications such as high-speed streaming of 4K or 8K UHD movies and self-driving cars will be supported.

According to the release of specifications by The 3rd Generation Partnership Project (3GPP), a standards organization developing the protocols for mobile telephony, the following bands as tabulated in Table 1.5 [53] have been defined for fifth-generation (5G) new radio (NR) networks. The allocated bandwidths reach up to 3 GHz each while the total frequency ranges from 24.25 to 40.00 GHz.

Table 1.6 compares the features of mobile communication network operating at 5G NR sub-6 GHz and mmW bands. Not surprisingly, the systems operating at mmW bands suffer from the higher path-loss and blocking due to weaker refraction and reflection in propagation in denser environments.

Based on the previous analysis and initial feedback from field trials, the mmW in 5G NR networks will have the typical applications as shown in Figure 1.11. It is estimated that for 5G NR the mmW systems can be used with the similar scenarios at sub-6 GHz bands, but coverage will be significantly reduced due to the higher path-loss and blocking of mmW propagation. The higher

Table 1.5 mmW frequency ranges for 5G NR (TDD).

NR operating band	$F_{UL/DL\_low} - F_{UL/DL\_high}$
n257	26.50–29.50 GHz
n258	24.25–27.50 GHz
n260	37.00–40.00 GHz

- Duplex Mode: Time-division duplexing (TDD)
- Uplink (UL): base station (BS) receive and user equipment (UE) transmit
- Downlink (DL): base station (BS) transmit and user equipment (UE) receive.

Table 1.6 Comparison of features of network at 5G NR sub-6 GHz and mmW bands.

	Sub-6 GHz	mmW
Bandwidth	~100 MHz	500 MHz @28 GHz/>2 GHz @E-band
BS/UE antenna configuration	Single/sectorized arrays	Very high-gain directional arrays
Network deployment	Low base-station density	Very high base-station density
Small-scale fading	Correlated with high rank	Correlated with low rank, varies with line-of-sight (LOS) or Non-LOS (NLOS)
Large-scale fading	Distance dependent path-loss	Distance dependent with random blockage model and total outage
No of users served simultaneously	High (>10)	Low (<10)

path-loss is caused by the reduced physical aperture of receive antennas with the same gain as the antennas operating at lower frequencies. Therefore, the electrical apertures or gain of the mmW antennas should be increased to keep the gain unchanged or link budget compared with the antennas operating at lower frequencies.

Besides the mmW frequency bands released by 3GPP for 5G NR, the extension of mmW bands from the lower edge frequencies around 30 to 60 GHz even higher has been proposed and investigated for mobile communication networks for years. However, due to even higher path-loss and severer blocking at upper mmW bands, the mmW systems may have different applications as shown in Figure 1.11. The systems operating at 60 GHz bands may be a good option for indoor hotspots or hot-region coverage for opportunistic connections and/or pointing connections for high data transmission. Systems operating at frequencies higher than typically 110 GHz have been used for long distance backhauls to replace higher cost optical fiber or cabled connections. The new application may be the backhauls between macro BSs and small cell BSs for low-cost and flexible network setup.

In short, it can be foreseen that the applications of mmW systems in the 5G NR are just a start and still present many technical challenges. In the 5G and even beyond 5G, the antenna technologies will play increasingly important roles. The small or compact multiple functional massive MIMO, beamforming/scanning, multi-beam, and reconfigurable antenna systems will be greatly developed to meet the challenging system demands.

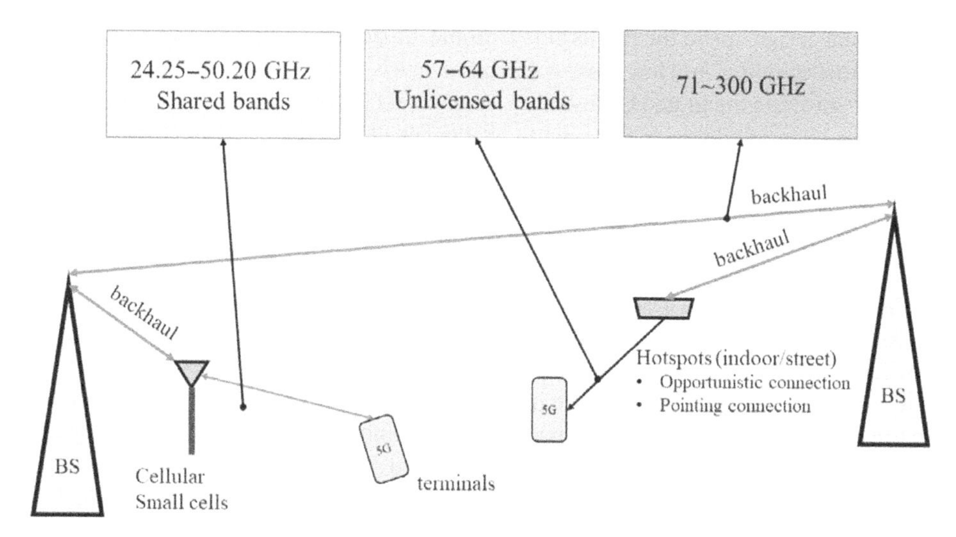

Figure 1.11 Potential mmW applications in 5G NR and future networks.

1.9 Organization of the book

This book is organized in the following way:

Chapter 1, "Introduction to Millimeter Wave Antennas" by Zhi Ning Chen, is an introductory chapter of this book. First the relevant concepts of mmW technology are introduced. Then the unique propagation characteristics of mmW are reviewed, associated with the existing and promising applications of mmW technologies. After that, the unique design challenges of antennas at mmW bands are addressed, followed by a brief overview of state-of-the-art mmW antenna designs. The last part briefly discusses the challenges in the fabrication of mmW antennas from the selection of materials to processes. The latest developments and applications of the mmW systems for 5G new radio and beyond are briefed for future research development of antenna technologies.

Chapter 2, entitled "Measurement Methods and Setups of Antennas at 60–325 GHz Bands" by Xianming Qing and Zhi Ning Chen introduces the testing setups of mmW antennas by addressing expensive testing setup, limited system dynamic range, complicated, and tedious calibration, as well as measurement procedures. This chapter deals with the measurement issues of the mmW antennas from 60 to 300 GHz. First, the state-of-the-art mmW antenna measurement systems are presented. Then the key considerations of configuring the measurement systems are addressed. In the last part, the detailed setup configurations for achieving the maximum system dynamic range with the available commercial accessories are described, wherein the measurement of reflection coefficient, gain, and radiation pattern of a number of antennas at 60, 140, and 270 GHz bands with different feeding connections (coax, waveguide, and probe) are exemplified.

In Chapter 3, "Substrate Integrated mmW Antennas on LTCC" Zhi Ning Chen and Xianming Qing introduce the basic concepts of SIW antennas, in particular, SIW slot antennas. Then, design examples in LTCC are elaborated. The examples include planar high-gain arrays operating at 60, 140, and 270 GHz. In particular, the three-dimensional corporate feeding network is introduced by taking advantage of the LTCC process.

In Chapter 4, "Broadband Metamaterial-Mushroom Antenna Array at 60 GHz Bands," by Wei Liu and Zhi Ning Chen, the techniques for enhancing the bandwidth of patch antennas are reviewed first. Then, the bandwidth enhancement techniques are evaluated for substrates of high dielectric constant. In particular, the metamaterial mushroom antenna technique are introduced for the

LTCC mmW antenna design due to the merits of low profile, broadband, high gain, high radiation efficiency, low mutual coupling, and low cross-polarization levels.

Narrow-wall-fed Substrate integrated Cavity (SIC) Antenna at 60 GHz is discussed in Chapter 5, by Yan Zhang. This chapter addresses the unique challenges of mmW SIC antenna design. At first, the mmW cavity antennas are reviewed. Then, the selected state-of-the-art SIC antennas are introduced and discussed. In particular, the technique to excite an SIC using a narrow-wall slot is elaborated, and furthermore a 2 × 2 array with narrow-wall-slot fed SIW at 60 GHz is presented as an example.

Chapter 6, entitled "Cavity-Backed SIW Slot Antennas at 60 GHz," by Ke Gong, introduces the history and milestones of the cavity-backed antennas (CBAs) first, together with some challenges for mmW applications. Then, the low-profile design methods and fabrication techniques about CBAs are analyzed, especially for the substrate integrated CBAs. After that, the low-profile SIW CBAs are reviewed. Their operating mechanisms are discussed, and methods for improving the performance, such as bandwidth enhancement, size reduction, and gain improvement, are presented. At last, a type of cavity-backed SIW slot antennas with big-aperture is presented as an example, with design details as a study case, including the antenna element. This type of antennas retains the advantage of conventional metallic CBAs, including high gain, high front-to-back ratio, and low cross-polarization level, and also keep the advantages of the planar antenna including low profile, light weight, low fabrication cost, and easy integration with the planar circuit.

In Chapter 7, "Circularly Polarized SIW Slot LTCC Antennas at 60 GHz," Yue Li first reviews the selected state-of-the-art techniques for mmW circularly polarized (CP) antennas. Second, as a feasible example to achieve both wide impedance and axial ration (AR) bandwidths, an SIW-fed slot antenna array with strip loading is introduced at 60 GHz using an LTCC substrate. The AR bandwidth enhancement property is systematically described with the potential adoption in various mmW CP applications.

Chapter 8, entitled "Gain Enhancement of LTCC Microstrip Patch Antenna by Suppressing Surface Waves," by Zhi Ning Chen and Xianming Qing, introduces the technology to suppress the surface wave losses caused by the thick and high-permittivity dielectric substrate in mmW antennas. First, the mechanism of generation of surface wave losses is discussed. Then the method to suppress the surface waves is addressed with an example of patch antenna. After that, the planar and via-less antenna array operating at 60 GHz is exemplified for gain enhancement by suppressing the surface wave losses. The method to suppress the surface wave is to cut open-air cavities around their radiating patches. The open-air cavities reduce the losses caused by severe surface waves and dielectric substrate at mmW bands. The arrays are excited through either a microstrip-line or stripline feed network with a grounded coplanar-waveguide (GCPW) transition. The GCPW transition is designed so that the antenna can be measured with the patch array facing free space, therefore reducing the effect of the probe station on the measurement. The proposed antenna arrays with the open-air cavities achieve gain enhancement of 1–2 dB compared with the conventional antenna array without any open-air cavity across the impedance bandwidth of about 7 GHz at the 60 GHz band.

In Chapter 9, "Substrate Integrated Antennas for Millimeter Wave Automotive Radars," Xianming Qing and Zhi Ning Chen introduce the PCB-based high-gain substrate integrated mmW antennas for car radar sensors. First, the general aspects of automotive radar are addressed including the classification, the frequency band regulation, system requirements, and antenna design considerations. Second, the selected state-of-the-art antenna designs for 24-GHz and 77-GHz automotive radars are reviewed. After that, two types of antenna arrays are introduced. A compact co-planar waveguide (CPW) center-fed SIW slot antenna array is elaborated to achieve narrow H-plane beamwidth and low sidelobe levels for 24-GHz automotive radars. A transmit-array on dual-layer PCB is introduced for automotive 77-GHz radar applications. With four SIW slot antennas as the primary feeds, the transmit-array is able to generate four switched beams. The coplanar structure significantly simplifies the transmit-array design and eases the fabrication, in particular, at mmW frequencies.

Chapter 10 is entitled "Sidelobe Reduction of Substrate Integrated Antenna Arrays at Ka-Band." Teng Li first introduces the synthesis technologies of low sidelobe array factors and the optimization methods. To accurately get the desired pattern, a brief analysis of mutual coupling is presented. Then, the selected state-of-the-art feeding technology for SIW array antenna are reviewed. After that, the examples of the small array, monopulse array, and shaped beam array with sidelobe reduction and different feeding technologies are introduced for SIW array antenna at Ka-band.

In Chapter 11, "Substrate Edge Antennas," Lei Wang and Xiaoxing Yin introduce substrate edge antennas (SEAs), which radiate from the edges of the PCBs. To diminish the mismatch between the PCB edge and the free space, two types of planar strips are printed in front of the SEA aperture. With the printed strips, both the impedance bandwidth and the front-to-back ratio are improved. Aiming at increasing the aperture efficiency, two kinds of substrate-integrated lenses are embedded in the SEAs. The phase-correcting lenses are integrated into the SEAs, maintaining the compact profiles of SEAs. Moreover, a leaky-wave SEA loaded with a prism lens is presented with a fixed-beam over a wide frequency band. The prism lens is implemented by utilizing a dispersive metasurface. By compensating for the dispersion of the leaky-wave SEA and the prism lens, fixed radiation beams are achieved over a 20% fraction bandwidth at Ka-band.

1.10 Summary

The research, development, and applications of mmW antennas have a long history. With the fast progress in device technologies and rapid deployment of system for a variety of commercial applications, theory, and technologies related to mmW antennas have been extensively investigated and developed [54–64]. This book will address the critical design challenges of mmW antennas for wireless communications and radar systems.

References

1 ITU-R Recommendation (2015). V.431: "Nomenclature of the frequency and wavelength bands used in telecommunications (Table I)." Geneva: International Telecommunication Union. https://www.itu.int/dms_pubrec/itu-r/rec/v/R-REC-V.431-8-201508-I!!PDF-E.pdf (accessed 19 December 2020).

2 Seybold, J.S. (2005). *Introduction to RF Propagation*, 3–10. Wiley.

3 Petty, K.R. and Mahoney, W.P. III, (2007). Weather applications and products enabled through vehicle infrastructure integration (VII). (Section 5) United States Department of Transportation – Federal Highway Administration Report No. FHWA-HOP-07-084. https://ops.fhwa.dot.gov/publications/viirpt/viirpt.pdf (accessed 19 December 2020).

4 Shannon, C.E. (1949). Communication in the presence of noise. *Proc. Inst. Rad. Eng.* 37 (1): 10–21.

5 Wiltse, J.C. (1984). History of millimeter and submillimeter waves. *IEEE Trans. Microwave Theory Tech.* 32 (9): 1118–1127.

6 Nichols, E.F. and Tear, J.D. (1923). Short electric waves. *Phys. Rev.* 21: 587–610.

7 Nichols, E.F. and Tear, J.D. (1923). Joining the infra-red and electric wave spectra. *Proc. Nat. Acad. Sci.* 9: 211–214.
8 Tear, J.D. (1923). The optical constants of certain liquids for short electric waves. *Phys. Rev.* 21: 611–622.
9 Cleeton, C.E. and Williams, N.H. (1934). Electromagnetic waves of 1.1cm wave-length and the absorption spectrum of ammonia. *Phys. Rev.* 45: 234–237.
10 Boot, H.A.H. and Randall, J.T. (1976). Historical notes on the cavity magnetron. *IEEE Trans. Electron Dev.* 23: 724–729.
11 Bennger, R. (1946). The absorption of one-half centimeter electromagnetic waves in oxygen. *Phys. Rev.* 70: 53–57.
12 Warters, W.D. (1977). WT4 millimeter waveguide system: introduction. *Bell Syst. Tech. J.* 56: 1925–1928.
13 Button, K.J. and Wiltse, J.C. (eds.) (1981). *Millimeter Systems*, vol. 4, (series on Infrared and Millimeter Waves). NY: Academic.
14 Schwartz, R.F. (1954). Bibliography on directional couplers. *IRE Trans. Microwave Theory Tech.* 2: 58–63.
15 Convert, G., Yeou, T., and Pasty, B. (1959). Millimeter-wave O-carcinotron. In: *Proceedings of Symposium on Millimeter Waves*, vol. IX, 313–339.
16 Wiltse, J.C. (1959). Some characteristics of dielectric image lines at millimeter wavelengths. *IRE Trans.* 7: 63–69.
17 Taub, J.J., Hindin, H.J., Hinckelmann, O.F., and Wright, M.L. (1963). Submillimeter components using oversize quasi-optical waveguide. *IEEE Trans. Microwave Theory Tech.* 11 (9): 338–345.
18 Richer, K.A. (1974). Near earth millimeter-wave radar and radiometry. *Proc. IEEE Int. Symp. Microw. Theorv Tech.*: 470–474.
19 Wiltse, J.C. (1979). Millimeter wave technology and applications. *Microw. J.* 22: 39–42.
20 Chang, K. and Sun, C. (1983). Millimeter-wave power-combining techniques. *IEEE Trans. Microwave Theory Tech.* 31 (2): 91–107.
21 Chen, Z.N. (ed.) (2016). *Handbook of Antenna Technologies*. Springer.
22 Balanis, C.A. (2016). *Antenna Theory: Analysis and Design*, 4e. Wiley.
23 Uchimura, H., Takenoshita, T., and Fujii, M. (1998). Development of a "laminated waveguide". *IEEE Trans. Microwave Theory Tech.* 46 (12): 2438–2443.
24 Takenoshita, T. and Uchimura, H. (1999). Laminated aperture antenna and multilayered wiring board comprising the same. EP20030026894 (Application number in 1998) and EP0893842B1(Grant number in 2004). https://patentimages.storage.googleapis.com/74/5e/eb/c2c2e031af31c7/EP0893842A2.pdf (accessed 19 December 2020).
25 Hirokawa, J. and Ando, M. (1998). Single-layer feed waveguide consisting of posts for plane TEM wave excitation in parallel plates. *IEEE Trans. Antennas Propag.* 46 (5): 625–630.
26 Deslandes, D. and Wu, K. (2001). Integrated microstrip and rectangular waveguide in planar form. *IEEE Microwave Wirel. Compon. Lett.* 11 (2): 68–70.
27 Yan, L., Hong, W., Hua, G. et al. (2004). Simulation and experiment on SIW slot array antennas. *IEEE Microwave Wirel. Compon. Lett.* 14 (9): 446–448.
28 Sebastian, M.T., Ubic, R., and Jantunen, H. (eds.) (2017). *Microwave Materials and Applications*. Wiley.
29 Sebastian, M. and Jantunen, H. (2008). Low loss dielectric materials for LTCC applications. *Int. Mat. Rev.* 53 (2): 57–90.
30 Ullah, U., Ain, M.F., Mahyuddin, N.M. et al. (2015). Antenna in LTCC technologies: a review and the current state of the art. *IEEE Antennas Propag. Mag.* 57 (2): 241–260.

31 Deslandes, D. and Wu, K. (2006). Accurate modeling, wave mechanisms, and design considerations of a substrate integrated waveguide. *IEEE Trans. Microwave Theory Tech.* 54 (6): 2516–2526.
32 She, Y., Tran, T.H., Hashimoto, K. et al. (2011). Loss of post-wall waveguides and efficiency estimation of parallel-plate slot arrays fed by the post-wall waveguide in the millimeter-wave band. *IEICE Trans. Electron.* E94-C (3): 312–320.
33 Gatti, F., Bozzi, M., Perregrini, L. et al. (2006). A novel substrate integrated coaxial line (SICL) for wideband applications. In: *Proceedings of the 36th European Microwave Conference*, 1614–1617.
34 Shao, Y., Li, X.-C., Wu, L.-S., and Mao, J.-F. (2017). A wideband millimeter-wave substrate integrated coaxial line array for high-speed data transmission. *IEEE Trans. Microwave Theory Tech.* 65 (8): 2789–2800.
35 Zhu, F., Hong, W., Chen, J.-X., and Wu, K. (2012). Ultra-wideband single and dual baluns based on substrate integrated coaxial line technology. *IEEE Trans. Microwave Theory Tech.* 60 (10): 3062–3070.
36 Yang, T.Y., Hong, W., and Zhang, Y. (2016). An SICL-excited wideband circularly polarized cavity-backed patch antenna for IEEE 802.11aj (45 GHz) applications. *IEEE Antennas Wirel. Propag. Lett.* 15: 1265–1268.
37 Liu, B., Xing, K.J., Wu, L. et al. (2017). A novel slot array antenna with substrate integrated coaxial line technique. *IEEE Antennas Wirel. Propag. Lett.* 16: 1743–1746.
38 Miao, Z.-W. and Hao, Z.-C. (2017). A wideband reflectarray antenna using substrate integrated coaxial true-time delay lines for QLink-pan applications. *IEEE Antennas Wirel. Propag. Lett.* 16: 2582–2585.
39 Xing, K., Liu, B., Guo, Z. et al. (2017). Backlobe and sidelobe suppression of a Q-band patch antenna array by using substrate integrated coaxial line feeding technique. *IEEE Antennas Wirel. Propag. Lett.* 16: 3043–3046.
40 Liang, W. and Hong, W. (2012). Substrate integrated coaxial line 3 dB coupler. *IET Electron. Lett.* 48 (1): 35–36.
41 Chu, P. et al. (2014). Wide stopband bandpass filter implemented with spur stepped impedance resonator and substrate integrated coaxial line technology. *IEEE Microwave Wirel. Compon. Lett.* 24 (4): 218–220.
42 Zhang, J., Zhang, X., and Shen, D. (2016). Design of substrate integrated gap waveguide. *IEEE MTT-S Int. Microwave Symp. Dig.*: 1–4.
43 Zhang, J., Zhang, X., Shen, D., and Kishk, A.A. (2017). Packaged microstrip line: a new quasi-TEM line for microwave and millimeter-wave applications. *IEEE Trans. Microwave Theory Tech.* 65 (3): 707–718.
44 Cao, B., Wang, H., Huang, Y., and Zheng, J. (2015). High-gain L-probe excited substrate integrated cavity antenna array with LTCC-based gap waveguide feeding network for W-band application. *IEEE Trans. Antennas Propag.* 63 (12): 5465–5474.
45 Cao, B., Wang, H., and Huang, Y. (2016). W-band high-gain TE220 -mode slot antenna array with gap waveguide feeding network. *IEEE Antennas Wirel. Propag. Lett.* 15: 988–991.
46 Dadgarpour, A., Sorkherizi, M.S., and Kishk, A.A. (2016). Wideband low-loss magnetoelectric dipole antenna for 5G wireless network with gain enhancement using meta lens and gap waveguide technology feeding. *IEEE Trans. Antennas Propag.* 64 (12): 5094–5101.
47 Sorkherizi, M.S., Dadgarpour, A., and Kishk, A.A. (2017). Planar high-efficiency antenna array using new printed ridge gap waveguide technology. *IEEE Trans. Antennas Propag.* 65 (7): 3772–3776.

48 Bayat-Makou, N. and Kishk, A. (2017). Millimeter-wave substrate integrated dual level gap waveguide horn antenna. *IEEE Trans. Antennas Propag.* 65 (12): 6847–6855.

49 Dadgarpour, A., Sorkherizi, M.S., Denidni, T.A., and Kishk, A.A. (2017). Passive beam switching and dual-beam radiation slot antenna loaded with ENZ medium and excited through ridge gap waveguide at millimeter-waves. *IEEE Trans. Antennas Propag.* 65 (1): 92–102.

50 Zhang, J., Zhang, X., and Kishk, A.A. (2018). Broadband 60 GHz antennas fed by substrate integrated gap waveguides. *IEEE Trans. Antennas Propag.* 66 (7): 3261–3270.

51 Shen, D., Ma, C., Ren, W. et al. (2018). A low-profile substrate-integrated-gap-waveguide-fed magnetoelectric dipole. *IEEE Antennas Wirel. Propag. Lett.* 17: 1373–1376.

52 Yeap, S.B., Chen, Z.N., Li, R. et al. (2012). 135-GHz co-planar patch array on BCB/silicon with polymer-filled cavity. *Int. Workshop Antennas Tech.*: 1–4.

53 3GPP TS 38.101–2 v15.2, online available: https://www.3gpp.org (accessed 19 December 2020).

54 Chen, Z.N., Chia, M.Y.W., Gong, Y. et al. (2011). Microwave, millimeter wave, and Terahertz technologies in Singapore. In: *Proceedings of the 41st European Microwave Conference*, 1–4.

55 Chen, Z.N. et al. (2012). Research and development of microwave & millimeter-wave technology in Singapore. In: *Proceedings of the 42nd European Microwave Conference*, 1–4.

56 Li, T., Meng, H.F., and Dou, W.B. (2014). Design and implementation of dual-frequency dual-polarization slotted waveguide antenna array for Ka-band application. *IEEE Antennas Wirel. Propag. Lett.* 13: 1317–1320.

57 Mao, C.-X., Gao, S., Luo, Q. et al. (2017). Low-cost X/Ku/Ka-band dual-polarized array with shared-aperture. *IEEE Trans. Antennas Propag.* 65 (7): 3520–3527.

58 Wang, Z., Xiao, L., Fang, L., and Meng, H. (2014). A design of E/Ka dual-band patch antenna with shared aperture. In: *Proceedings of the Asia-Pacific Microwave Conference*, 333–335.

59 Han, C., Huang, J., and Chang, K. (2005). A high efficiency offset-fed X/Ka-dual-band reflectarray using thin membranes. *IEEE Trans. Antennas Propag.* 53 (9): 2792–2798.

60 Hsu, S.-H., Han, C., Huang, J., and Chang, K. (2007). An offset linear-array-fed Ku/Ka dual-band reflectarray for planet cloud. *IEEE Trans. Antennas Propag.* 55 (11): 3114–3122.

61 Chaharmir, M. and Shaker, J. (2015). Design of a multilayer X-/Ka-band frequency-selective surface-backed reflectarray for satellite applications. *IEEE Trans. Antennas Propag.* 63 (4): 1255–1262.

62 Attia, H., Abdelghani, M.L., and Denidni, T.A. (2017). Wideband and high-gain millimeter-wave antenna based on FSS Fabry–Perot cavity. *IEEE Trans. Antennas Propag.* 65 (10): 5589–5594.

63 Li, T. and Chen, Z.N. (2020). Wideband Sidelobe-level reduced Ka-band Metasurface antenna Array fed by substrate integrated gap waveguide using characteristic mode analysis. *IEEE Trans. Antennas Propag.* 68 (3): 1356–1365.

64 Hong, W. et al. (2017). Multibeam antenna technologies for 5G wireless communications. *IEEE Trans. Antennas Propag.* 65 (12): 6231–6249.

2

Measurement Methods and Setups of Antennas at 60–325 GHz Bands

Xianming Qing[1] *and Zhi Ning Chen*[2]

[1] *Signal Processing, RF, and Optical Department, Institute for Infocomm Research, Singapore 138632, Republic of Singapore*
[2] *Department of Electrical and Computer Engineering, National University of Singapore, Singapore 117583, Republic of Singapore*

2.1 Introduction

Measurement is basic and crucial for characterizing an antenna to ensure the design to meet the specifications, to validate simulation or design, and to verify the construction and fabrication process of the designed antenna. Antenna metrology requires not only sound background knowledge in both antenna theory and ample experience in antenna engineering but also sophisticated and expensive equipment capable of providing the acceptable accuracy and purity of measured data. Many antenna measurement methods have been developed even before World War II. Commercial systems specifically designed for a variety of antenna measurement have become available since the 1960s, in particular, due to the strong demand of the aerospace, space, and defense industries. New approaches and advanced technologies of antenna measurement have continuously emerged with the rapid growth of modern mobile communications in the last decades [1–5].

The modern wireless systems such as 5G wireless communications systems and Internet of Things (IoT) are bringing a growing need for electrically small and large, low- and large-gain antennas, also at mmW bands. Frequencies of 24, 28, 38, 58, 60, 71–86, 94, 120–150 GHz, and some even higher "window" frequencies have attracted increasing interest in radar, imaging, and communication systems because of their wide bandwidths and short wavelengths for potential high-resolution and high-data-rates applications [6–9]. Compared with the measurement at lower microwave frequencies, the mmW antenna measurement is much more complicated, expensive, and tedious [10–14]. First, the much larger loss of the transmission lines such as coaxial cables and waveguides at mmW bands and limited output power as well as the sensitivity of the mmW modules result in a limited dynamic range, which is critical for the measurement of the antennas with low gain. Second, the cable, connector and/or on-wafer probe used to connect the antenna under test (AUT) are of comparable size or even larger than the AUT, which shows non-ignorable effect on the measurement results, in particular, the radiation performance. Third, the mmW measurement setups are always expensive. In addition, the commercially available mmW high performance devices/components such as high-power amplifier, low-noise power amplifier, joint rotator, and so on, are very limited, which makes the system configuration more challenging.

The mmW antenna measurement techniques can be categorized as far- and near-field methods. In the far-field antenna measurement system, the AUT is illuminated by a uniform plane wave, and the antenna parameters are measured directly. In a near-field antenna measurement system, the

Substrate-Integrated Millimeter-Wave Antennas for Next-Generation Communication and Radar Systems, First Edition.
Edited by Zhi Ning Chen and Xianming Qing.

distribution of near-field from the AUT is measured using a probe, and the far-field characteristics are obtained through near-field to far-field transformation.

2.1.1 Far-Field Antenna Measurement Setup

The far-field measurement can be performed in either outdoor or indoor ranges. In general, there are two typical types of far-field antenna ranges, that is, reflection range and free space range. In the ground reflection range, the specular reflection by the ground is used to obtain a uniform phase and amplitude distribution over the AUT. In the free space ranges, the reflections by the environment are minimized. The typical free space ranges are elevated range, slant range, anechoic chamber, and compact range, wherein the setups using anechoic chambers [15] and compact ranges are the most applicable ones for mmW antenna measurement.

2.1.1.1 Free-Space Range Using Anechoic Chamber

For far-field antenna measurement, the AUT is required to be illuminated by a uniform plane wave so that a signal source and a transmit antenna with desired radiation characteristics are needed. The block diagram of the far-field antenna measurement system is shown in Figure 2.1; the main parts of the systems include a signal source, a transmitting antenna, a receiver, a positioning system (turntable), and a recording system/data processing system. Apart from the separate signal source and receiver configuration, the vector network analyzer (VNA) has been applied in antenna measurement systems for many years since it exhibits excellent phase stability, spectral purity, sensitivity, and dynamic range (well in excess of 100 dB), which makes it well suited for antenna measurement.

The anechoic chambers are the most popular antenna measurement sites. Compared with the outdoor ranges, the anechoic chamber is able to provide a controlled environment, all-weather

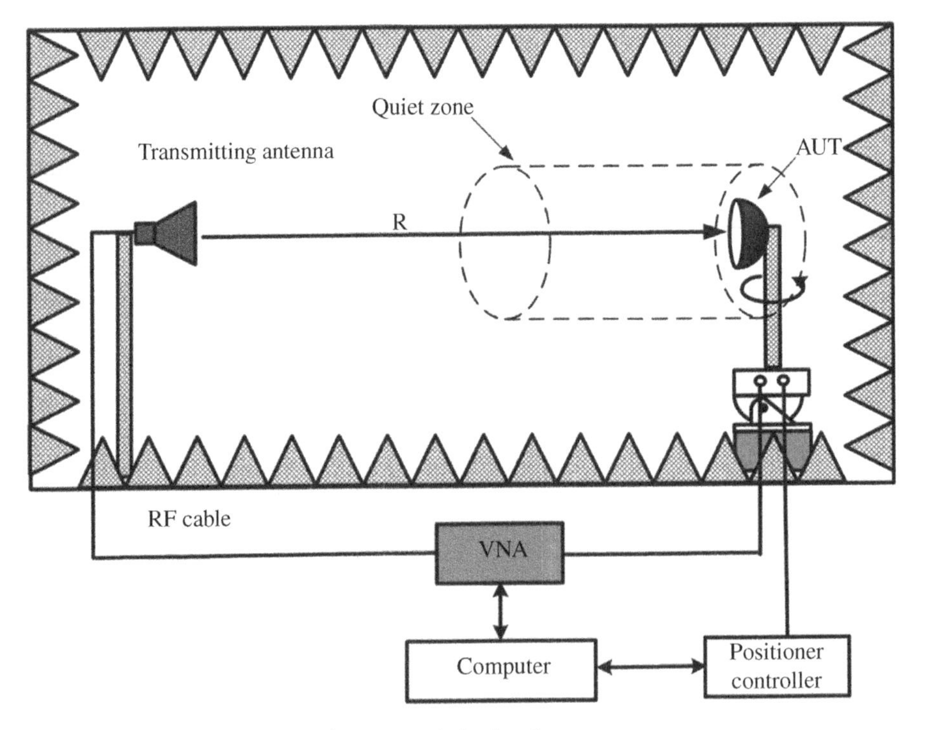

Figure 2.1 Free-space range using an anechoic chamber.

capability, and security and minimize possible electromagnetic interference for antenna measurement. In the test zone, the amplitude and phase of the incident plane wave from the transmitting antenna exhibit minimum variation, such as 0.2 dB and 0.5°, respectively. In general, The RF absorber used for microwave frequencies possesses improved performance at mmW bands. As the operating frequency is increased, thinner RF absorbing materials are able to maintain a given level of reflectivity performance [15].

In case of electrically large antennas, the classical far-field method has some major obstacles. The separation d between the AUT and the transmitting antenna must meet the far-field requirement, namely, $d \geq 2D^2/\lambda$, where D is the maximum dimension of the AUT and λ is the measurement wavelength. The far-field method is readily applicable for mmW antennas with gain of less than 30 dBi wherein the far-field distance is less than about 1 m; the measurement can be done in a controlled environment such as an anechoic chamber. For an antenna of higher gain, for example, a parabolic reflector antenna with diameter of 0.5 m at 60 GHz with gain of about 49 dBi (aperture efficiency of 80%), the required far-field distance is 100 m, carrying out the desired measurement is extremely challenging considering the large free space transmission attenuation and the loss from the RF cables.

2.1.1.2 Compact Range

The alternative far-field solution for electrically large antenna measurement is the compact antenna test range (CATRs), which is a collimating setup that generates nearly planar wave fronts in a very short distance [16, 17]. As exhibited in Figure 2.2, in the compact test range, a transmitting antenna is used as an offset primary feed to illuminate a paraboloidal reflector, which converts the impinging spherical waves into plane waves. This is very similar to the principle on which a dish antenna operates. Compact range configurations are often designated according to their analogous reflector antenna configurations: parabolic, Cassegrain, Gregorian, and so on [18–20]. The AUT is placed into the quiet-zone (QZ). Typically, the maximum amplitude and phase deviations have to be less than ±0.5 dB and ±5°, respectively. The major problem is the very stringent surface accuracy requirement at mmW frequencies; for example, the root-mean-square (rms) surface error should be less than 0.01λ, for instance, 40 μm at 75 GHz [11].

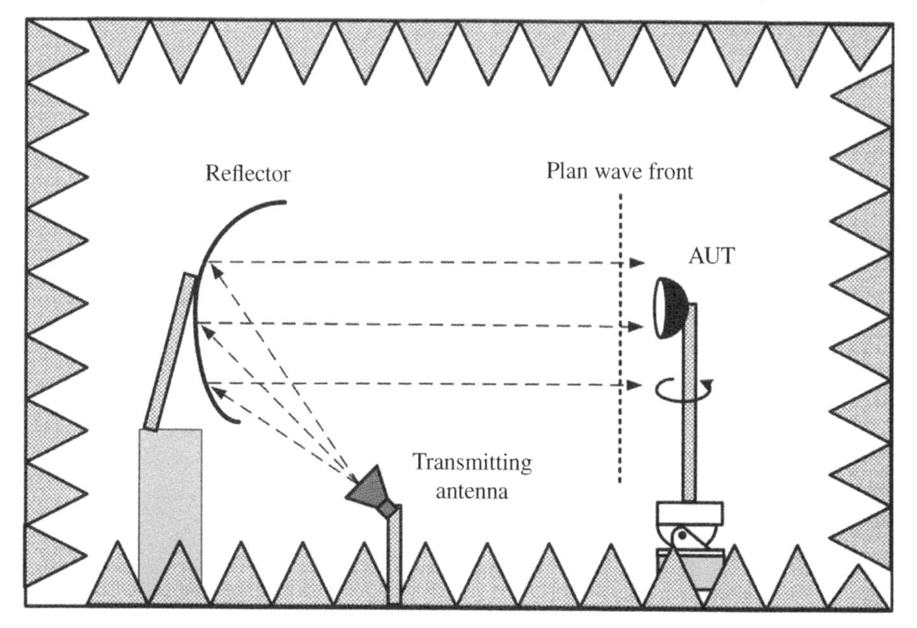

Figure 2.2 Compact antenna test range.

In general, the AUT in the conventional far-field measurement setups is positioned on a turntable and rotated for receiving the signal from the transmitting antenna that is kept stationary. However, such traditional setups are with the feeding mechanism limitations, in particular, not suitable for the on-wafer antenna or antenna in/on package, where the AUT is fed by an on-wafer probe and mounted on a probe station and thus cannot be rotated. For such cases, the transmitting antenna is required to be installed on an arm and rotated around the AUT instead.

2.1.2 Near-Field Antenna Measurement Setup

The antenna radiation performance can be characterized by measuring the near-field distribution of the AUT. The amplitude and phase of the AUT near field are sampled with a probe, and the far-field radiation characteristics are calculated using numerical transformation methods [21–24]. The small probe scans over a surface surrounding the AUT. Typically, the separation between the probe and the AUT is in the order of 4–10 wavelengths. During the measurement, the phase and amplitude of the near field is collected over a discrete matrix of points. The data is then transformed to the far field using the Fourier transform. In near-field testing, the AUT is usually aligned to the scanner's coordinate system and then either the probe or the test antenna is moved. In practice, it is easier and more cost effective to scan the RF probe over linear axes or the AUT over angular axes. In principle, the measurement can be done over a surface, which can be defined in any of the six orthogonal coordinate systems such as rectangular, cylindrical (circular-cylindrical), spherical, elliptic-cylindrical, parabolic-cylindrical, and conical. However, only the first three are deemed convenient for data acquisition, and of them, the complexity of the analytical transformations increases from the planar to the cylindrical and then to the spherical surfaces.

In case of mmW antennas, the sampling of the near-field is normally accomplished over a plane as shown in Figure 2.3. The sampling area has to be comprehensive so that all the significant energy transmitted by the AUT can be captured. The sampling has to be dense enough for satisfying the Nyquist sampling criterion, that is, the distance between two sampling points has to be smaller than one half of the wavelength. Major error sources in the near-field scanning are the phase errors in

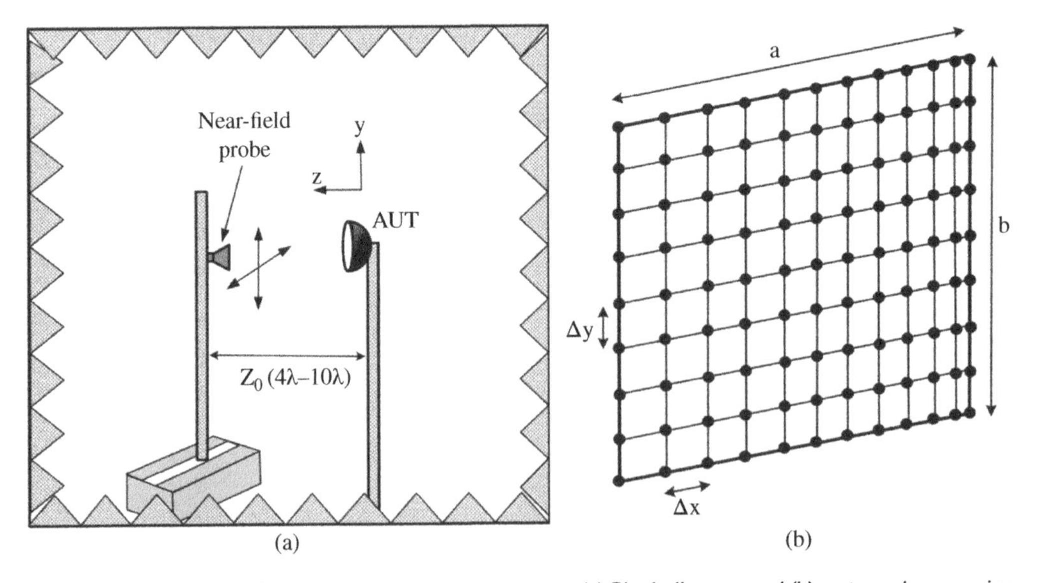

Figure 2.3 Planar near-field antenna measurement setup. (a) Block diagram and (b) rectangular scanning surface.

bending and twisting cables, inaccurate positioning of the probe, and the reflections between the probe and the AUT.

2.2 State-of-the-Art mmW Measurement Systems

2.2.1 Commercially Available mmW Measurement Systems

ETH-MMW-1000 (18–75 GHz) [25]: The ETH-MMW-1000 from Ethertronics Inc. is a fully anechoic mmW far-field measurement system capable of testing antennas from 18 to 75 GHz. The system is self-contained, moveable, and compact to fit into a laboratory or production environment. As shown in Figure 2.4, the ETH-MMW-1000 includes a distributed axis positioning system, consisting of an azimuth mast rotator for rotating the AUT about the phi axis and a theta ring positioner for elevating the horns around the AUT. Different mmW bandwidths of 18–26.5, 26.5–40, 33–50, 40–60, 50–67, and 50–75 GHz are applicable. Each measurement bandwidth is covered by dedicated paths (RF cables, rectangular waveguides, measurement horns), associated with a common amplification stage. The full anechoic enclosure provides a shielded environment and ensures stable measurement results. The system is suitable for the AUT with a coaxial or waveguide feeding structure but has a co-planar waveguide (CPW) structure for on-wafer measurement. The key parameters of the ETH-MMW-1000 are summarized in Table 2.1.

μ-Lab (50–110 GHz) [26]: As shown in Figure 2.5, the μ-Lab is developed by Microwave Vision Group (MVG) for collecting far- and near-field electromagnetic data of on-wafer antennas and miniature antennas at the mmW range from 50 to 110 GHz. The system is designed for convenient manual changeover and applicable for on-wafer antenna measurement. The positioning subsystem consists of a lightweight precision gantry arm assembly mounted on an azimuth positioner.

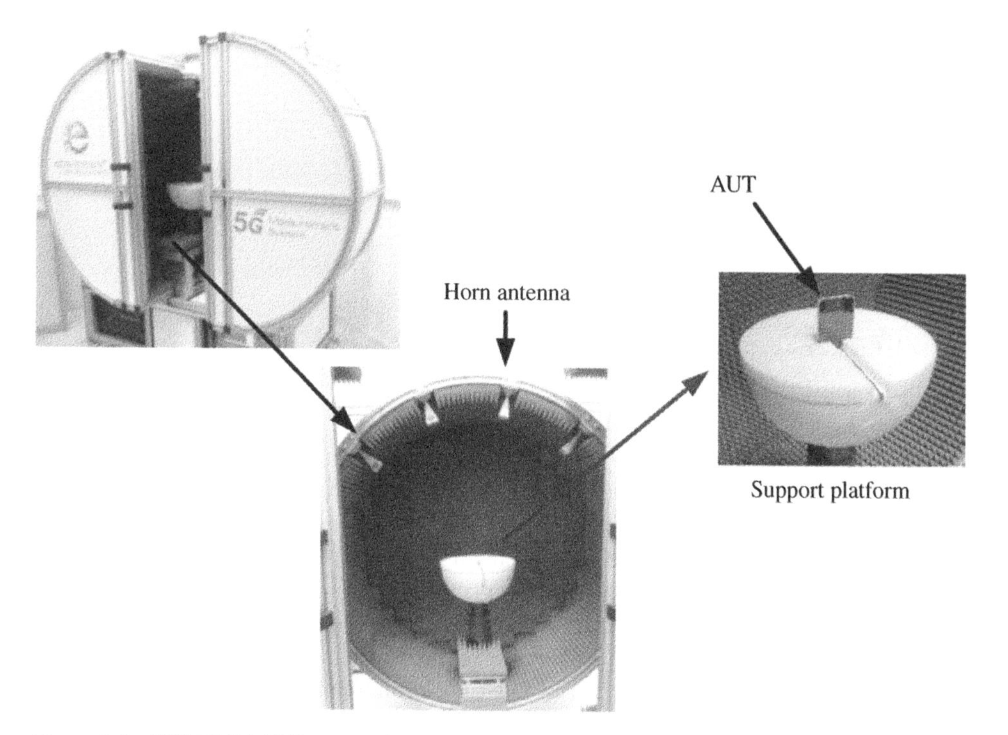

Figure 2.4 ETH-MMW-1000 system (courtesy of Ethertronics Inc.).

Table 2.1 Main features of the ETH-MMW-1000.

Technology	Far-field / Spherical with oversampling
Frequency Range	18–75 GHz
Measurement Radius	NA
Gain Accuracy	+/− 0.9 dB typical
Typical Dynamic Range	>50 dB, 40–67 GHz >55 dB, 20–40 GHz
Dimensions	1.56 m (L) × 1.25 m (W) × 2.12 m (H)
Maximum Size of DUT	45 cm
Maximum Mass of DUT	10 kg

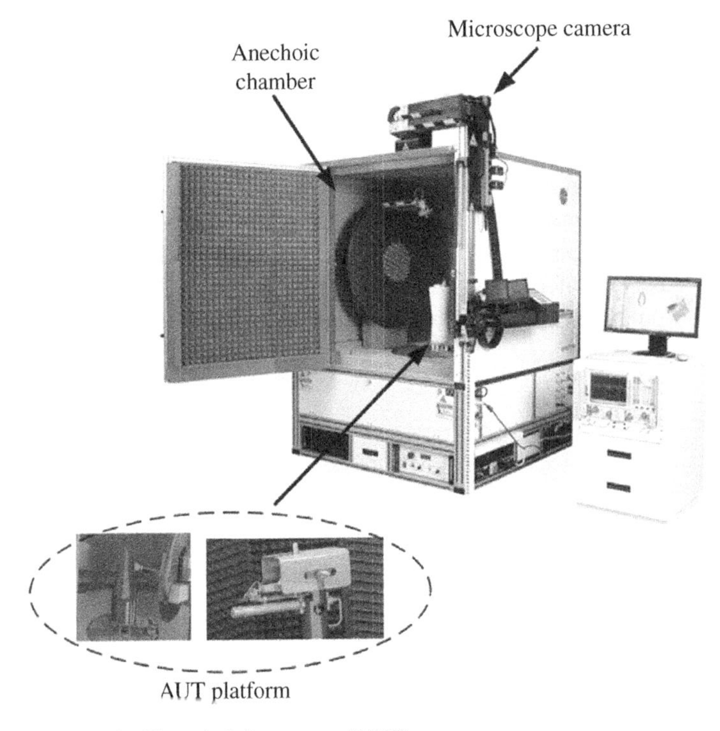

Figure 2.5 The μ-Lab (courtesy of MVG).

The near-field probe, mounted on the gantry arm, can be rotated to change polarization. The gantry arm assembly rotates in azimuth to cover all the longitudinal cuts on the measurement sphere. The AUT remains fixed on a stationary disk while the probe rotates in elevation and azimuth around the AUT to cover the measurement sphere. Measurement bands are reconfigurable to allow wide bandwidth operation of the system. The key parameters of the μ-Lab are summarized in Table 2.2.

Mini-Compact Range (4 GHz – 110 GHz) [27]: As shown in Figure 2.6, the mini-compact range and chamber assembly developed by MVG enable cost effective testing of microwave and mmW antennas, with a quiet zone diameter up to 0.5 m. The basic configuration allows full 3-D patterns to be collected using standard vector network analyzers. A compact range feed polarization rotator enables the transmit polarization to be changed during a single test or in between tests. Linked

Table 2.2 Main features of the μ-Lab.

Technology	Near-field/Spherical Far-field/Spherical
Frequency Range	50–110 GHz
Measurement Radius	15 in. (38.1 cm) nominal
Gain Accuracy	+/− 0.5 dB
Typical Dynamic Range	>60 dB
Dimensions	1.52 m (L) × 1.52 m (W) × 2.13 m (H)
Maximum Size of DUT	On centered support column: as large as a standard laptop On offset column for chip measurements: 5 cm × 5 cm (chipset)
Maximum Mass of DUT	10 kg on the mast

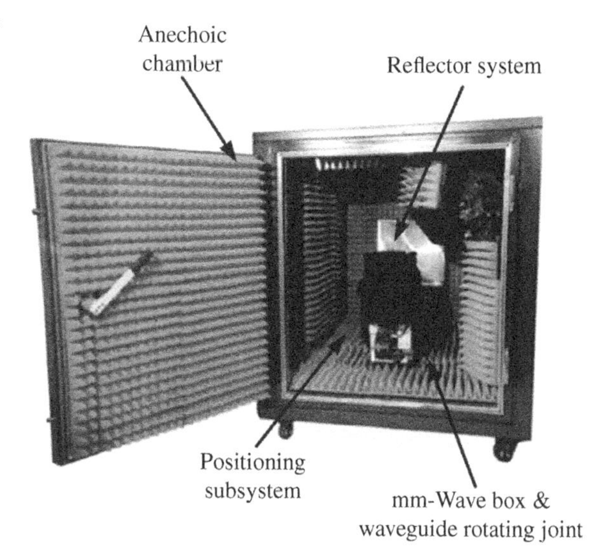

Figure 2.6 Mini-compact range (courtesy of MVG).

axis motion of the transmit rotator and roll axis allows for automatic acquisition of E- and H-plane patterns in a single test. A squint (elevation) axis allows E- and H-plane patterns through the peak of the beam in case electrical and mechanical boresight do not coincide. The system is not applicable for on-wafer antenna measurement. The key parameters of the Mini-Compact Range are summarized in Table 2.3.

2.2.2 Customized mmW Measurement Systems

Current commercial measurement setups are desired for mmW antenna characterization up to 110 GHz, while they are not widely utilized, in particular in universities and research organizations. This is partially attributed to the lack of the on-wafer antenna measurement capability, in particular, at the early (late 1990s). Numerous customized mmW antenna measurement setups have been reported for various purposes at frequencies up to 950 GHz. In this section, the typical mmW antenna measurement setups are overviewed, noting that most of the setups are applicable for the on-wafer antenna measurement.

Table 2.3 Main features of the mini-compact range.

Technology	Compact range
Frequency Range	4–110 GHz
Quiet zone diameter	0.5 m
Gain Accuracy	+/− 0.5 dB
Typical Dynamic Range	>80 dB
Dimensions	2.9 m (L) × 2.4 m (W) × 4.0 m (H)
Maximum Size of DUT	Up to 0.5 m diameter
Maximum Mass of DUT	Up to 45 kg for Azimuth (AZ) Only Up to 13 kg for Roll/AZ (according to EL axis) Up to 23 kg for Roll/AZ with AL-161-1P

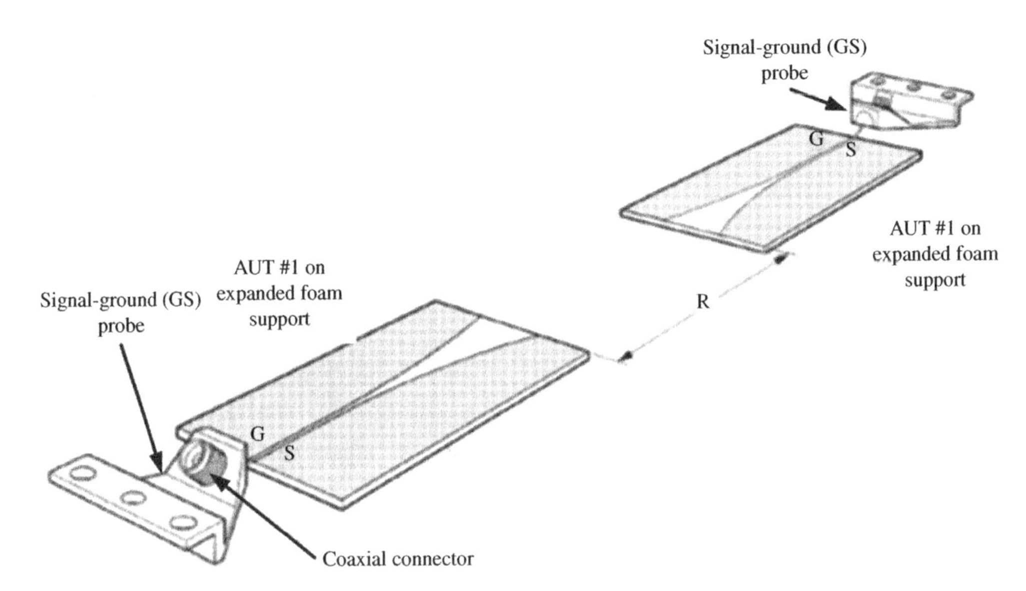

Figure 2.7 Gain measurement of two identical antennas on a probe station [28].

In 1999, Simons and Lee demonstrated the first experimental setup for mmW antenna measurement using on-wafer probes [28]. Figure 2.7 shows the configuration of the gain measurement for tapered slot antennas. In this setup, the metallic chuck has been replaced by a block of foam to avoid unwanted metallic reflections, and all the surrounding unavoidable metallic parts have been covered with microwave absorbers. Two identical antennas are placed at a distance *R* from each other, and the gain is computed from the measured S_{21} using the Friis equation. The antenna-to-antenna distance is larger than the far-field distance. Using a calibration substrate, the feeding probe of the AUT is calibrated to the tip of the probe. With a pair of ground-signal (GS) microwave probes (Picoprobe Model 40 A, pitch = 10 mil), a wafer probe station (Cascade Model 42), and a vector network analyzer (HP8510C), the system is able to measure the input impedance matching and the gain of the antenna accurately. However, the first setup is only applicable for gain measurement of the end-fire antenna and unable to measure the radiation pattern because the AUTs are all stationary.

To carry out on-wafer antenna radiation pattern measurement, several simple far-field measurement setups are proposed as shown in Figure 2.8. In these setups [29–31], the AUT is positioned

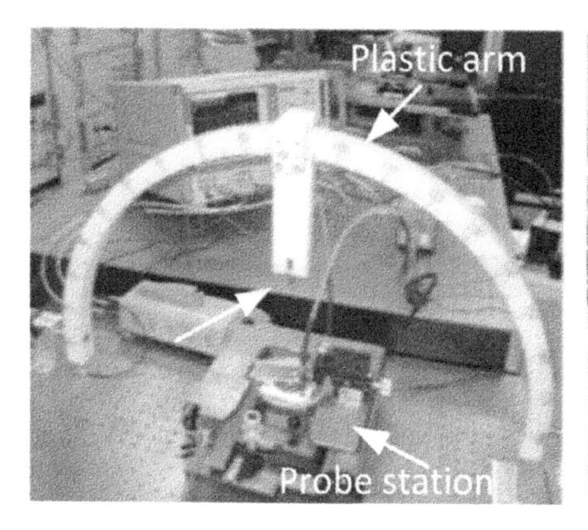

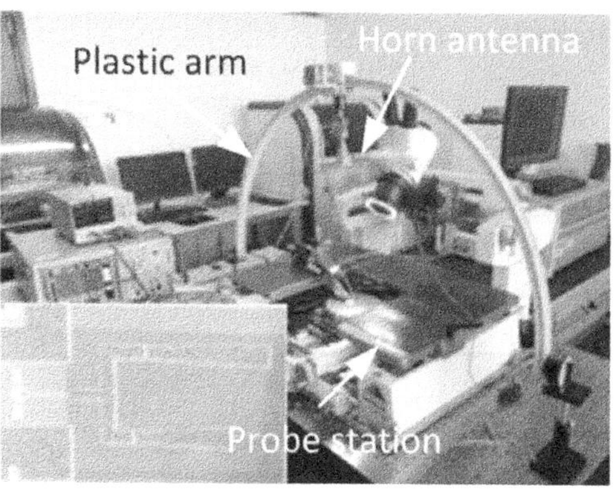

Figure 2.8 Simplified manual on-wafer antenna measurement setup [29–31].

on a metallic chuck and fed by a ground-signal-ground (GSG) probe, which is connected to the VNA through a coaxial cable. A horn antenna is mounted on a plastic holder and rotates along a plastic arc manually at a distance away from the AUT. The hemispherical radiation patterns can be measured by moving the horn along the θ-direction. The drawbacks of the setups are obviously the reflections by the environment including the metallic chuck, the probe positioner, and the mmW module for frequencies above 67 GHz, which seriously distort the measured patterns. A full hemispherical pattern is unable to be measured, only a few cuts are applicable since the AUT is stationary. In addition, the blockage of the probe/module positioner limits the measurement angles as well.

Pilard et al. reported an improved measurement setup as shown in Figure 2.9 [32]. The AUT is placed on a rotating Rexolite chuck. A 3-axis micromanipulator allows the connection between the AUT and the network analyzer, such as Agilent PNA E8361A, through an on-wafer probe mounted on a probe holder. A standard-gain horn is moved along a Rexolite arch with a two-degree step to depict a quarter of a circle around the AUT, leading to the determination of the elevation radiation pattern. The radius R of the arch is chosen such that the AUT is placed at a constant far-field distance from the horn. The horn and the probe are connected to the network analyzer by coaxial cables. With the rotation of the chuck, the pattern measurements in the hemisphere above the AUT can be carried out. The cross-polarization levels in each plane can be measured by a 0°–90° rotation of the horn. Furthermore, the allocation of the system in an anechoic chamber and covering the micromanipulator and the chuck with RF absorbers makes the so-called parasitic reflections minimized.

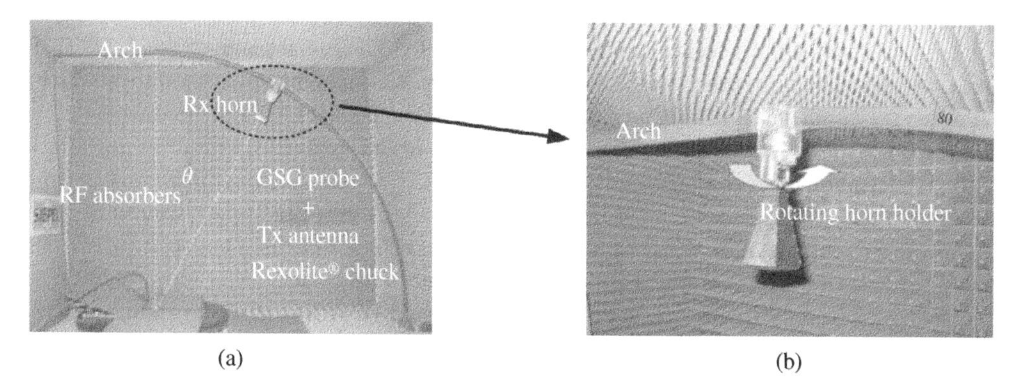

Figure 2.9 (a) Overview of the system configuration and (b) standard-gain horn on the arch [32].

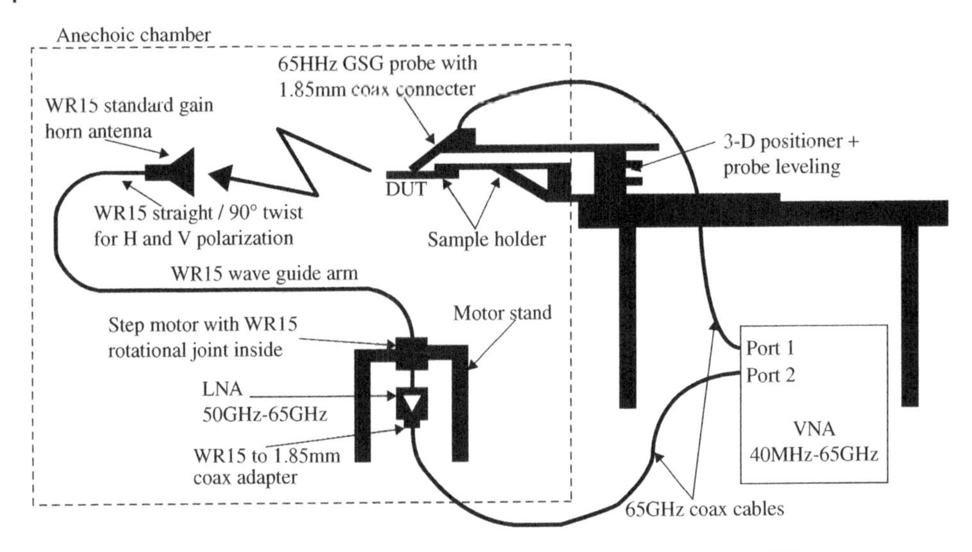

Figure 2.10 Automatic on-wafer mmW antenna measurement setup diagram [33].

Antenna radiation pattern measurement is a tedious and time-consuming task. Therefore, an automatic antenna measurement setup is always preferable for antenna researcher/engineers. In 2004, Thomas et al. demonstrated an automatic on-wafer mmW antenna test setup that is able to measure radiations patterns with a ±90° range in three different planes [33]. As shown in Figure 2.10, The AUT is held by a customized sample holder that reaches into the anechoic chamber from an anti-vibration table. A second arm, with the microwave probe, is mounted on a probe positioner with additional probe leveling, which also sits on the anti-vibration table. An arm with WR15 waveguide sections and a standard horn antenna rotates around the AUT at a distance of 38 cm to ensure a far-field condition. A WR15 rotational joint with a low-noise amplifier (LNA) is mounted in the center of the stepper motor to avoid any cable bending due to motor movement and to enhance the sensitivity for the gain measurement.

To further enlarge the measurement angle range of radiation pattern, the on-wafer-based setups for nearly 3-D radiation pattern are proposed [34, 35]. As exhibited in Figure 2.11, the AUT holder is attached to a plastic arm whose other end is attached to an anti-vibration table. The stationary AUT that is placed upward into free space transmits waves upward, and the receiving horn mounted on the two-axis turntable receives waves at various observation angles for the radiation pattern measurements. The two-axis turntable and two arms are used to rotate the receiving horn around the AUT, and a bigger rotary stage was positioned onto the floor, exactly in the axis below the AUT, and thus the second, smaller rotary stage is positioned in the horizontal plane of the AUT. The two rotary stages, which are consistent with the azimuth and elevation angles, allow it to measure either nearly the complete 3-D radiation pattern or three sectional planes without circumstantially rearranging the setup. The attachment of the measurement horn is rotatable around 90° to realize the pattern measurement of two orthogonal polarizations. The rotating angle of the arms is partly blocked by the table such that the horizontal plane can only be measured within 270° and one vertical plane only within 255°. A 60 cm measurement distance is achieved for both setups, which is sufficiently large for most of antenna arrays at 60 GHz and above.

In general, a probe station with an anti-vibration table is necessary for an on-wafer antenna measurement setup. However, the huge metallic platform causes severe effects on measured antenna impedance matching and radiation characteristics. To overcome this issue, some on-wafer antenna measurement setups without using probe stations are presented [36, 37]. Titz et al. presented a

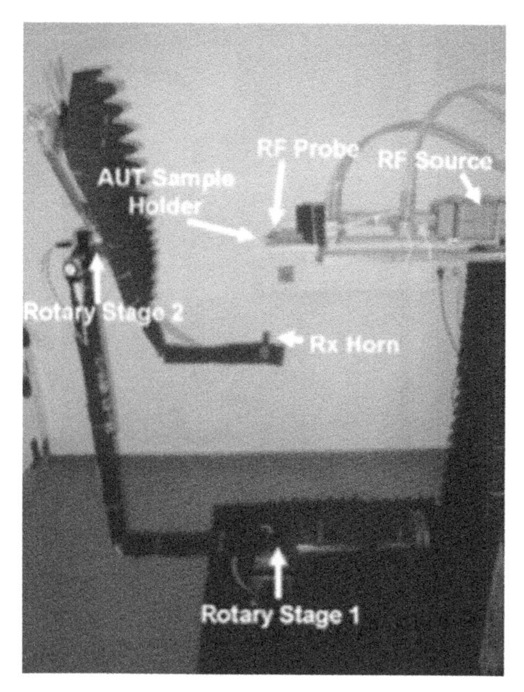

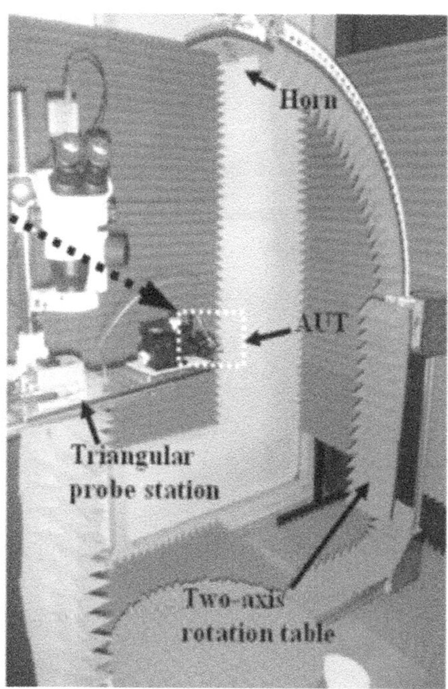

Figure 2.11 Nearly 3-D radiation pattern measurement setups [34, 35].

60-GHz on-wafer antenna measurement system as shown in Figure 2.12. Different from conventional on-wafer antenna measurement setups, the AUT is not maintained on the metallic chuck but on a specific foam holder with $\varepsilon_r \approx 1$. This foam is 1 cm thick and sufficiently rigid and horizontally planar, and a cavity is cut into the foam to ensure fitting the AUT properly. The holder is screwed to a special carrier fabricated in a rigid polyurethane material, and the carrier is attached to a metallic three-dimensional positioner, classically used in a probe station. This carrier greatly helps in leaving the space around the AUT free of any objects. The mechanical parts fit in less than 1 m^3. A very heavy and stable table is used to avoid any vibration during the measurement and thus any damage to the probe. The table is purposely cut to let the rotating arms turn around the AUT. In addition, the table is covered with the absorbers to avoid any reflections.

Mosalanejad et al. proposed an antenna measurement setup with a backside probing technique as shown in Figure 2.13. At the center of this setup, there is a metal table, which is used as a reflector and also as a holder for the AUT, micro positioner, and the probe. Four commercial vibration dampers are mounted around the table to isolate it from any vibration that comes from the fixed rail or the holding frame. A scanning arm fabricated from non-conducting material is utilized to carry the horn antenna and rotate it around the AUT. This arm is motorized with a step motor and rotated in elevation to measure the radiation pattern in a specified plane cut. This elevation scanning (along θ) can be done over the range from $-90°$ to $90°$. The selection of the desired plane cut can be done manually, by pushing/pulling the carriage along the fixed rail. This fixed rail covers more than 180° in azimuth (along the φ direction) to ensure complete scanning of the bottom hemisphere. The setup is able to measure the 3-D radiation pattern in any cut of a full hemisphere. In addition, compared with a conventional setup, it is claimed a 20 dB increment in measurement dynamic range can be achieved by shielding the probe from the measurement hemisphere.

Because of the limited measurement distance, the far-field mmW antenna measurement setup is generally applicable for low-gain antenna elements or smaller antenna arrays. For large antenna

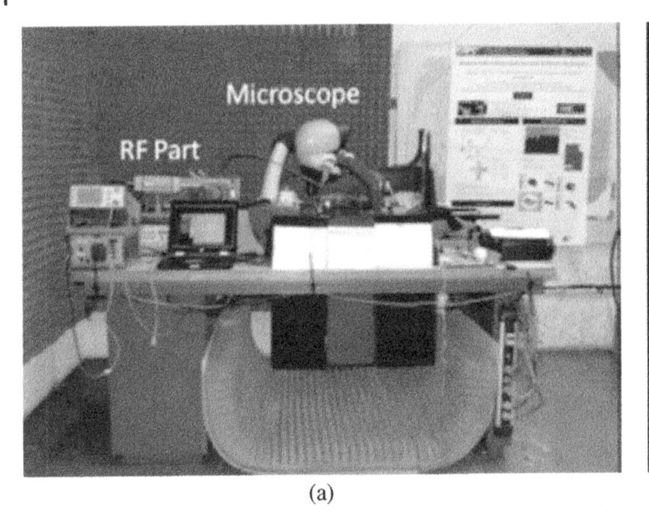

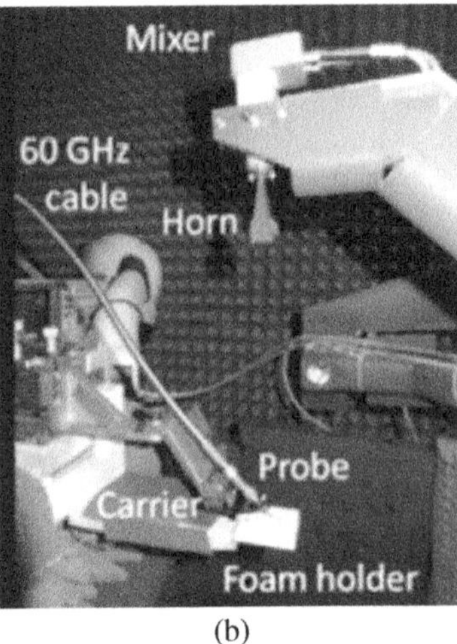

(a) (b)

Figure 2.12 (a) A photo of the measurement setup and (b) a closer view of the measurement setup [36].

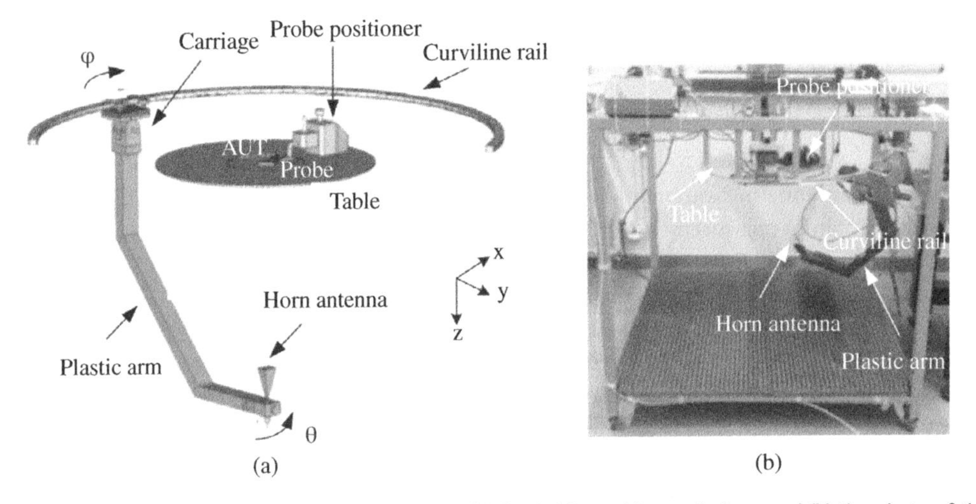

Figure 2.13 (a) Antenna measurement setup with backside probing technique and (b) the photo of the antenna measurement system [37].

arrays or reflector type antennas with higher gain, for example 40 dBi and above, the near-field measurement setup is more preferable and applicable.

Rensburg et al. from Nearfield Systems Inc. (NSI, now NSI-MI Technologies) reported a sub-mmW planar near-field antenna test system for earth observation and radio astronomy antenna characterization [38]. Figure 2.14 shows the scanner that supports the Jet Propulsion Laboratory (JPL) earth Observation System Microwave Limb Sounder at 650–660 GHz. This sub-mmW planar near-field scanner was designated the NSI-905V-8 × 8 (2.4 m × 2.4 m) and has a planarity of 5 μm root mean square (RMS). This scanner uses a granite base for thermal stability and extensive air cooling over the vertical tower to minimize structural deformation due to ambient temperature variation. The scanner construction consists of two parallel granite beams forming the *x*-axis with a vertical granite tower forming the *y*-axis. This construction provides precise surfaces with low-frequency spatial errors.

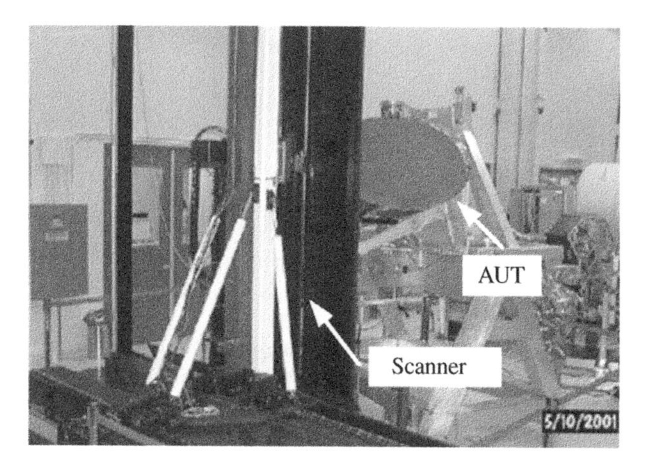

Figure 2.14 NSI planar near-field scanner model NSI-905 V-8x8 shown during testing JPL earth Observation System Microwave Limb Sounder at 660 GHz [38].

Figure 2.15 NSI tiltable planar near-field scanner model NSI-906HT-3 × 3 built for testing ALMA receiver modules up to 950 GHz [38].

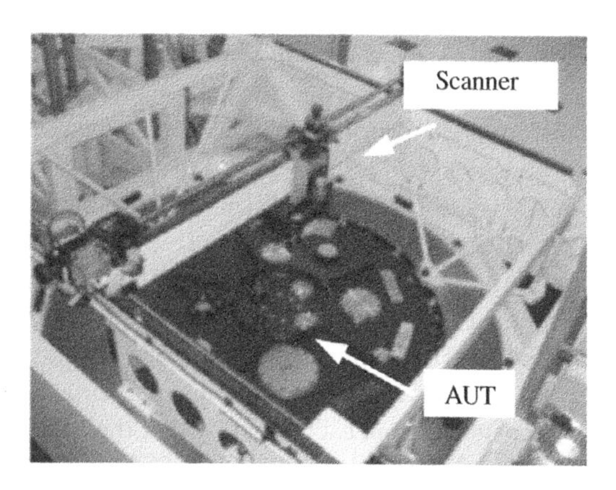

The other reported sub-mmW scanner was built in support of the Atacama Large Millimeter Array (ALMA) program at 950 GHz [38]. This sub-millimeter wave tiltable planar near-field scanner, shown in Figure 2.15, was designated the NSI-906HT-3 × 3 (0.9 m × 0.9 m) and has a planarity of 20 μm RMS in any scan plane from vertical to horizontal. An active structure correction is employed to enhance the scanner planarity, wherein no laser is being used for structure monitoring, but a predetermined structural data set is used for planarity correction during acquisition by using a z-directed linear actuator.

There are several possibilities to obtain the near-field data. One way is to measure the radiated field on a Huygens surface, that is, a plane perpendicular to the main beam propagation direction, and computing equivalent sources in the antenna region. From these sources, the radiation fields can be computed at any point in the near-field and also in the far-field. A second way is to directly measure the volumetric near-field in the desired observation points in the 3-D space in front of the aperture. This direct method is certainly desirable. It provides precise measurement results in the region of interest, and the obtained data can directly be used as reference or calibration data without the influence of processing errors. Besides the presented application, a volumetric near-field scanner can be used for fast and accurate planar measurements on multiple parallel planes. Such measurements help to analyze and remove the effect of multiple reflections between the AUT and the probe antenna [39] Also, investigations toward phaseless near-field to far-field

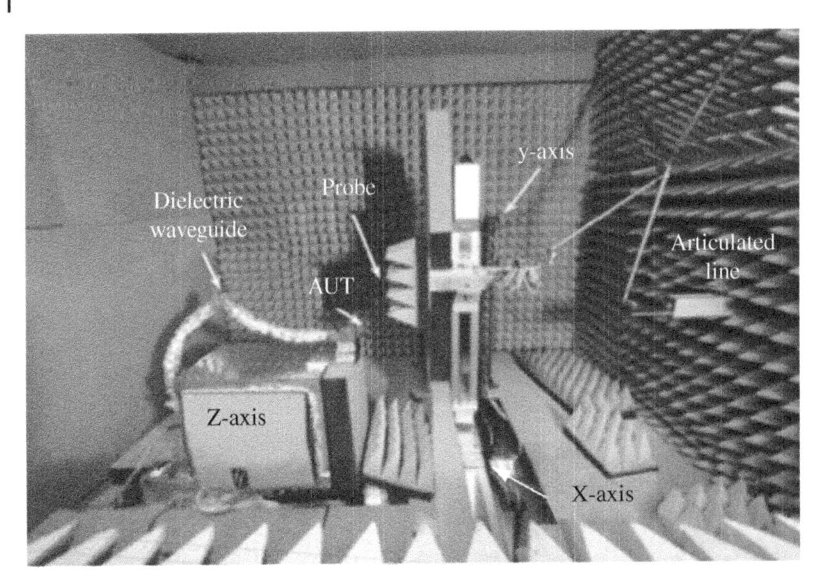

Figure 2.16 Image of the volumetric mmW measurement facility [40].

transformation algorithms can take advantage of the possibility to accurately measure amplitudes and phases on two planes, where the phases can serve as reference.

A 3-D volumetric near-field setup for mmW antenna measurements ranging from 40 to 110 GHz is presented [40]. As shown in Figure 2.16, the volumetric near-field setup consists of an RF acquisition unit in the form of an HP8510C network analyzer with mmW extension and a mechanical positioning unit with three axes, one for the AUT and two for the probe antenna. The z-axis, which carries the AUT, is aligned perpendicular to the xy-plane, on which the probe antenna moves. The xy-axis system is a fixed combination of a horizontal and a vertical axis. The positioning of the axes is independent from each other so that the complete grid within Δx, Δy, and Δz can be reached with a resolution and repeatability of $\pm 20\,\mu m$. The probe antenna is mounted on a sledge on the y-axis. A flexible dielectric waveguide is used to connect the AUT at a 110 GHz band to reduce the transmission loss and thus enhance the system dynamic range. The utilization of the "articulated lines" that comprise rigid line sections and rotary joints ensure a phase stability of $\pm 0.5°$.

In addition to the traditional near-field measurement setups where the scanners are utilized to position the near-field probe in the x-, y-, and z-direction for different scan geometries, some robotically controlled near-field mmW antenna measurement setups have been reported. In such setups, a six degrees of freedom (DOF) industrial robot is utilized to measure the radiation pattern in the frequency up to 550 GHz [41–43]. Boehm et al. reported a robotic-based near field antenna measurement setup (60–330 GHz) as shown in Figure 2.17. A horn antenna is mounted on the robotic arm for receiving the signal from the AUT, which is connected to the Tx module and the VNA. With the six DOF, the Rx position can be adjusted in the x-, y-, and z-directions, and the orientations can be changed by rotation around the three different axes (Rotx, Roty, Rotz), which offers high flexibility in terms of scan geometries such as sphere and planar scans. The position repeatability of 50 μm guarantees highly repeatable measurements. The computer controls the whole measurement process, and the data are sent for post processing and far-field transformations. Besides the waveguide- and coax-fed antenna, the system is able to measure the on-wafer antenna using a probe positioner.

Novotny et al. reported another robotically controlled near-field pattern setup ranging from 50 to 500 GHz. As shown in Figure 2.18, the system incorporates a precision industrial six-axes robot, six-axes parallel kinematic hexapod, and high precision rotation stage. A laser tracker is used to

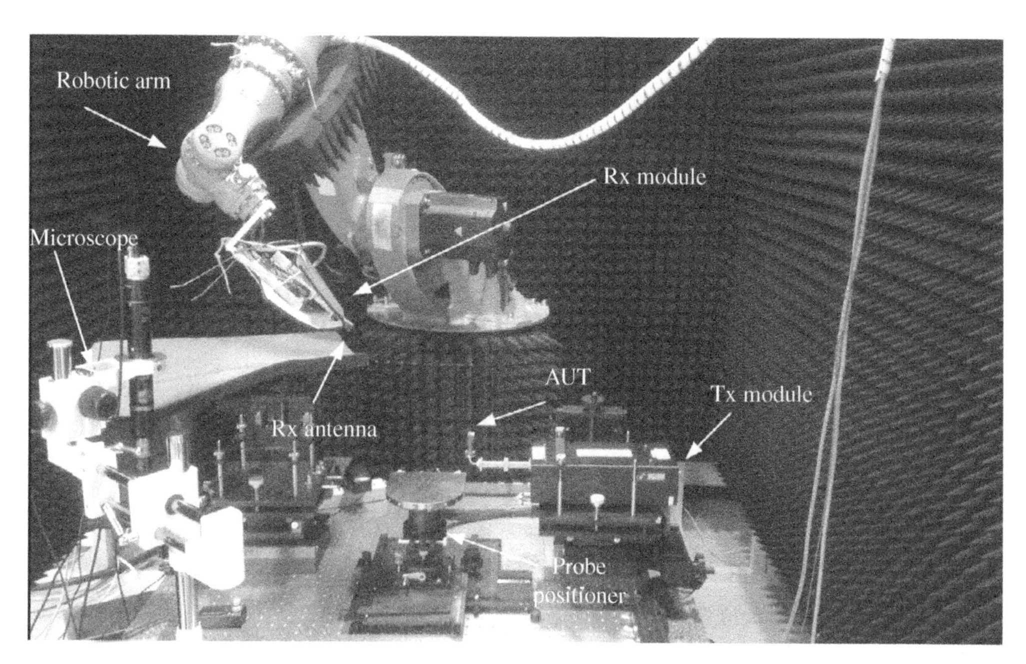

Figure 2.17 Robotic based near-field antenna measurement setup [41].

Figure 2.18 The layout of the mechanical positioning and measurement systems [43].

determine position and to calibrate the robot. The robotic positioning arm is programmable and allows scanning in a variety of geometries including spherical, planar, and cylindrical and performing in-situ extrapolation measurements, as well as other user-defined geometries. For the planar geometry, the coverage is a rectangle 1.25 m × 2 m. For spherical geometry, radii from 2 cm to 2 m are possible, while the coverage in θ is $\pm 120°$ and in φ is $\pm 180°$. Robot positioning repeatability has been evaluated and determined to be about 30 μm, and absolute positioning determination via the laser tracker is ~15 μm. This setup does not support on-wafer antenna measurement.

2.3 Considerations for Measurement Setup Configuration

As briefed in previous sections, the configuration of an antenna measurement setup is a difficult task where many factors must be taken into account. The capability for on-wafer antenna measurement makes it more challenging. The key issues for configuring a mmW antenna measurement setup are addressed below [10, 11, 44]

2.3.1 Far-Field versus Near-Field versus Compact Range

2.3.1.1 Far-Field Measurement

The natural choice for antenna measurement is the far-field method. However, in case of electrically large antennas, the classical far-field method has some major obstacles. The separation between the AUT and the source antenna will be too large to be applicable. For example, the required separation is 100 m for a parabolic reflector antenna with a diameter of 0.5 m at 60 GHz (gain of ~48.5 dBi with 80% aperture efficiency); a much larger separation may be needed when very low sidelobe levels are measured. The large attenuation and the distortions due to change of temperature and humidity of the atmosphere are also major problems. The far-field method is readily applicable for mmW antennas with a gain of less than 30 dBi, wherein the measurement can be done in a controlled environment such as an anechoic chamber with an antenna separation of less than 1 m. On the other hand, for an antenna with lower gain of a few dBi (for example, a single microstrip patch antenna), the far-field distance is not an issue while the measurement interface such as coax/waveguide adapter/connector the on-wager probe tends to disturb the measurement severely.

2.3.1.2 Near-Field Measurement

Near-field range measurement can be carried out indoors and in a relatively small space. The humidity and temperature of the measurement environment can be controlled, and thus the distortions due to the atmosphere become small. In case of a mmW antenna, the sampling of the near-field is normally accomplished over a plane that has to be comprehensive so that all the significant energy transmitted by the AUT can be captured. The sampling has to be dense enough for satisfying the Nyquist sampling criterion, i.e., the distance between two sampling points has to be smaller than one half of the wavelength ($D < \lambda/2$). Major error sources in the near-field scanning are the phase errors in bending and twisting cables, inaccurate positioning of the probe, and the reflections between the probe and the AUT.

The near-field measurement setup features a big challenge for on-wafer antenna measurement: the existence of the on-wafer probe as well as the probe positioner limit the sampling area of the AUT, so the energy transmitted by the AUT can only be partially captured and the far-field performance is unable to be calculated accurately.

2.3.1.3 Compact Antenna Test Range

CATRs enable the antenna measurements indoors in plane wave conditions and the radiation pattern can be directly measured by rotating the AUT. Most of the CATRs are based on reflectors, so the major problem is the very stringent surface accuracy requirement at mmW frequencies; for example, the root-mean-square (rms) surface error should be less than 0.01λ, (e.g., 40 μm at 75 GHz). The CATRs are not applicable for on-wafer antenna measurement since the AUT is required to be rotated.

2.3.2 RF System

The mmW frequency introduces challenges for the RF instrumentation in any antenna measurement. Minimum requirement is to have a frequency stable transmitter with sufficient output power and a sensitive receiver at the given mmW frequency. Amplitude measurements may suffice for most parameters such as gain, efficiency, radiation patterns, cross polarization level, and so on. A spectrum analyzer is an excellent incoherent receiver (when equipped with external down-conversion units). In most antenna measurements, however, a vector measurement is desired. Today, there are commercially available VNAs with extension units up to terahertz frequencies based on frequency multiplication and harmonic mixing.

The loss of coaxial cable above 50 GHz is significant. To achieve good system sensitivity and dynamic range, the use of frequency up/down conversion, such as port extension modules (PEMs), located close to the transmitting and receiving antennas is critical. Standard PEMs from 75 GHz and above are readily available from various test equipment manufacturers [45–48], as shown in Figure 2.19. Using the PEMs, the coax cables between the VNAs and the PEMs carry frequencies less than 26 GHz. Calibrated noise floors in excess of 40 dBi and dynamic ranges in excess of 70 dB from 50 GHz to 110 GHz are achievable.

2.3.3 Interface Between the RF Instrument and AUT

Note that the current RF instruments such as spectrum analyzer and vector network analyzer are all with coaxial interface. In the microwave frequency range, the AUT can be directly connected to the spectrum analyzer/vector network analyzer using a coaxial cable via the coaxial connector or coax/waveguide adapter, as shown in Figure 2.20.

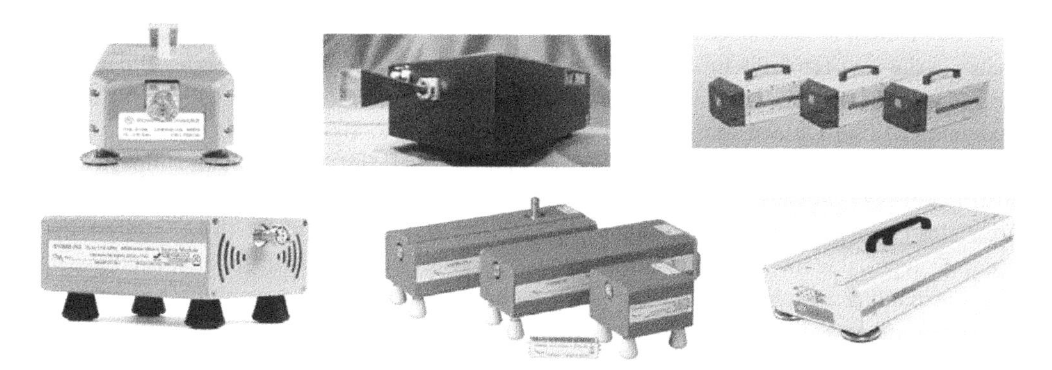

Figure 2.19 Commercially available port extension modules.

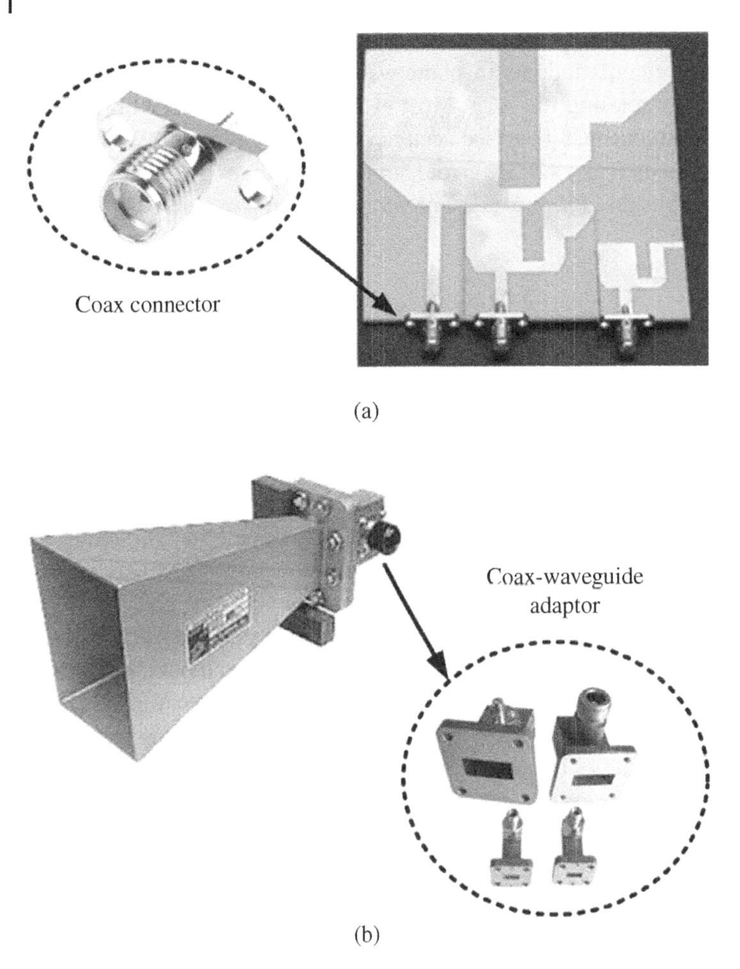

Figure 2.20 Interfaces for microwave antenna measurements. (a) Coax connector and (b) coax-waveguide adapter.

At the mmW frequency range of 30–60 GHz, the utilization of coax-waveguide adapter is still applicable for the antenna with waveguide feeding structure. For antenna on PCB, the normal coaxial connector cannot be used since its tiny inner conductor is difficult to solder to the PCB. Instead, a special coaxial connector, the End Launch Connector as shown in Figure 2.21a, is used to install the connection between the cable and the antenna with microstrip or GCPW feeding structures. The commercial End Launch Connector is claimed to operate to 100 GHz while it is suggested to be used up to the 60 GHz band because of the high loss from the coaxial cables connected to RF instruments. For on-wafer antenna measurement, a coax-probe is normally used, as shown in Figure 2.21c. The coaxial air coplanar probe (ACP) is a rugged mmW probe with a compliant tip for accurate, repeatable measurements for both on-wafer as well as signal integrity applications. It delivers outstanding compliance for probing non-planar surfaces and offers stable and repeatable over-temperature measurements, with a typical probe life of 500 000 contacts on gold pads. Configurations for both single and dual signal applications are available. The commercially available probes are with typical pitches of 100, 125, 150, 200, and 250 μm.

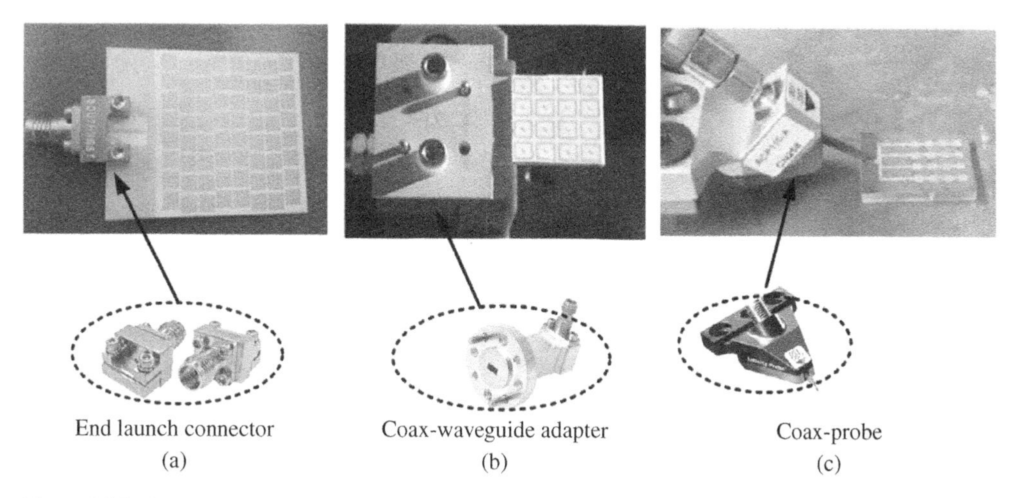

Figure 2.21 Interfaces for mmW antenna measurements up to 60 GHz. (a) End Launch connector, (b) coax-waveguide adapter, and (c) coax-probe.

At the frequencies above 75 GHz, the PEMs have to be used, and thus only waveguide interface can be used. The AUT must be with a waveguide feeding structure so that it can be connected to the PEM directly or with a CPW feeding structure and therefore can be connected to the PEM through a waveguide-probe, as shown in Figure 2.22.

2.3.4 On-Wafer Antenna Measurement

Micro-probing of the on-wafer antenna brings new challenges to antenna measurements. Micro-probe stations are physically/electrically large and heavy due to anti-shock and anti-vibration techniques implemented to protect the probe and improve measurement integrity. The metallic environment makes the probe station an undesired choice for radiated measurements. In addition, the existence of the micro-probe and associated probe positioner not only causes physical blockage for spherical coverage but also results in radiation interference with the intended signal. Compared with a conventional antenna measurement setup, the on-wafer antenna measurement setup is with several restrictions as discussed below.

2.3.4.1 Feeding and Movement Limitations

In most of the conventional radiation pattern testing setups, the AUT rotates and the reference horn antenna is kept stationary. However, the rotation of an AUT is impossible in on-wafer antenna measurement since the AUT is positioned on the heavy and huge probe station associated with a probe positioner. Instead of a three-axis positioner of conventional measurement setups, this source horn antenna requires a modified rotating positioner that can rotate around the AUT. Similarly the problem arises with near-field scanning measurements where the near-field probe must rotate around the AUT to capture the radiating fields.

2.3.4.2 Reflection Caused by Probes/Metallic Environment

A typical commercially available probe station is with an anti-vibration table, a chuck, a microscope, a positioner, and a probe. All the metallic sections of a measurement setup are sources of reflections and scattering of electromagnetic fields that lead to erroneous measurement. The chuck

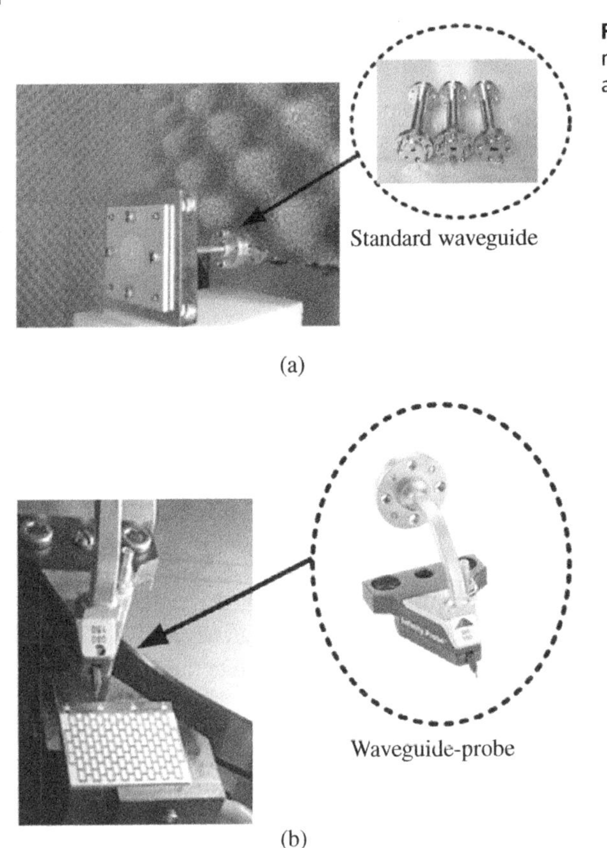

Figure 2.22 Interfaces for mmW antenna measurement above 75 GHz. (a) Waveguide and (b) waveguide-probe

itself acts as a ground plane and results in the generation of undesired modes. The back radiation from an AUT also gets reflected field from the chuck and adds to the direct signal with a certain phase delay. Alongside, the radiation from the standard reference antenna may also get bounced signals between the chuck and other metallic surfaces in the close proximity thus causing multiple reflections. In addition, for a smaller on-wafer AUT, the proximity of the physical/electrically large probe causes the environment reflections and distorts the antenna radiation patterns.

Another common problem associated with the on-wafer measurements is the precise alignment of probes with the feeding point of the AUT. The probes are aligned and skid precisely to make contact with the AUT. This alignment, especially with narrow pitch probes, requires a microscopic view of the feeding point and probes. Incorporating a microscope within the test chamber causes reflections that consequently affect the radiation pattern of the AUT, so a movable microscope is desired.

Figure 2.23 demonstrates four selected mmW measurement setups, which are not expected to obtain accurate measurement results because of the rich reflections from the messy metallic environment.

2.3.4.3 Undesired Coupling Effects Caused by Measurement Probes

The substrate residual current of an on-wafer antenna may interact with the measurement probes thus producing errors in the measurements. The unshielded tips of the ACP probe cause this coupling effect. The unintended coupling cannot be suppressed through the de-embedding process that is carried out with a calibration substrate. In addition, the conventional on-wafer GSG pads for measurement purposes introduces capacitive effects and consequently changes the resonant

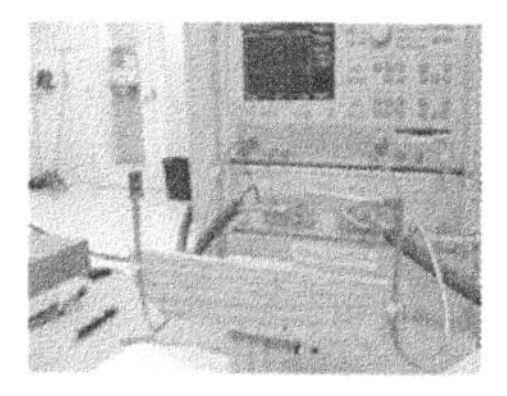

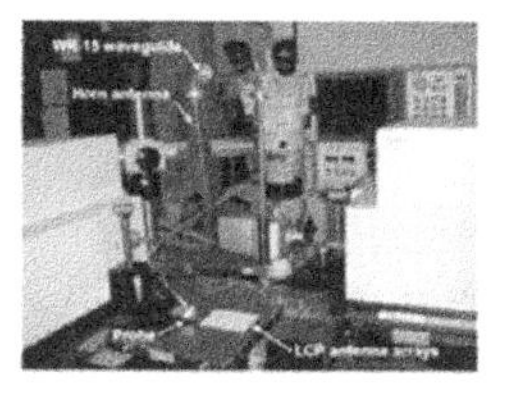

Figure 2.23 Examples of mmW antenna measurement setup with multiple reflection environment.

frequency of the AUT, while the location and placement of these pads may directly disturb the impedance matching of the AUT and result in resonant frequency shifting [49, 50]

2.4 mmW Measurement Setup Examples

In this section, three mmW antenna measurement setups at 60, 140, and 270 GHz bands, are demonstrated, respectively. The 60-GHz system is composed of a VNA, an LNA, and a power amplifier, which features the largest system dynamic range and is applicable to antennas with a coax, waveguide, or probe feeding structure. The 140-GHz system consists of a VNA, two mmW PEMs, and an LNA, which is with the desirable system dynamic range and applicable to antennas fed by either a waveguide or probe. The 270-GHz system is with a VNA and two mmW PEMs, which delivers an acceptable system dynamic range and is applicable to the antennas connected by either a waveguide or probe feeding structures. The systems are verified by measuring various mmW antennas with accurate and reliable results that agree very well with the simulations.

2.4.1 60-GHz Antenna Measurement Setup

Figure 2.24 shows the schematic diagram of the 60-GHz antenna measurement setup at the Institute for Infocomm Research, Singapore [14]. The AUT is positioned on a low-permittivity plastic holder extending from an anti-vibration table, which is essential for mounting the probe for on-wafer antenna measurement. The AUT is connected to one of the VNA ports through a coaxial line or waveguide. A standard horn antenna is mounted on the rotating arm (the distance between the horn and the AUT is adjustable, up to 1 m) and connected to the other port of the VNA. Waveguide sections instead of coaxial cable are positioned along the rotating arm to reduce the system loss. A power amplifier and an LNA are utilized to further enhance the system dynamic range, which is important to characterize the side-lobe and cross-polarization levels of the antenna, in particular, for low-gain antennas.

Figure 2.25 exhibits the photos of the practical measurement setup with AUT in a mini anechoic chamber. Figures 2.26–2.28 demonstrate the examples of the measured radiation patterns for three antennas operating at 60 GHz bands. The AUTs include a low-gain antipodal Fermi antenna fed through an end launch connector, a 12 dBi low-gain planar array fed by a probe, and a 20 dBi high-gain planar antenna array excited through a waveguide. Desired agreement has been achieved between the measured and simulated results in all tests.

2.4.2 140-GHz Antenna Measurement Setup

The configuration of the 140-GHz antenna measurement system is shown in Figure 2.29. Different from the setup for 60-GHz antennas, two mmW PEMs [46] are used here since without frequency

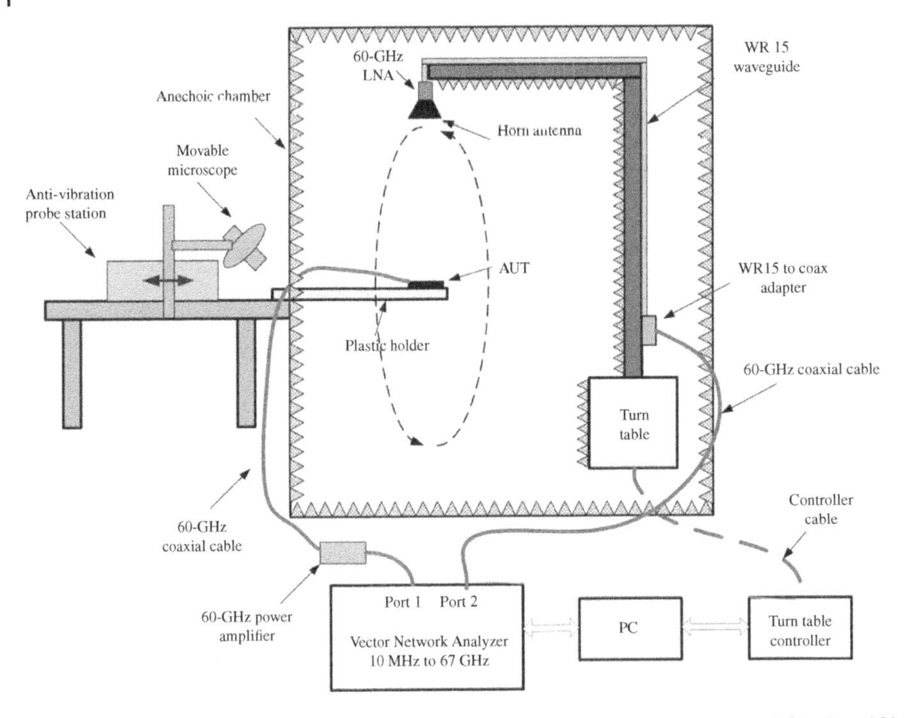

Figure 2.24 Schematic diagram of the measurement setup for antennas at a 60 GHz band [14].

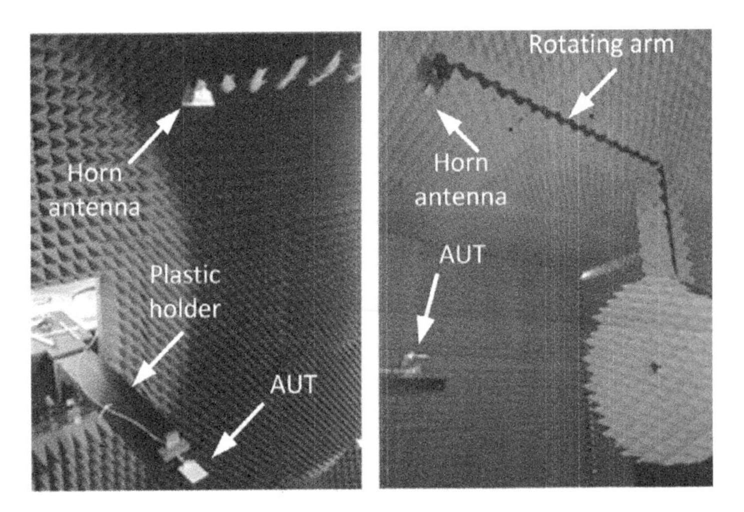

Figure 2.25 Photos of the measurement setup for antennas at a 60 GHz-band.

extension the used VNA system can operate only up to 67 GHz. The AUT is positioned on the anti-vibration table and connected to a PEM through a waveguide or a waveguide probe. The standard horn antenna is mounted on the rotating arm and connected to the other PEM through a waveguide. The two PEMs are connected to the ports of the VNA through longer microwave cables of about 10 m. To ensure the mmW PEMs are operating properly, several LNAs are utilized to compensate the cable loss of the LO/RF links.

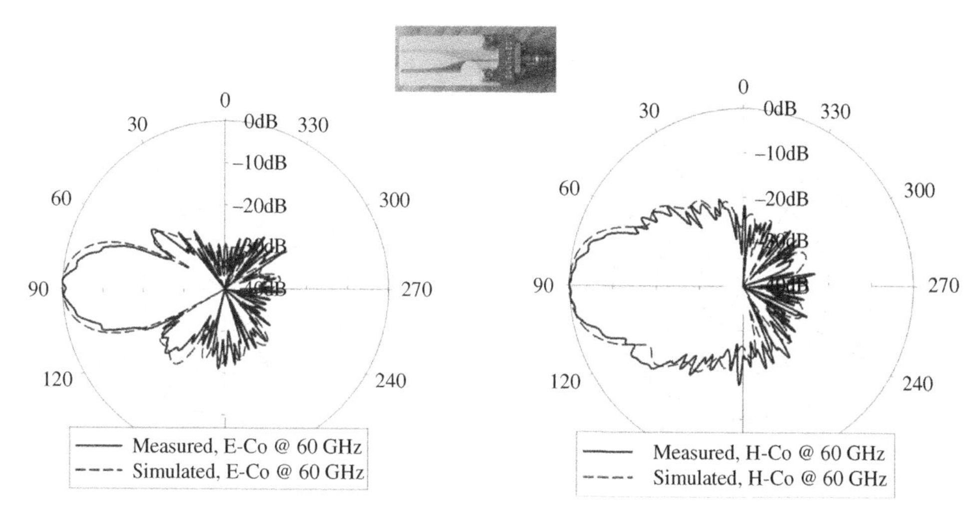

Figure 2.26 Measured radiation patterns of an antipodal Fermi antenna at 60 GHz [51].

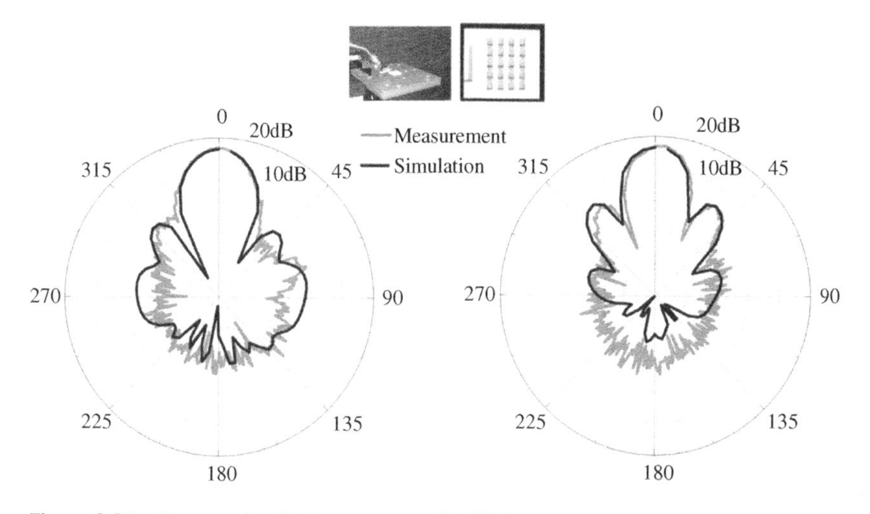

Figure 2.27 Measured radiation patterns of a LTCC antenna array with open-air cavities [52].

With the D-band (110–170 GHz) power amplifier and LNA, the system shown in Figure 2.29 achieves a desired system dynamic range. Figure 2.30 exhibits the photos of the practical measurement setup in a mini anechoic chamber at Institute for Infocomm Research, Singapore. To reduce the reflections, the major portions are covered with absorbers and a movable microscope is used for on-wafer antenna measurement. Figures 2.31 and 2.32 exhibit the radiation patterns of a 10 dBi low-gain on-chip antenna array and a 20-dB gain antenna array on LTCC, which are with probe and waveguide feeding structure, respectively. The measured radiation patterns agree very well with the simulated ones, and the side lobe levels of −40 dB can be characterized.

2.4.3 270-GHz Antenna Measurement Setup

Figures 2.33 and 2.34 show the configuration of the 270-GHz antenna measurement system built in the Institute for Infocomm Research, Singapore, wherein two 220–325 GHz mmW PEMs are

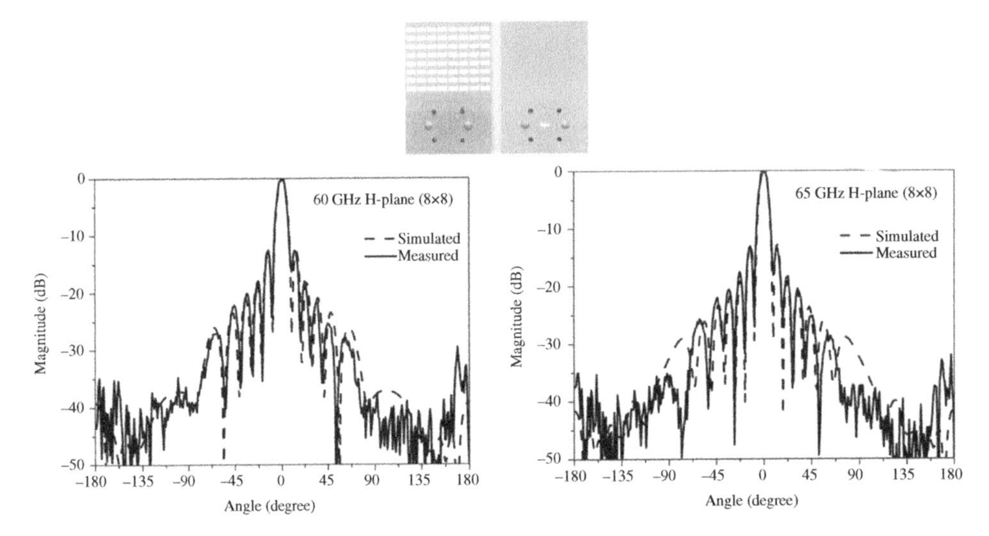

Figure 2.28 Measured radiation patterns of a 60-GHz substrate-integrated waveguide fed cavity array antenna on LTCC [53].

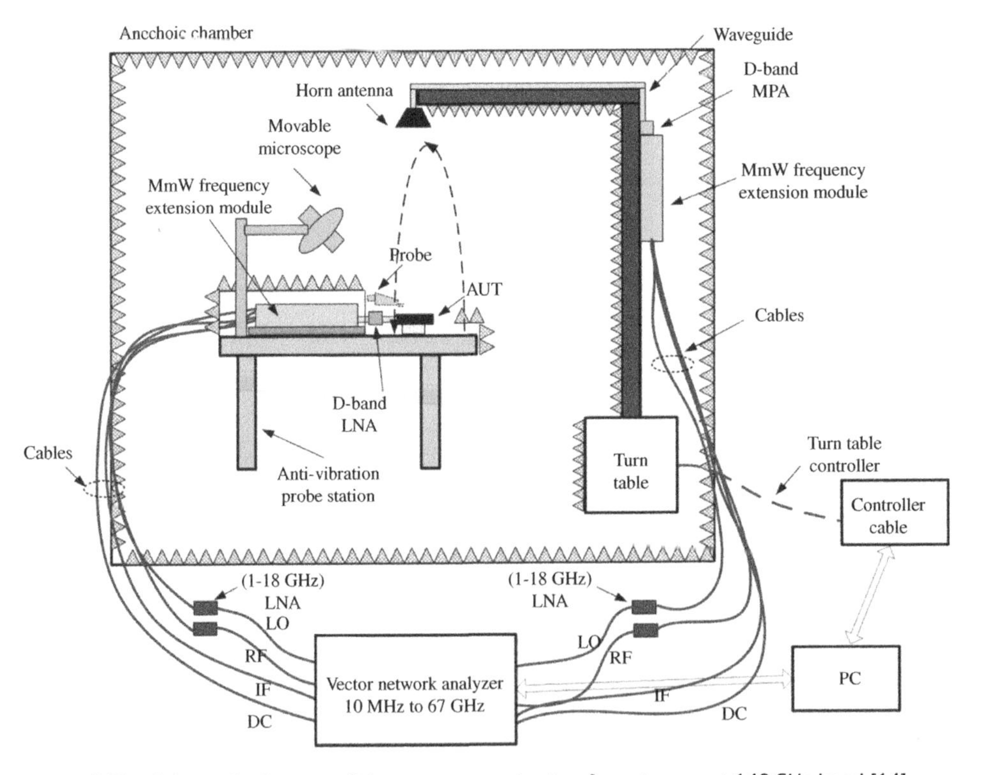

Figure 2.29 Schematic diagram of the measurement setup for antennas at 140 GHz band [14].

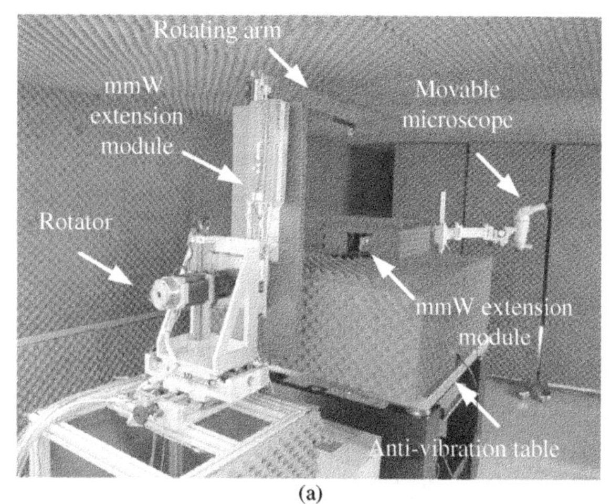

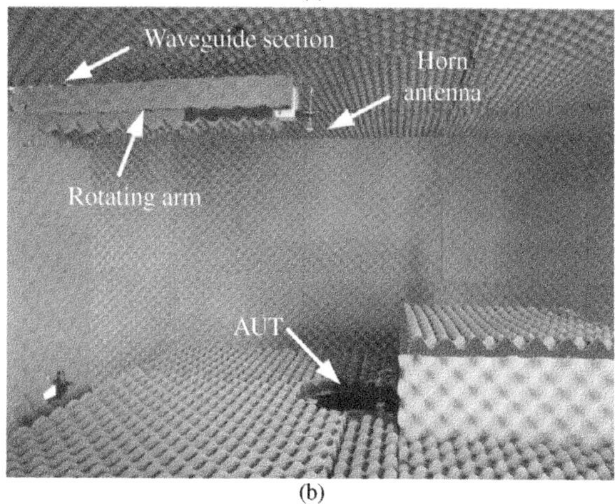

Figure 2.30 Photos of the measurement setup for antennas at a 140 GHz band. (a) Overall setup and (b) detailed view of AUT/horn positioning.

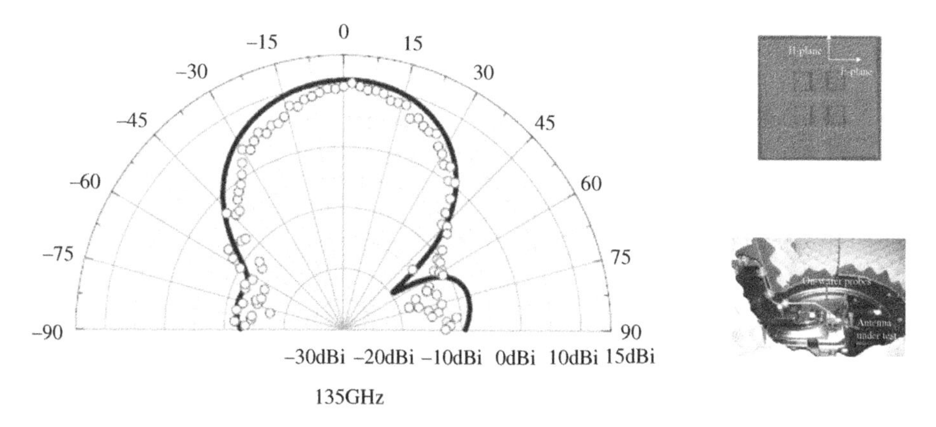

Figure 2.31 Measured radiation pattern of 135-GHz antenna array on Benzocyclobutene (BCB) [54].

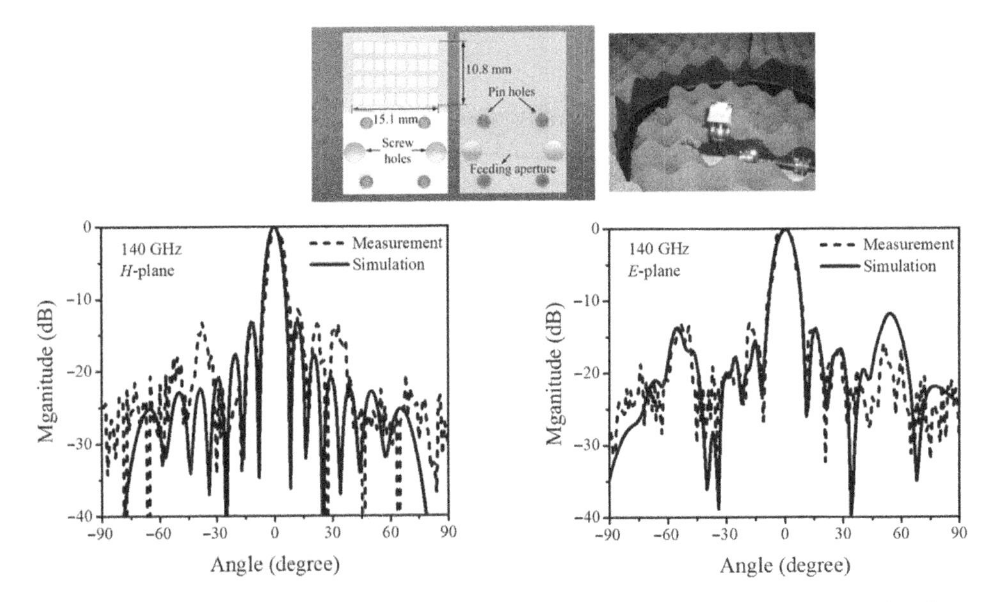

Figure 2.32 Measured radiation patterns of a TE_{20}-mode substrate integrated waveguide fed slot antenna array at 140 GHz [55].

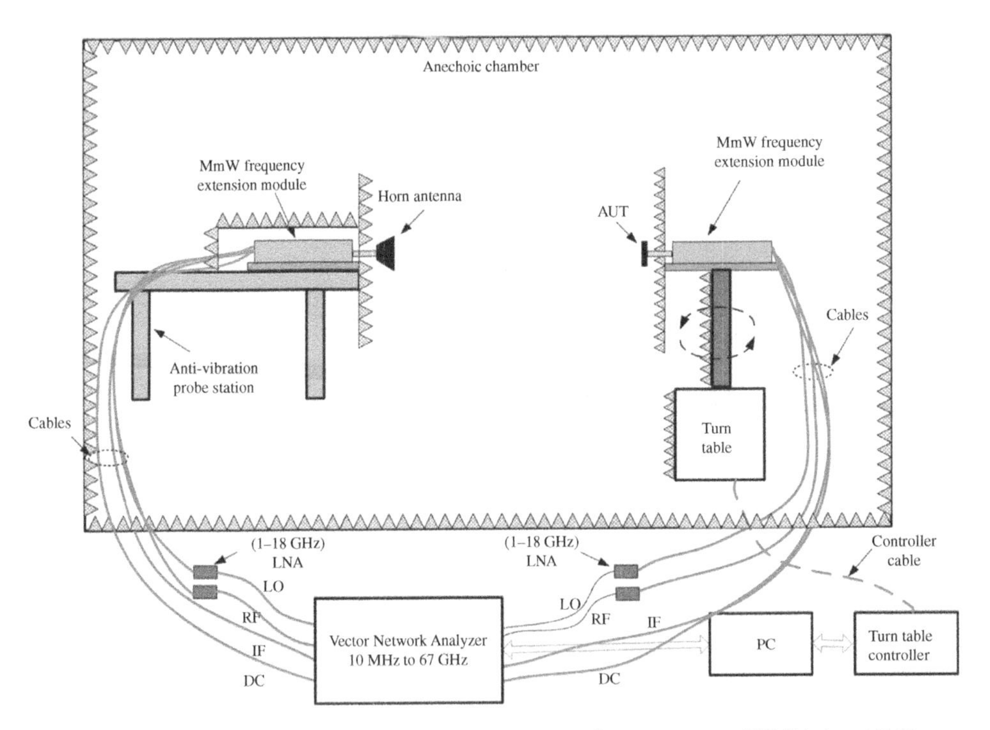

Figure 2.33 Schematic diagram of the measurement setup for antennas at 270 GHz band [14].

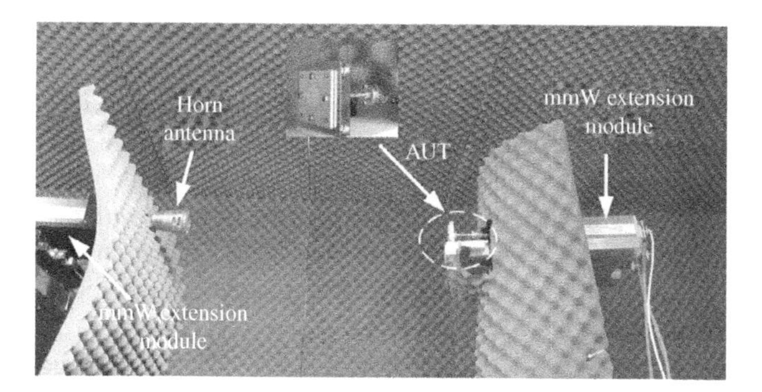

Figure 2.34 Photo of the measurement setup for antennas at a 270-GHz band.

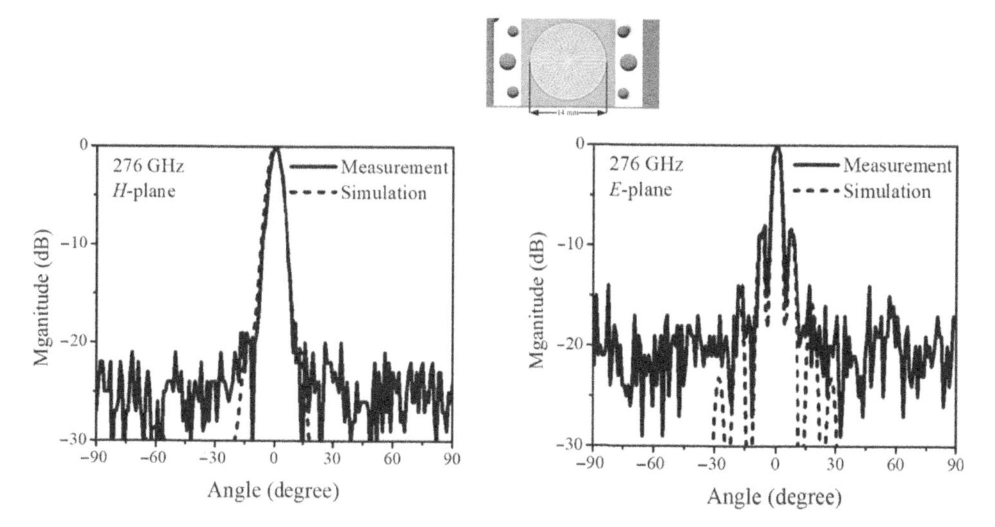

Figure 2.35 Measured radiation patterns of a 270-GHz LTCC-integrated strip-loaded linearly polarized radial line slot array antenna [56].

used. The standard horn antenna and the AUT are directly connected to the PMEs to minimize the system loss caused by the RF power transmission line and connections. The AUT positioned on the turntable can be rotated 360° horizontally. Due to the limited output power and sensitivity of the frequency extenders, the setup at the 270-GHz band suffers from a reduced system dynamic range.

Figure 2.35 exhibits the radiation patterns of a 270-GHz LTCC-integrated strip-loaded linearly polarized radial line slot array antenna with a waveguide fixture. The system is able to characterize the main beam and the 1st side-lobe with high accuracy but not other side-lobes. It should be noted that the calibration of all testing systems is also a time-consuming and challenging task. All the alignments between the standard antenna and the AUT as well as the turntables must be kept highly accurate to avoid unacceptable measurement errors.

2.5 Summary

Antenna development at mmW bands has many uniquely challenging issues wherein the measurement is one of the toughest ones because of the expensive testing setup, limited system dynamic range, and complicated and tedious calibration, as well as measurement procedures. Furthermore, the cable, connector, or probe used in the measurement shows non-ignorable effects on the measurement results of the AUTs, which must be considered carefully in the measurement system configuration. This chapter has dealt with the measurement issues of the mmW antennas operating from 60 to 300 GHz. First, the state-of-the-art mmW antenna measurement systems have been presented. Then the key considerations of configuring the measurement systems have been addressed. Last, the detailed setup configurations for achieving the maximum system dynamic range with the available commercial accessories have been described, wherein the measurement of radiation pattern of a number of antennas at 60, 140, and 270 GHz bands with different feeding connections (coax, waveguide, and probe) have been exemplified.

References

1 Kummer, W.H. and Gillespie, E.S. (1978). Antenna measurements. *Proc. IEEE* 66 (4): 483–507.

2 Hollis, J.S., Lyon, T.J., and Jr. Clayton, L. (1970). *Microwave Antenna Measurements*. Scientific-Atlanta.

3 IEEE, *IEEE Standard test procedures for antennas* (IEEE Std 149-1979), 1979.

4 Balanis, C.A. (2005). *Antenna Theory Analysis and Design*. Wiley.

5 Qing, X. and Chen, Z.N. (2016). *Handbook of Antenna Technologies: Antenna Measurement Setups-Introduction*. Springer.

6 Liu, D., Gaucher, B., Pfeiffer, U., and Grzyb, J. (2009). *Advanced Millimeter-Wave Technologies: Antennas, Packaging and Circuits*. Chichester, UK: Wiley.

7 Seki, T., Honma, N., Nishikawa, K., and Tsunekawa, K. (2005). A 60GHz multilayer parasitic microstrip array antenna on LTCC substrate for system-on-package. *IEEE Microw. Wireless Components Lett.* 15 (5): 339–341.

8 Pozar, D.M. (1983). Considerations for millimeter wave printed antennas. *IEEE Trans. Antennas Propag.* 31 (5): 740–747.

9 Lamminen, A., Saily, J., and Vimpari, A.R. (2008). 60-GHz patch antennas and arrays on LTCC with embedded-cavity substrates. *IEEE Trans. Antennas Propag.* 56 (9): 2865–2870.

10 Karim, M.R., Yang, X., and Shafique, M.F. (2018). On chip antenna measurement: a survey of challenges and recent trends. *IEEE Access* 6: 20320–23333.

11 Räisänen, A.V., Zheng, J., Ala-Laurinaho, J., and Viikari, V. (2018). Antenna measurements at millimeter wavelengths - overview. In: *12th European Conference on Antennas and Propagation (EuCAP 2018), London*, 1–3.

12 Gulana, H., Luxey, C., and Titzc, D. (2016). *Handbook of Antenna Technologies: Mm-Wave Sub-mm-Wave Antenna Measurement*. Springer.

13 Reniers, A.C.F. and Smolders, A.B. (2018). Guidelines for millimeter-wave antenna measurements. In: *2018 IEEE Conference on Antenna Measurements & Applications (CAMA), Vasteras*, 1–4.

14 Qing, X. and Chen, Z.N. (2014). Measurement setups for millimeter-wave antennas at 60/140/270 GHz bands. In: *2014 IEEE International Workshop on Antenna Technology: Small Antennas, Novel EM Structures and Materials, and Applications (iWAT), Sydney, NSW*, 281–284.

15 Chung, B.K. (2016). *Handbook of Antenna Technologies: Anechoic Chamber Design*. Springer.

16 Johnson, R.C., Ecker, H.A., and Moore, R.A. (1969). Compact range techniques and measurements. *IEEE Trans. Antennas Propag.* 17 (5): 568–576.

17 Johnson, R. (1986). Some design parameters for point-source compact ranges. *IEEE Trans. Antennas Propag.* 34 (6): 845–847.

18 Burnside, W.D., Gilreath, M., Kent, B.M., and Clerici, G. (1987). Curved edge modification of compact range reflector. *IEEE Trans. Antennas Propag.* 35 (2): 176–182.

19 Pistorius, C.W.I. and Burnside, W.D. (1987). An improved main reflector design for compact range applications. *IEEE Trans. Antennas Propag.* 35 (3): 342–347.

20 Sanad, M.S.A. and Shafai, L. (1990). Dual parabolic cylindrical reflectors employed as a compact range. *IEEE Trans. Antennas Propag.* 38 (6): 814–822.

21 Johnson, R.C., Ecker, H.A., and Hollis, J.S. (1973). Determination of far-field antenna patterns from near-field measurements. *Proc. IEEE* 61 (12): 1668–1694, 1973.

22 Paris, D.T., Jr. Leach, W.M., and Joy, E.B. (1978). Basic theory of probe compensated near-field measurements. *IEEE Trans. Antennas Propag.* 26 (3): 373–379, 1978.

23 Joy, E.B., Jr. Leach, W.M., Rodrigue, G.P., and Paris, D.T. (1978). Applications of probe compensated near-field measurements. *IEEE Trans. Antennas Propag.* 26 (3): 379–389.

24 Yaghjian, A.D. (1986). An overview of near-field antenna measurements. *IEEE Trans. Antennas Propag.* 34 (1): 30–45.

25 Online available: http://www.avx.com/products/antennas/mmwave-measurement-system

26 Online available: https://www.mvg-world.com/sites/default/files/2019-09/Datasheet_Antenna%20Measurement_%C2%B5-Lab_BD_0.pdf

27 Online available: https://www.mvg-world.com/sites/default/files/2019-09/Datasheet_Antenna%20Measurement_Mini-Compact%20Range_BD.pdf

28 Simons, R.N. and Lee, R.Q. (1999). On-wafer characterization of millimeter-wave antennas for wireless applications. *IEEE Trans. Microw. Theory Tech.* 47 (1): 92–95.

29 Liu, H., Guo, Y.X., Bao, X., and Xiao, S. (2012). 60-GHz LTCC integrated circularly polarized helical antenna array. *IEEE Trans. Antennas Propag.* 60 (3): 1329–1335.

30 Deng, X.D., Li, Y., Liu, C. et al. (2015). 340 GHz on-chip 3-D antenna with 10 dBi gain and 80% radiation efficiency. *IEEE Trans THz Sci. Technol.* 5 (4): 619–627.

31 Song, Y., Xu, Q., Tian, Y. et al. (2017). An on-chip frequency-reconfigurable antenna for Q-band broadband applications. *IEEE Antennas Wireless Propag. Lett.* 16: 2232–2235.

32 Pilard, R., Montusclat, S., Gloria, D. et al. (2009). Dedicated measurement setup for millimetre-wave silicon integrated antennas: BiCMOS and CMOS high resistivity SOI process characterization. In: *3rd European Conf. Antenna Propagat*, 2447–2451.

33 Zwick, T., Baks, C., Pfeiffer, U.R. et al. (2004). Probe based MMW antenna measurement setup. *Int. Symp. Antennas Propag. USNC/URSI Nat. Radio Sci. Meeting*: 747–750.

34 Beer, S. and Zwick, T. (2010). Probe based radiation pattern measurements for highly integrated millimeter-wave antennas. In: *Proceedings of the Fourth European Conference on Antennas and Propagation, Barcelona*, 1–5.

35 Chin, K.S., Jiang, W., Che, W. et al. (2014). Wideband LTCC 60-ghz antenna array with adual-resonant slot and patch structure. *IEEE Trans. Antennas Propag.* 62 (1): 174–182.

36 Titz, D., Ferrero, F., and Luxey, C. (2012). Development of a millimeter-wave measurement setup and dedicated techniques to characterize the matching and radiation performance of probe-fed antennas. *IEEE Antennas Propag. Mag.* 54 (4): 188–203.

37 Mosalanejad, M., Brebels, S., Ocket, I. et al. (2015). A complete measurement system for integrated antennas at millimeter wavelengths. In: *9th European Conf. Antenna Propagat*, 1–4.

38 Van Rensburg, D.J. and Hindman, G. (2008). An overview of near-field sub-millimeter wave antenna test applications. In: *2008 14th Conference on Microwave Techniques, Prague*, 1–4.

39 IEEE, "IEEE recommended practice for near-field antenna measurements," *IEEE Std 1720–2012*, Dec 2012.

40 Koenen, C., Hamberger, G., Siart, U., and Eibert, T.F. (2016). A volumetric near-field scanner for millimeter-wave antenna measurements. In: *2016 10th European Conferenceon Antennas and Propagation, Davos*, 1–4.

41 Boehm, L., Boegelsack, F., Hitzler, M., and Waldschmidt, C. (2015). An automated millimeter-wave antenna measurement setup using a robotic arm. *Int. Symp. Antennas Propag. USNC/URSI Nat. Radio Sci. Meeting*: 2109–2110.

42 Boehm, L., Foerstner, A., Hitzler, M., and Waldschmidt, C. (2017). Refection reduction through modal filtering for integrated antenna measurements above 100 GHz. *IEEE Trans. Antennas Propag.* 65 (7): 3712–3720.

43 Novotny, D., Gordon, J., Coder, J. et al. (2013). Performance evaluation of a robotically controlled millimeter-wave near-field pattern range at the NIST. In: *2013 7th European Conference on Antennas and Propagation, Gothenburg*, 4086–4089.

44 Lee, E., Soerens, R., Szpindor, E., and Iversen, P. (2015). Challenges of 60 GHz on-chip antenna measurements. *Int. Symp. Antennas Propag. USNC/URSI Nat. Radio Sci. Meeting*: 1538–1539.

45 Online available: Virginia Diodes, Inc., https://www.vadiodes.com/en/about-vdi

46 Online available: OML, Inc., https://www.omlinc.com

47 Online available: Rohde & Schwarz, https://www.rohde-schwarz.com/us/products

48 Online available: Farran Technology, Ltd, https://www.farran.com

49 Esfahlan, M.S. and Tekin, I. (2016). Radiation influence of ACP probe in S11 measurement. In: *2016 10th European Conference on Antennas and Propagation, Davos*, 1–4.

50 Esfahlan, M.S., Kaynak, M., Göttel, B., and Tekin, I. (2013). SiGe process, integrated on-chip dipole antenna on finite-size ground plane. *IEEE Antennas Wireless Propag. Lett.* 12: 1260–1263.

51 Sun, M., Qing, X., and Chen, Z.N. (2011). 60-GHz antipodal Fermi antenna on PCB. In: *Proceedings of the 5th European Conference on Antennas and Propagation (EUCAP), Rome*, 3109–3112.

52 Yeap, S.B., Chen, Z.N., and Qing, X. (2011). Gain-enhanced 60-GHz LTCC antenna array with open air cavities. *IEEE Trans. Antennas Propagat.* 59 (9): 3470–3473.

53 Xu, J., Chen, Z.N., Qing, X., and Hong, W. (2011). Bandwidth enhancement for a 60 GHz substrate integrated waveguide fed cavity array antenna on LTCC. *IEEE Trans. Antennas Propagat.* 59, 3o.3: 826–832.

54 Yeap, S.B., Chen, Z.N., and Qing, X. (2012). 135-GHz antenna array on BCB membrane backed by polymer-filled cavity. In: *2012 6th European Conference on Antennas and Propagation (EUCAP), Prague*, 1337–1340.

55 Xu, J., Chen, Z.N., Qing, X., and Hong, W. (2013). 140-GHz TE20-mode dielectric-loaded SIW slot antenna array in LTCC. *IEEE Trans. Antennas Propagat.* 61 (4): 1784–1793.

56 Xu, J.F., Chen, Z.N., and Qing, X. (2013). 270-GHz LTCC-integrated strip-loaded linearly polarized radial line slot array antenna. *IEEE Trans. Antennas Propagat.* 61 (4): 1794–1801.

3

Substrate Integrated mmW Antennas on LTCC

Zhi Ning Chen[1] *and Xianming Qing*[2]

[1] *Department of Electrical and Computer Engineering, National University of Singapore, Singapore 117583, Republic of Singapore*
[2] *Signal Processing, RF, and Optical Department, Institute for Infocomm Research, Singapore 138632, Republic of Singapore*

3.1 Introduction

In theory, the performance of antennas such as impedance matching and radiation patterns strongly depends on their electrical dimensions once their shapes are fixed. For instance, a half-wave dipole is usually used to operate at the frequency calculated from half a wavelength, wherein the operating frequency can be any value.

In engineering, the antennas must be fabricated using materials and processes. The materials include conductors and dielectrics. The former are usually used as radiators while the latter are substrate, supporters, loadings, fillings, or resonators. Unlike a perfect electric conductor (PEC), an idealized material featuring infinite electrical conductivity or, equivalently, zero resistivity and perfect dielectric exhibiting zero resistivity, all materials suffer from ohmic losses with frequency-dependent finite conductivities due to skin effects for metals and dielectric constants. For example, for direct currents at 20 °C the resistivity $\rho(\Omega\cdot m)$/conductivity $\sigma(S/m)$ of silver, copper, gold, and aluminum are respectively $1.59\times10^{-8}/6.30\times10^{7}$, $1.68\times10^{-8}/5.98\times10^{7}$, $2.44\times10^{-8}/4.52\times10^{7}$, and $2.82\times10^{-8}/3.5\times10^{7}$. There is a skin effect of metals when the conductors are imperfect. The resistivity of the conductor increases proportionally to the decreases in skin depth. The skin depth decreases as operating frequency increases. For example, the skin depth of copper decreases from 2.06 μm at 1 GHz to 0.38 μm at 30 GHz and 0.12 μm at 300 GHz. These frequency-dependent properties of conductors used as radiators definitely affect the performance of antennas, in particular the radiation efficiency, especially at mmW bands.

Another important issue for mmW antenna design is the fabrication processes. This chapter will discuss the substrate integration antenna design based on low temperature co-fired ceramic (LTCC) process at mmW bands and focus on the technologies developed for the substrate integrated antennas (SIAs), in particular, LTCC-based SIAs for the operation at 60–270 GHz bands.

3.1.1 Unique Design Challenges and Promising Solutions

As mentioned in Chapter 1 (Introduction), due to the shorter operating wavelengths, the design of antennas at mmW bands presents unique challenges, in particular, losses. The major loss is the ohmic loss from materials including dielectric and conductors, which is caused by the conversion of the electrical energy into heat. The other loss is caused by surface wave radiation from the antenna

Substrate-Integrated Millimeter-Wave Antennas for Next-Generation Communication and Radar Systems, First Edition.
Edited by Zhi Ning Chen and Xianming Qing.

structures. Due to the excitation of surface waves, part of the radiated energy is guided to other undesired directions so that this part of radiation "loses" at the desired radiation directions. Also part of the radiated energy contributes to the radiation in the desired directions but through different paths with uncontrollable phase so that such radiation distorts the radiation patterns in the desired directions. These losses are not considered as ohmic losses if the additional paths don't cause more additional loss. Moreover, part of the energy is coupled to other lossy circuits through surface waves so that the antenna radiation efficiency is reduced due to the ohmic losses and even the antenna radiation patterns are distorted.

The mmW systems suffer from large free space path-loss. For example, the free space path-loss of a radio frequency (RF) link operating at 60 GHz is 20 dB higher than that of a RF link operating at 6 GHz if the antenna gain is kept unchanged. Therefore, the large-scale high-gain mmW antenna arrays are necessary to compensate for the higher path-loss. The much more complicated feeding structures of the large-scale antenna arrays at mmW bands make the antenna array design much more challenging than single-element antennas. The long current/power paths of feeding structures of large-scale arrays inevitably introduce much insertion losses and degrade the antenna efficiency.

The promising solutions to the loss suppression include at least the reduction of insertion loss caused by materials used for the transmission systems of feeding networks and radiators. One way is to select the transmission systems with less insertion losses. The study, such as the work in Ref 32 of Chapter 1 has clearly suggested that laminated or post-wall waveguide or substrate integrated waveguide (SIW) are more suitable for designing the feeding network of antenna arrays at mmW bands. SIWs combine the merits of low ohmic losses such as a waveguide with the easy fabrication process such as a PCB or LTCC (Ref 23–27 of Chapter 1). If considering the SIW as a waveguide structure, the SIWs can naturally be used to design a variety of waveguide-based antennas, in particular at mmW bands (Ref 25 in Chapter 1) [1, 2]. More generally, besides the waveguide-based antennas, all the antennas based on substrate integrated or laminated technology are regarded as SIAs. So far, such antennas can be designed and fabricated by PCB and LTCC processes.

Furthermore, shortening the current or power paths of feeding structures is another effective way to reduce the insertion loss caused by feeding networks. At mmW bands, the shorter operating wavelengths provide the opportunities to integrate the antennas and event arrays into packages of integrated circuits (ICs), the case preventing physical damage and corrosion of ICs and supporting the electrical contacts that connect the device to a circuit board [3–6]. Based on packaging technology, antennas can be integrated in, below, above, and even by package as mentioned in Ref. [6]. Furthermore, the package itself, usually a low-loss ceramic or dielectric, can be used as a dielectric antenna directly [7]. All such package technology–based antennas are categorized as *package integrated antennas*, or PIAs in short. PIAs allow a great reduction of losses caused by the lossy substrate such as semiconductor used in antenna on chip. Also, the larger volume of package provides much more space to host antennas, even arrays for acceptable performance. Such flexibility offers the industry big room to configurate the entire system design onto the package [8, 9].

3.1.2 SIW Slot Antennas and Arrays on LTCC

This subsection demonstrates the procedure to design one SIW antenna array at 24 GHz bands. As an example, an antenna is designed on the substrate of RO4003C with thickness of 0.8128 mm.

Step 1: Microstrip-SIW-microstrip transition is designed to connect the SIW and the feeding structures.

As shown in Figure 3.1a, the light gray rectangular portion connected to two stubs is metal and the two arrays of small circles are the top of the metalized through vias. The vias connect the ground

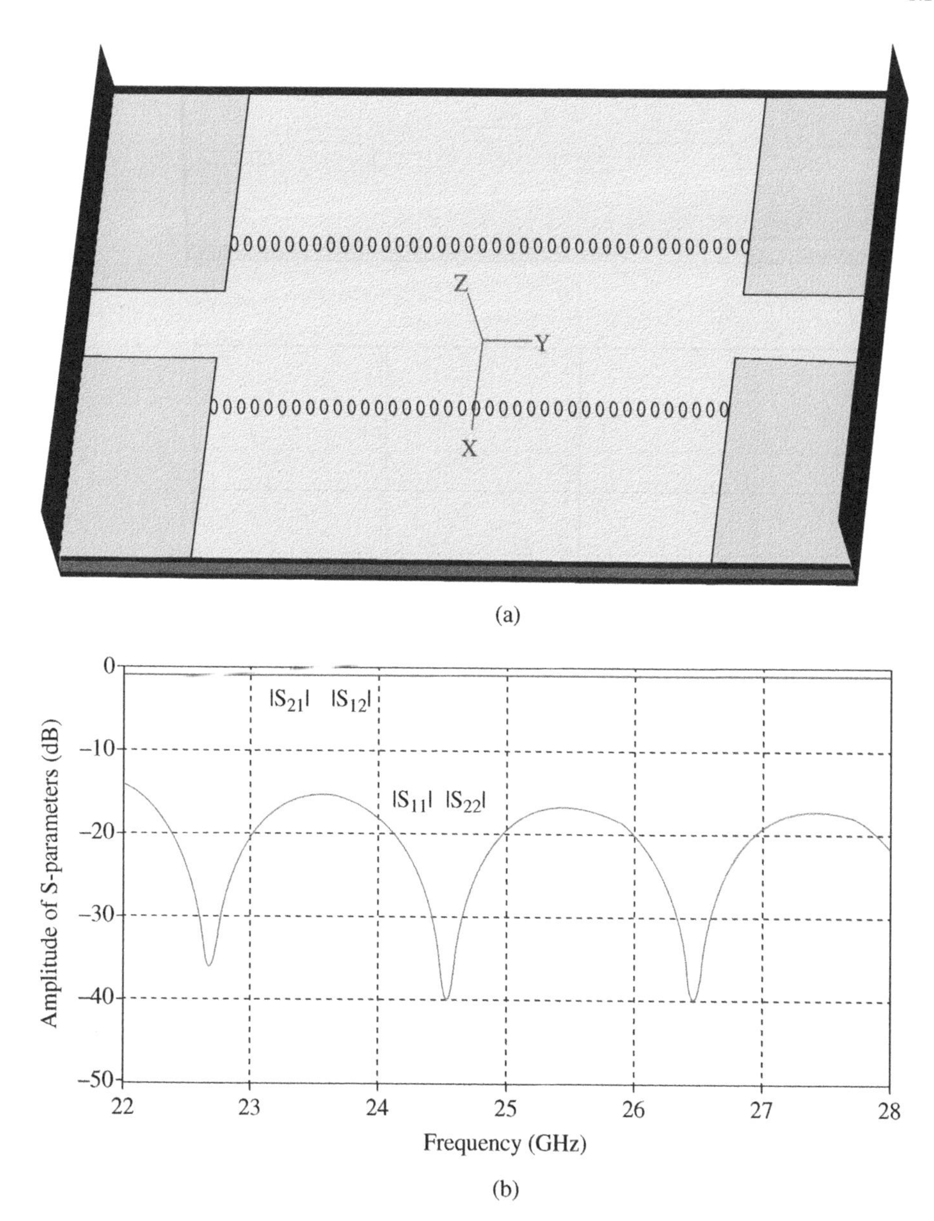

Figure 3.1 (a) Geometry of the SIW simulation model and (b) the amplitude of the simulated S-parameters.

plane fully covered the bottom of the substrate and the tope metal layer. At the two ends, two 50-Ω microstrip lines directly and centrally connected to SIW ends. The brown walls at the two ends of substrate board are feeding ports in simulation. Figure 3.1b shows the amplitude of the simulated S-parameters. It is seen that over the bandwidth of 22–28 GHz, the simulated $|S_{11}|$ and $|S_{22}|$ are identical and less than −15 dB while the transmission $|S_{21}|$ and $|S_{12}|$ are identical as well and less than −1 dB.

Therefore, the design of microstrip-SIW-microstrip transition is acceptable.

Step 2: A 1 × 8 SIW linear slot array is designed as a subarray.

Figure 3.2a shows the configuration of the sub-array. The detailed design procedure of the SIW slot array can be found in many literatures [10, 11]. Figure 3.2b shows the simulated impedance

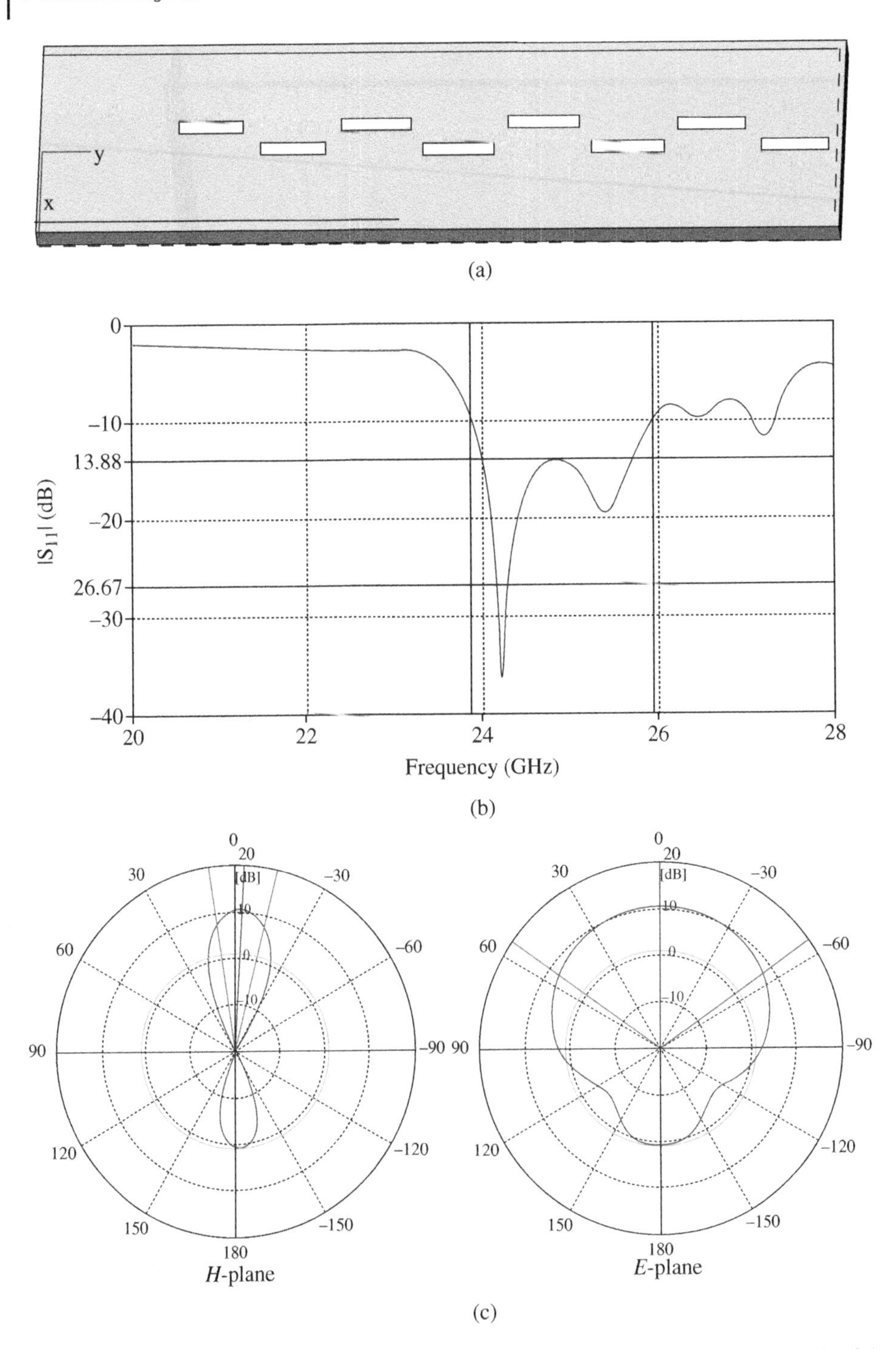

Figure 3.2 (a) The configuration of a 1 × 8 SIW linear array in simulation, (b) the amplitude of the simulated S_{11}, and (c) the simulated radiation patterns in *H*- and *E*-plane at 24 GHz.

matching; $|S_{11}| \leq -10$ dB is observed over the frequency range of 24–26 GHz. Figure 3.2c shows the simulated H- and E-plane radiation patterns at 24 GHz. The array achieves a gain of 10.3 dBi at 24 GHz. The E-plane beamwidth (110°) is much wider than that of the H-plane (24°) because of the smaller ground plane size in the E-plane. In addition, the front-to-back radiation ratio is 9.2 dB at 24 GHz.

Step 3: With eight 1×8 SIW linear slot subarrays, an 8×8 SIW planar slot array is designed for high gain.

Figure 3.3a shows the antenna array configuration and simulated results. An SIW power divider is designed to feed all eight subarrays uniformly in amplitude and in phase. The power divider is directly connected to the microstrip-SIW transition designed above. Figure 3.3b shows that the $|S_{11}|$ is less than −10 dB over the frequency range of 23.7–25.9 GHz. Figure 3.3c demonstrates the H- and E-plane radiation patterns at 24 GHz. The planar array achieves a gain of 18.1 dBi at 24 GHz, which is 7.8 dB higher than those of the 1×8 SIW linear slot subarray. The beamwidths of radiation patterns in the H- and E-plane are 22.2° and 18.3° at 24 GHz, respectively. The first side lobe levels (SLLs) are −29.1 dB at 24 GHz.

Therefore, the design of the SIW slot antenna array with linearly polarized radiation at mmW bands is similar to conventional slotted waveguide arrays at lower microwave bands. However, special attention should be paid to the design of the microstrip-SIW transition and the amplitude-uniform and in-phase SIW power divider.

3.2 High-Gain mmW SIW Slot Antenna Arrays on LTCC

As mentioned in Chapter 1, the most critical design consideration of the mmW antennas in general is how to achieve high antenna efficiency. In particular, the antenna efficiency issue becomes much severer in the design of large-scale mmW antenna arrays for high gain. The large-scale array must be fed by complicated and large-scale feeding networks in planar designs. The long power path of feeding network significantly increases the losses caused by materials. As mentioned in Section 1.6.2, the SIW is a good option for low-loss transmission at mmW bands.

Another challenge is how to keep the direction of maximum radiation consistent over the wide operating bandwidth, for example, 57–64 GHz. Thus, corporate feeding structures are selected to achieve symmetrical and in-phase feeding power division over a wide operating bandwidth.

3.2.1 SIW Three-Dimensional Corporate Feed

As mentioned in Chapter 1, one of the features of the LTCC process is its flexibility to form the metalized vias on each layer so that it is easy to achieve the connection across layers. With such flexibility, it is feasible to form waveguide structures in any layers even cross layers in LTCC to form three-dimensional waveguide structures. Thus, it is feasible to form the three-dimensional power dividers in LTCC. This is very important to form a symmetrically parallel feeding structure rather than the series feeding structure for a large antenna array operating over a wide bandwidth. Using the parallel feeding structure, the maximum radiation direction is consistent or independent of the frequency over the operating bandwidth [12].

Figure 3.4a shows a side view of a 60 GHz multilayered SIW fed cavity array antennas (on the left-hand side) with a transition (on the right-hand side) in LTCC [13]. There are a total of 20 layers of LTCC sheets. The SIW functional blocks are situated in the gray region. The feeding structure of the array is a three-dimensional symmetrically parallel configuration implemented in multilayered LTCC. The SIW is formed with metallic broadwalls (top and bottom walls) and closely

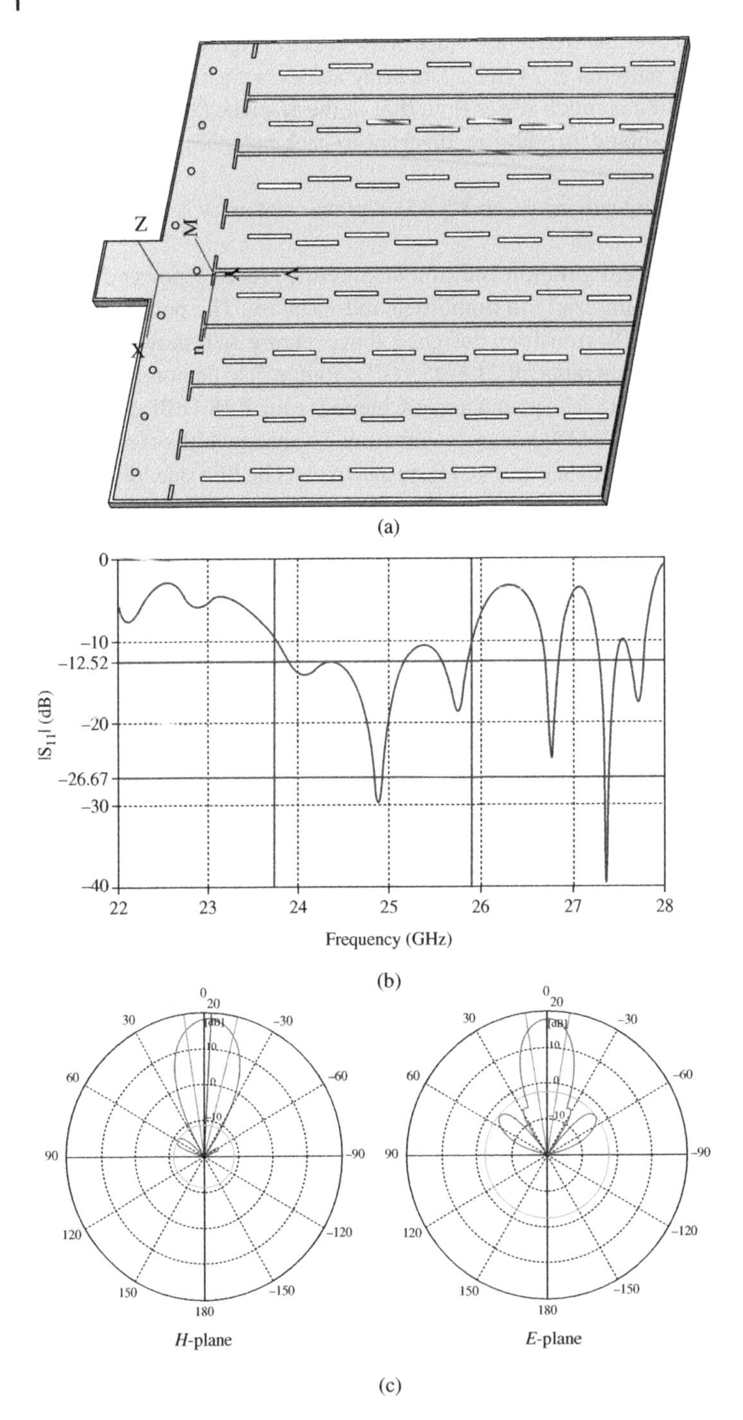

Figure 3.3 (a) The configuration of an 8 × 8 SIW linear array in simulation, (b) the amplitude of the simulated S_{11}, and (c) the simulated radiation patterns in *H*- and *E*-plane at 24 GHz.

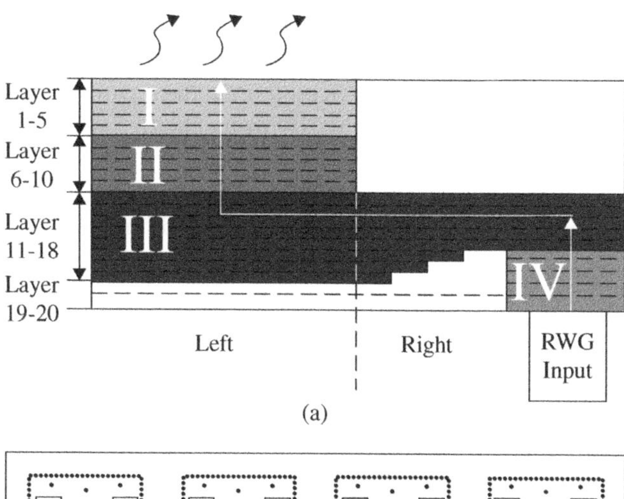

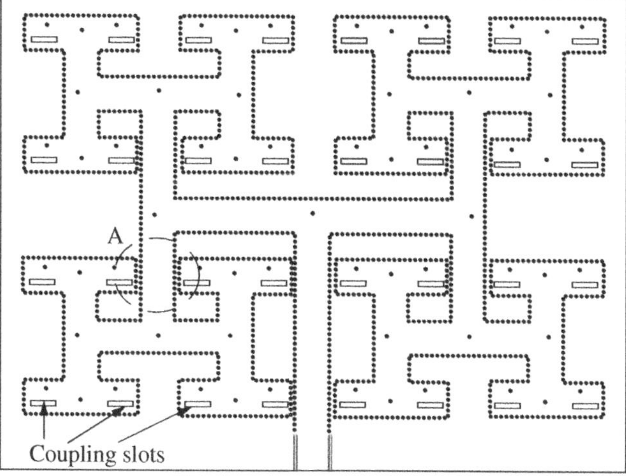

Figure 3.4 (a) Side view of an SIW fed cavity array antenna in LTCC substrate and (b) The top view (Layer 11) of three-dimensional power divider beneath the subarray layer.

aligned metallic via arrays functioning as sidewalls, which are electrically connected to the top and bottom metallic layers. The via-arrays stop radio frequency (RF) power leaking out from the waveguide structure. The arrows indicate the path of the RF power flowing through the SIW regions from the input of the RF power (Region IV) to the array of radiators (substrate integrated cavities and subarrays in Regions I and II) through the feeding structure (Region III). Each LTCC layer is with a co-fired thickness of $b_1 = 0.095$ mm. The material of LTCC substrate is Ferro A6-M with $\varepsilon_r = 5.9 \pm 0.20$, $\tan\delta = 0.002$ at 60 GHz [14].

Figure 3.4a shows that the power coupled from Region IV flows into Region III. Region III homes a 4×8-way power divider. The top view of Region III is shown in Figure 3.4b. The arrays of dots indicate metallized vias of the side walls of SIWs. The output of the power divider goes through the 4×8 coupling slots (broadwall couplers). Each slot couples the power upward into a two-element subarray in Region II. The thickness of Region III affects the resonant frequency of the couplers. The eight-layer thickness is selected for the resonant frequency at 60 GHz.

As part of the feeding structure, the broadband transition between the SIW and the rectangular waveguide (RWG) are designed in Region IV for measurement. The RF power is coupled through the feeding aperture on the bottom of the substrate integrated cavity (SIC). On the top of the SIC (top of 16th layer), there is a coupling slot on the broadwall of the SIW to couple RF power into Region III.

With the high dielectric constant of the LTCC substrate, a five-layer SIC is formed to achieve the broadband transition between the SIW and RWG as shown in Figure 3.5. The dashed-line rectangle in the top view indicates the feeding aperture connected with the RWG as shown in Figure 3.5a. The simulation model includes the RWG as port 1, the transition, and the SIW with width W3 as port 2, as shown in Figure 3.5b. The cavity thickness greatly affects the impedance bandwidth. Figure 3.5c shows that the transit achieved the wide bandwidth for $|S_{11}| < -20$ dB of about 18% with the maximum insertion loss of 0.8 dB.

3.2.2 Substrate Integrated Cavity Antenna Array at 60 GHz

With the feeding structure, an SIC antenna array operating at 60 GHz is designed in LTCC. The design includes the SIC array in Region I and the SIW slot subarrays in Region II. The design follows the following steps:

Step 1: Design of the radiation element.

Figure 3.6 shows the top view of the open-ended cavities (layers 1–5), the apertures on the top of layer 1 in LTCC. Figure 3.6a shows the radiating element formed by a five-layered rectangular open-ended SIC with via array sidewalls. The open radiating aperture (dashed-line rectangle) is cut onto the top surface of the LTCC substrate in the screen-printing process. The cavity is fed by a transverse feeding slot cut onto the center of the cavity bottom. Figure 3.6b,c show the three-dimensional configuration of a radiating element fed by a transverse feeding slot in simulation as well as the detailed rectangular radiating open-ended SIC and its transverse feeding slot. By changing the dimensions of the rectangular feeding slot, the impedance matching can be achieved over a 23% bandwidth for $|S_{11}| < -10$ dB as shown in Figure 3.7. The simulated mutual coupling for two configurations, namely, two slots aligned in the co-*H*- and co-*E*-planes, is exhibited in Figure 3.7 as well, which is lower than −22 and −16 dB over the bandwidth of 50–70 GHz, respectively. The simulated gain of the single radiating element is higher than 6.7 dBi.

Step 2: Design of the SIW slot coupler.

The broadwall coupler is the key component to achieve the vertical RF power transmission in a multilayered LTCC structure. Figure 3.8a,b shows two types of the cross broadwall coupler between Region II and Region III, namely, the two crossed SIW and an offset longitudinal coupling slot with three ports. In the design of array, Port 2 and Port 3 are connected with the feeding slots to form a two-element subarray as shown in Figure 3.9.

Figure 3.10 shows the simulated S-parameters of the parallel and cross couplers. Both the parallel and cross couplers with the optimized dimensions of eight-layered structure achieve $|S_{21}|$ and $|S_{31}|$ of −3.38 and −3.49 dB, respectively, at 60 GHz. The insertion loss of the coupler is less than 1.12 dB over 55–65 GHz.

Step 3: Design of the two-element subarray.

As shown in Figure 3.11, the two-element subarray are the elements of the array. With the design of the subarrays, then a large-scale array can be configured accordingly as shown Figure 3.9. With the symmetry of parallel-feeding structure, it is flexible to select the element spacing, even larger than one guided wavelength to decrease the mutual coupling caused by the high dielectric constant of the ceramic. However, it should be noted that the element-to-element spacing does not exceed

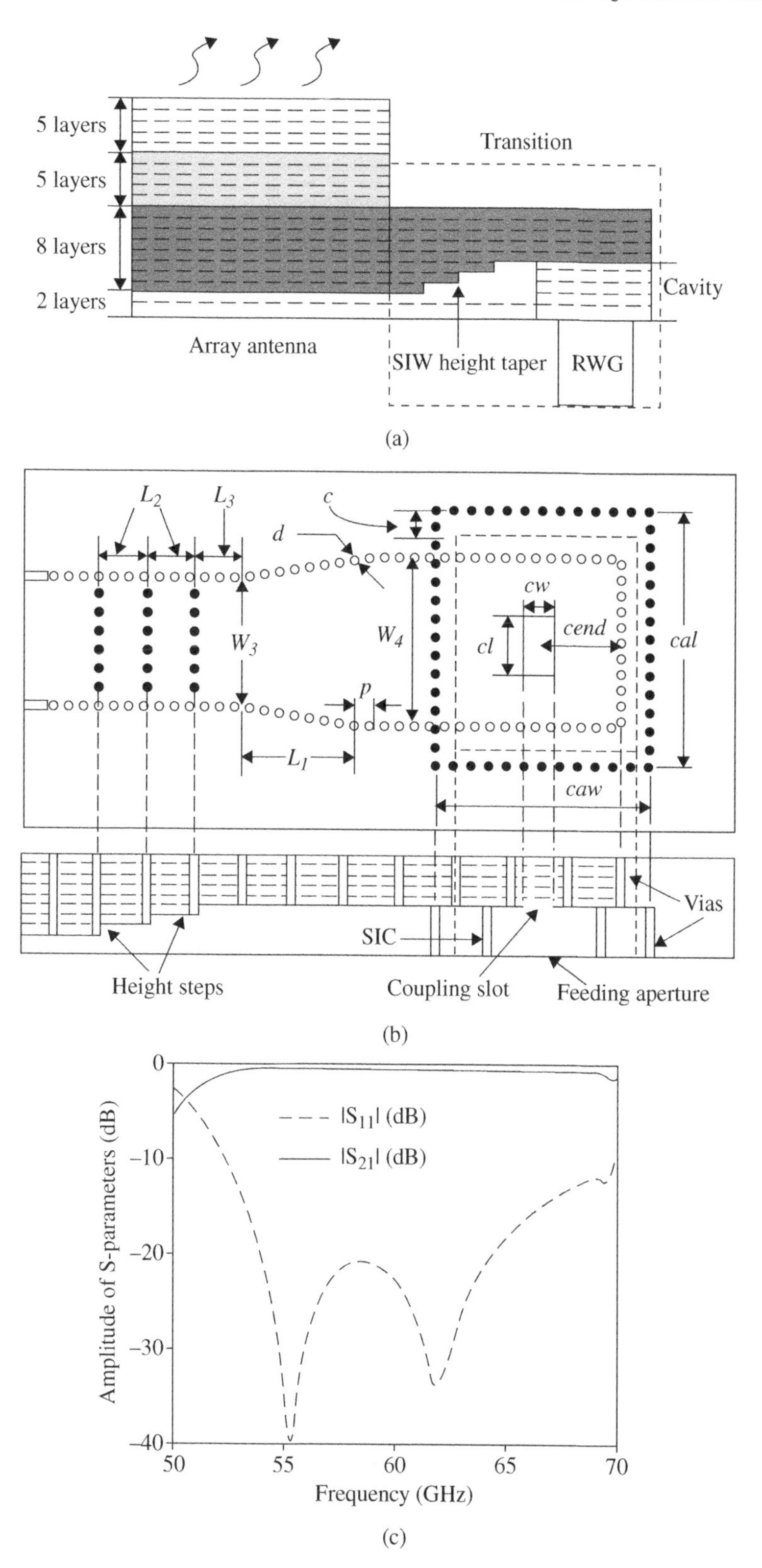

Figure 3.5 (a) Side view of the multilayer SIW-RWG transition in LTCC substrate, (b) top view of the multilayer, and (c) The simulated $|S_{21}|$ of the transition between the output of RWG and input of SIC as well as reflection at the input of RWG.

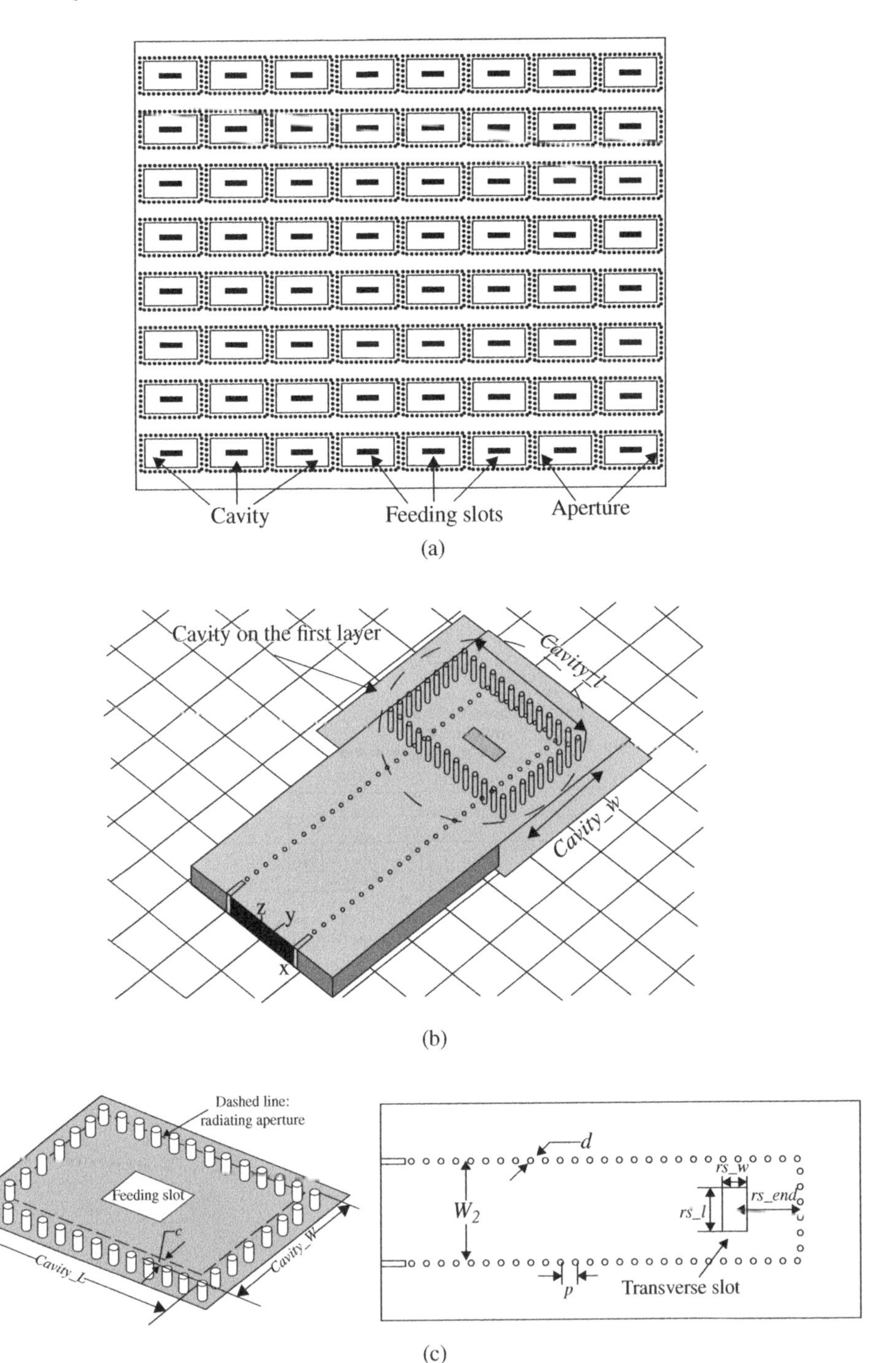

Figure 3.6 (a) Top view of apertures on the top of Layer 1 of open-ended cavity (Layer 1–5), (b) three-dimensional configuration of radiating element fed by a transverse feeding slot in simulation, and (c) the detailed rectangular radiating open-ended SIC and its transverse feeding slot.

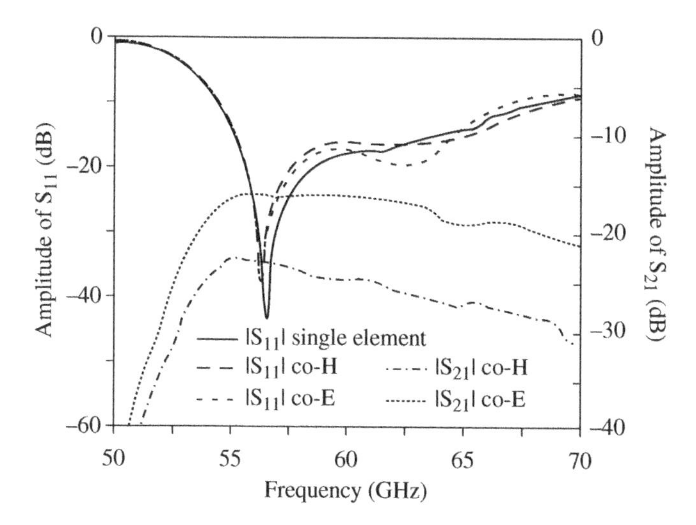

Figure 3.7 $|S_{11}|$ of single element and the mutual couplings in two configurations.

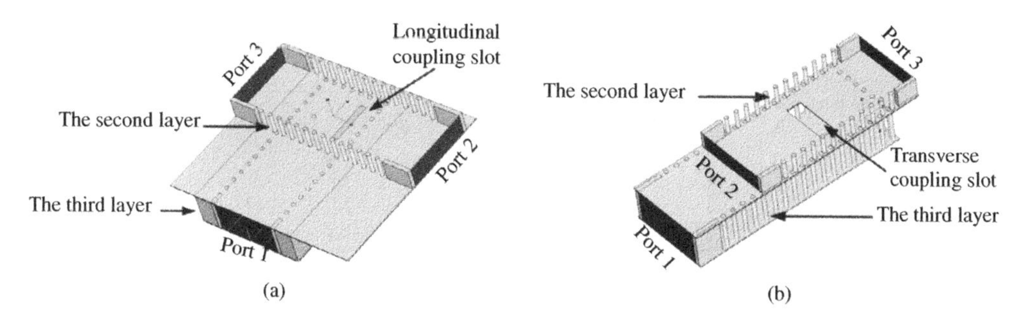

Figure 3.8 (a) The model of parallel coupler in simulation and (b) the model of cross coupler in simulation.

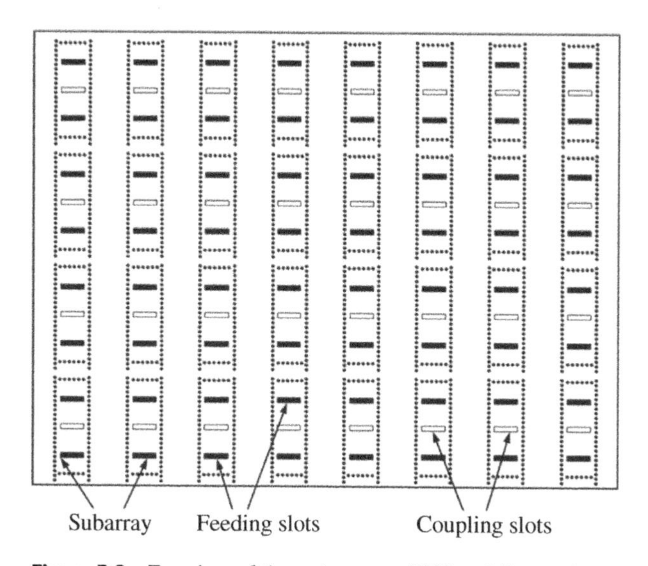

Figure 3.9 Top view of the antenna on LTCC multilayered structure of two-element subarray (Layer 6–10), feeding slots on the top of Layer 6.

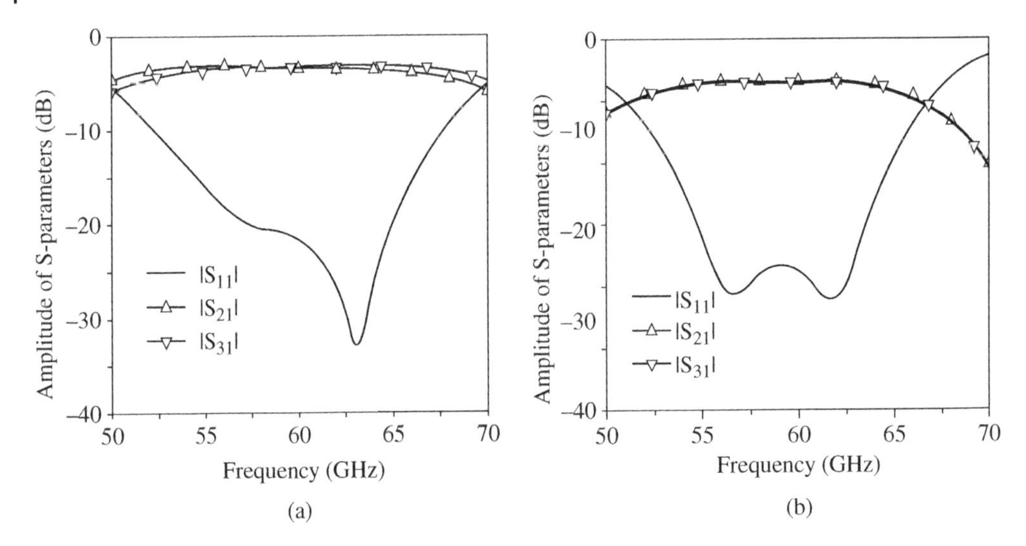

Figure 3.10 S-parameters of the eight-layered couplers. (a) Parallel coupler and (b) cross coupler.

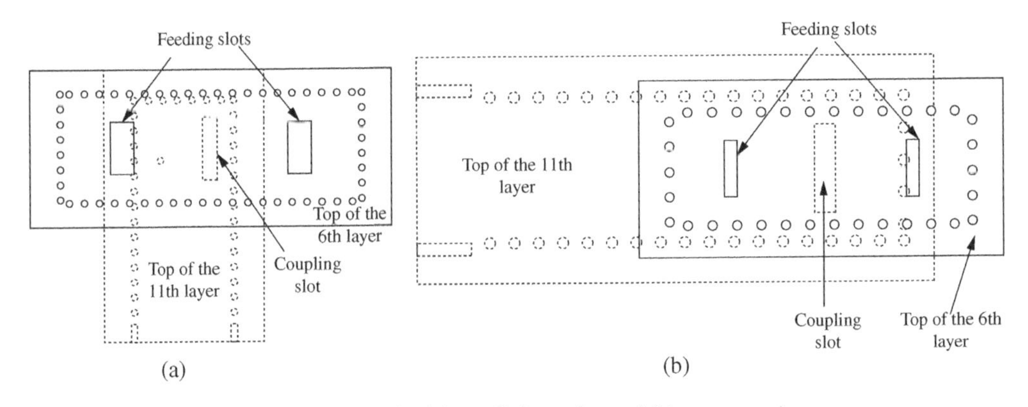

Figure 3.11 Two-element subarrays with (a) parallel coupler and (b) cross coupler.

one wavelength in free space to avoid grating lobes according to the array theory. Both the subarrays achieve the impedance matching over the bandwidth of 55–65 GHz.

Step 4: Design and verification of the 8×8 array.

As shown in Figure 3.9, the two-element subarrays are used to form an 8×8 array. The LTCC prototype of the array is shown in Figure 3.12a. At the bottom, the white rectangular aperture is the feeding input portion. The other portion is the metal ground. The overall size of the 8×8 array with transition is 47×31 mm. The measured and simulated $|S_{11}|$ of the 8×8 array are compared in Figure 3.12b. It is seen that the bandwidth for $|S_{11}| < -10$ dB is 17.1% or 54.86–65.12 GHz.

It is important to verify the radiation performance of the array. Besides the fabrication, the measurement is another challenge in the mmW antenna design, In particular, for high-gain designs. Figure 3.13 compares the measured and simulated radiation patterns of the array antenna in both the *E*- and *H*-plane. The measurement was conducted in a mini anechoic chamber as mentioned in Chapter 2. It is seen that the direction of the main lobe keeps consistently pointing to boresight and the SLLs are nearly −13 dB below the main beam levels over the whole operating bandwidth of 17% because of the symmetrically parallel-feeding network. More importantly, Figure 3.14 shows a consistent gain response with a gain variation smaller than 2.5 dB over the

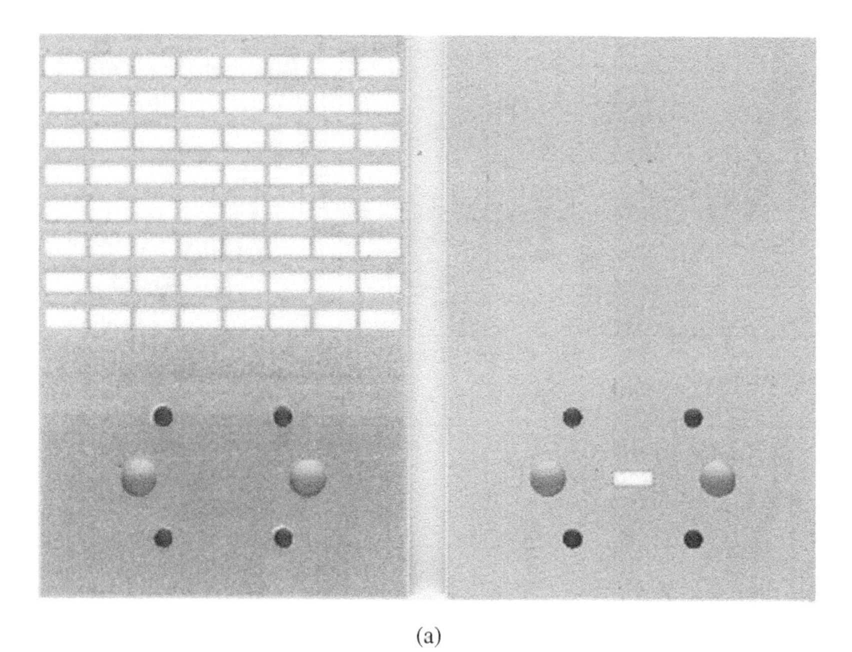

(a)

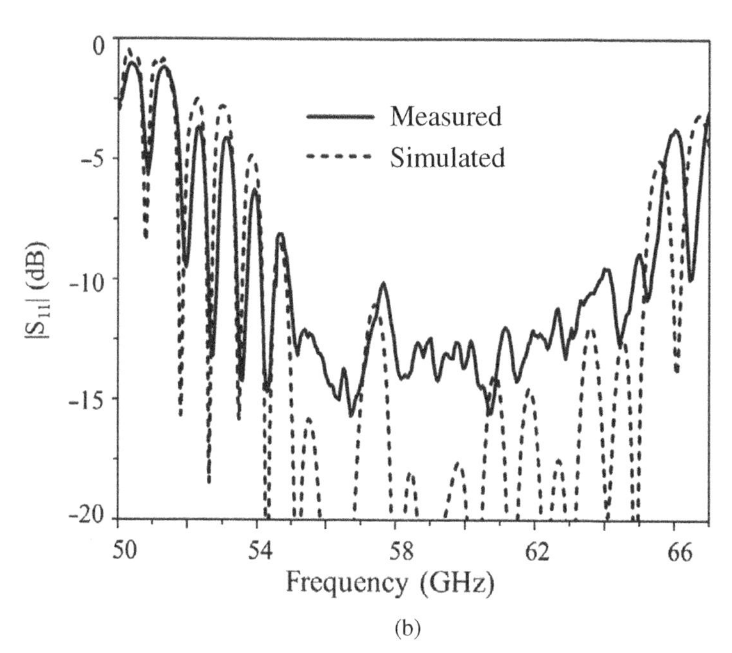

(b)

Figure 3.12 (a) Photograph of antenna array (left: top view, right: bottom view) and (b) $|S_{11}|$ of the array antenna.

operating bandwidth of 17.1%. The measured and simulated gains reach up to 22.1 and 23.0 dBi at 60 GHz, respectively as tabulated in Table 3.1.

Step 5: Loss analysis of the array.

Different from the antenna design at lower frequencies, the loss analysis in mmW antenna design is very important and necessary because the information is helpful in the suppression of loss in the design. The total array loss of 2.6 dB at 60 GHz is estimated by simulation. The insertion loss caused

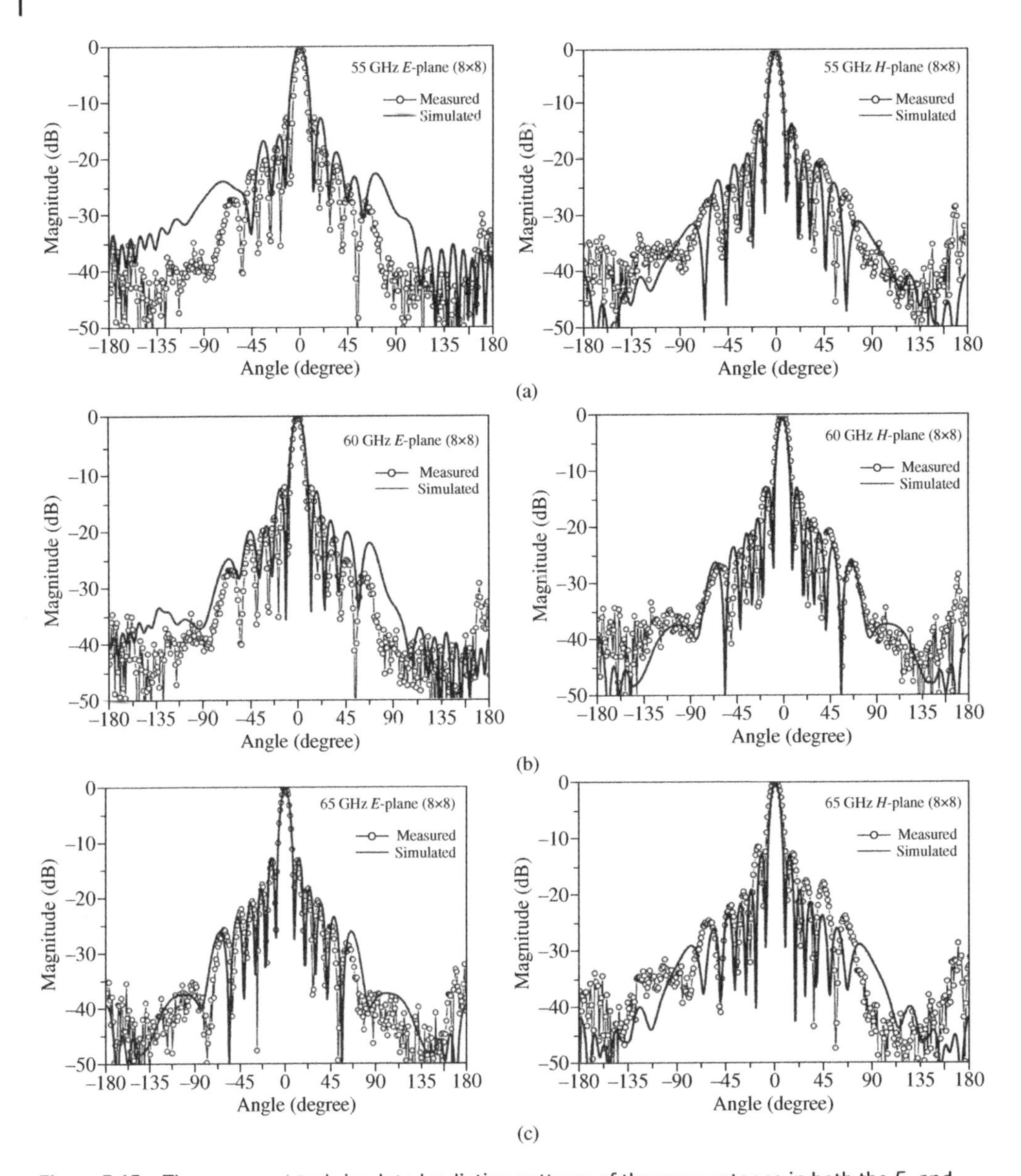

Figure 3.13 The measured and simulated radiation patterns of the array antenna in both the *E*- and *H*-plane at (a) 55 GHz, (b) 60 GHz, and (c) 65 GHz.

by the SIW-RWG transition and the 12-mm-long SIW feed line is 0.9 dB. The insertion loss of the broadwall coupler is 0.5 dB. The mismatch loss is 0.2 dB. The dielectric loss and the conductor loss in the feeding network are 1 dB. The measured and simulated efficiency are, respectively, 44.4% and 54.7% at 60 GHz.

In summary, based on the unique multilayered LTCC process a symmetrical parallel-feed structure can be realized for achieving consistent radiation performance. Also, the antenna array benefits from the high efficiency by the usage of electromagnetically closed and low-loss SIW structure. The loss reduction is another important design consideration of mmW antenna design, in particular for large-scale arrays because of usual high losses caused by materials and surface waves. The latter will be discussed in Chapter 8.

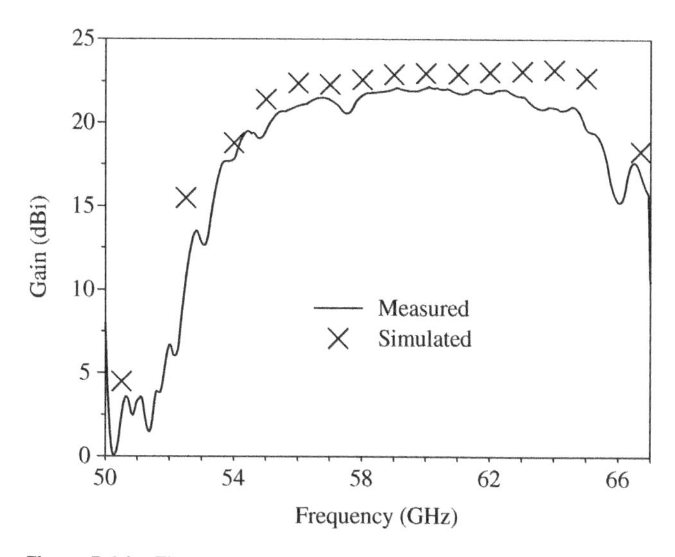

Figure 3.14 The measured and simulated gains of the 8 × 8 array.

Table 3.1 Gain and efficiency of the 8 × 8 array.

Frequency	55 GHz	60 GHz	65 GHz
Gain (dBi)	21.6	23.1	22.9
Efficiency	47.6%	55.2%	46%

3.2.3 Simplified Designs and High-Order-Mode Antenna Array at 140 GHz

D-band or 140 GHz band (110–170 GHz) has been allocated by the Federal Communications Commission (FCC) for radio astronomy, satellite communications, and applications in the allocated industry, scientific, and medical (ISM) band around 122 GHz [15]. Recently D-band applications such as imaging and radar systems have been proposed [16, 17]. Conventionally the antennas operating at 140 GHz include horns, reflectors, hollow-waveguide-fed antennas, quasi-optical antennas, antenna array integrated into a benzocyclobutene (BCB) membrane backed by low-loss polymer-filled cavities, and so on [18–22]. Compared with the antenna operating at 60 GHz bands, the integration of antennas operating at 140 GHz bands with other circuits in substrates is more important and preferred because the antenna is preferred to be integrated into package or even the system in package [23, 24] for the reduction of interconnection loss and fabrication cost as well as the enhancement of the fabrication reliability.

With the success in the design of three-dimensional feeding structure and large-scale array in LTCC at 60 GHz [13], the design of planar arrays at 140 GHz are carried out [25–29]. The study has shown that the fabrication tolerances severely affects the performance of the antenna at 140 GHz bands, and the shrinkage in LTCC process significantly increases the tolerance so that the antenna performance is significantly further degraded. Therefore, three solutions to 140 GHz SIW fed slot antenna arrays in LTCC have been proposed, namely, introduction of a large and simplified dielectric loading, the improved dielectric loading with a large via fence and apertures, and the higher-order mode instead of the dominant mode of the waveguide [25–28].

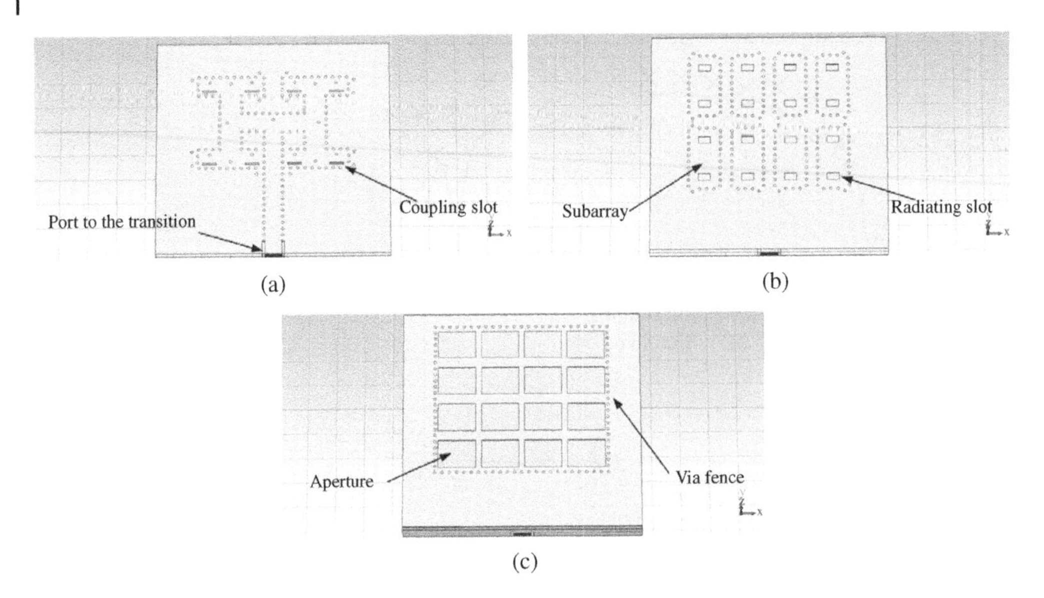

Figure 3.15 Configuration of the antenna array. (a) Power divider, (b) subarrays, and (c) dielectric loading.

3.2.3.1 140-GHz Slot Antenna Array with a Large-Via-Fence Dielectric Loading

In substrate IC design, the via arrays are used to form the electric walls. Not only the cost and tolerance of the via but also their pitch limits in LTCC become very critical at 140 GHz bands. Therefore, it is desirable to simplify the structure of the array but maintain the performance. Based on the technologies previously presented, the substrate integrated based antenna array design operating at 60 GHz bands is simplified to operating at 140 GHz bands by reducing the via fences.

Figure 3.15 shows the 4 × 4 SIW slot antenna array comprising an SIW-waveguide transition (layers 7, 8), power divider (layers 5, 6), subarrays (layers 3, 4), and dielectric loading (layers 1, 2) in an LTCC substrate. The input port in Figure 3.15a is linked to a transition for measurement through a transition while the incident power is distributed by the power divider and coupled upward into the two-element subarrays in Figure 3.15b. Replacing the dielectric-loaded SIC each enclosed by individual via fence with a large fence, a complete dielectric slab (multilayered LTCC substrate sheets) with the apertures above the radiating slots are formed as shown in Figure 3.15c. The other parts are the same as the design presented in the previous sub-section.

Figure 3.16a shows the cross-section view of an array in LTCC while Figure 3.16b shows the top view of the 4 × 4 dielectric loaded radiating slot array enclosed by a large fence. The antenna array consists of the dielectric loading in layers 1–2, the subarrays in layers 3–4, the power divider in layers 5–6, and the transition from the SIW to an external waveguide. The detailed view shows that there is a large via fence instead of individual via fence for each element as was explained in the previous sub-section. Removing all the internal vias will make fabrication easier and cost lower. The number of vias in this design is reduced by 60% if the 4 × 4 array design in the configuration used in [13].

Besides the large reduction of the numbers of vias, the performance of the array with a large-via-fence dielectric loading becomes insensitive against the fabrication tolerances. For instance, the effects of varying the *y*-direction strip width or spacing between two adjacent apertures, w_m, on the $|S_{11}|$ are shown in Figure 3.17. It is observed that the $|S_{11}|$ is consistent against w_m. A robust design of the antenna is important because fabrication tolerances are big issues at such a high frequency, and the shrinkage in LTCC technology is somewhat unpredictable, which increases the fabrication tolerances.

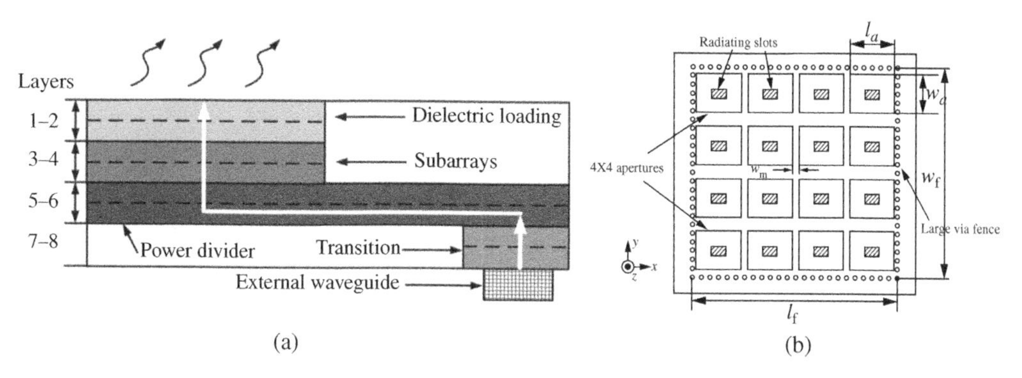

Figure 3.16 (a) The cross-section view of array in LTCC and (b) top view of a 4 × 4 dielectric loaded slot array enclosed by a large fence.

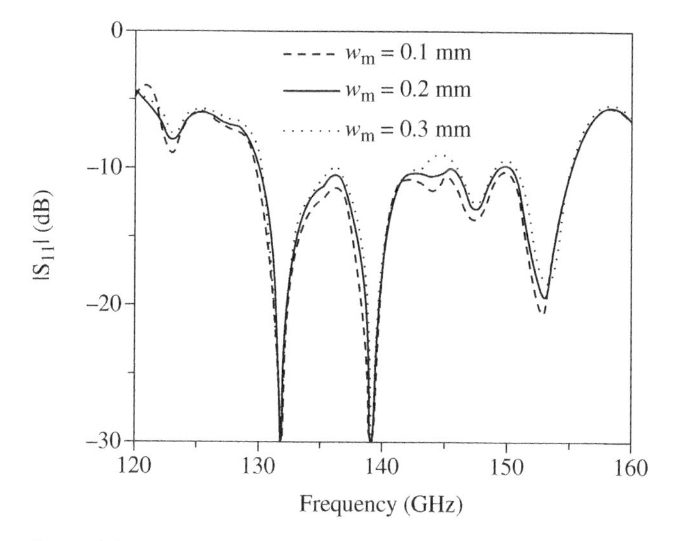

Figure 3.17 The simulated $|S_{11}|$ of the antenna array versus w_m.

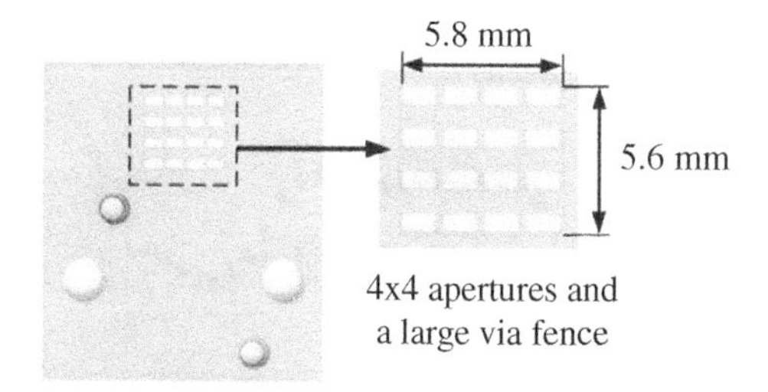

Figure 3.18 Photograph of the 4 × 4 antenna array prototype.

The antenna array was prototyped and fabricated using an LTCC multilayer process. The material of the LTCC substrate is Ferro A6-M with $\varepsilon_r = 5.9 \pm 0.20$, $\tan\delta = 0.002$ at 100 GHz. The co-fired thickness of each layer is 0.095 mm. The conductor used for metallization is Au, whose conductivity is 4.56×10^7 S m^{-1}. The photograph of the antenna prototype is shown in Figure 3.18. The size of the effective radiating aperture is 5.8 mm × 5.6 mm. The overall size of the antenna is 23 mm × 20 mm × 0.76 mm. The holes of screws and positioning are used to fix the standard flange (UG387/U) of WR-6 waveguide on the bottom of the antenna.

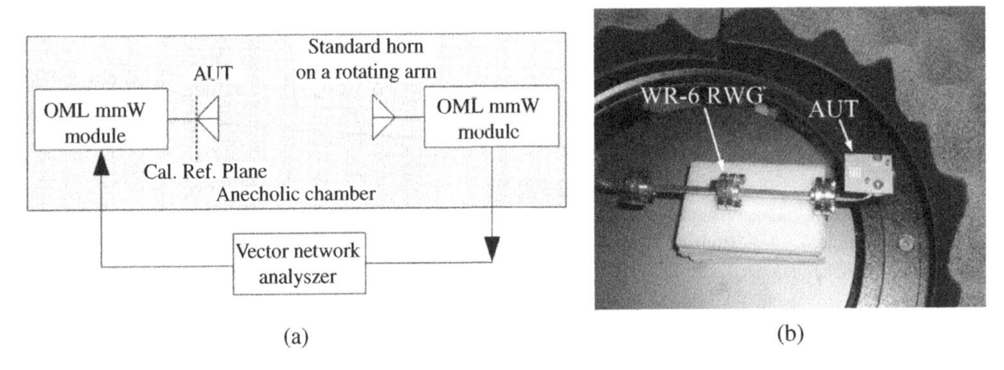

Figure 3.19 (a) Schematic of the measurement setup and (b) photograph of AUT in measurement setup.

The radiation patterns of antenna array prototype were measured in an anechoic chamber using a self-built far-field mmW antenna measurement system. The schematic of the measurement setup is shown in Figure 3.19. This measurement system can cover the D-band (110–170 GHz) antenna test in terms of return loss, gain, and radiation pattern. The calibration reference plane for return loss measurement is shown in Figure 3.19a. The horn antenna on an arm is rotated to measure the radiation pattern. The photograph of the antenna under test (AUT) is shown in Figure 3.19b. The WR-6 RWGs are used as the basic transmission lines in the test system.

Figure 3.20a shows that the measured and simulated $|S_{11}|$ is less than −7 and −9 dB over the bandwidth of 130–150 GHz, respectively. Figure 3.20a compares the measured and simulated boresight gains of the array. The measured gain reaches 15.6 dBi at 140 GHz and above 13.7 dBi in 130–150 GHz. The measured and simulated radiation patterns in both the *H*- and *E*-plane are shown in Figure 3.21, respectively. The measured patterns at 130 and 150 GHz are in good agreement with the simulated results. For the patterns at 140 GHz, the measured SLL is about 6 and 5 dB higher than the simulated results in the *H*- and *E*-plane, respectively because of the relatively lower gain of the amplifier used in the measurement system at 140 GHz.

In summary, the 4 × 4 SIW antenna array with a dielectric loading including a large via fence simplifies the fabrication and reduces the fabrication cost.

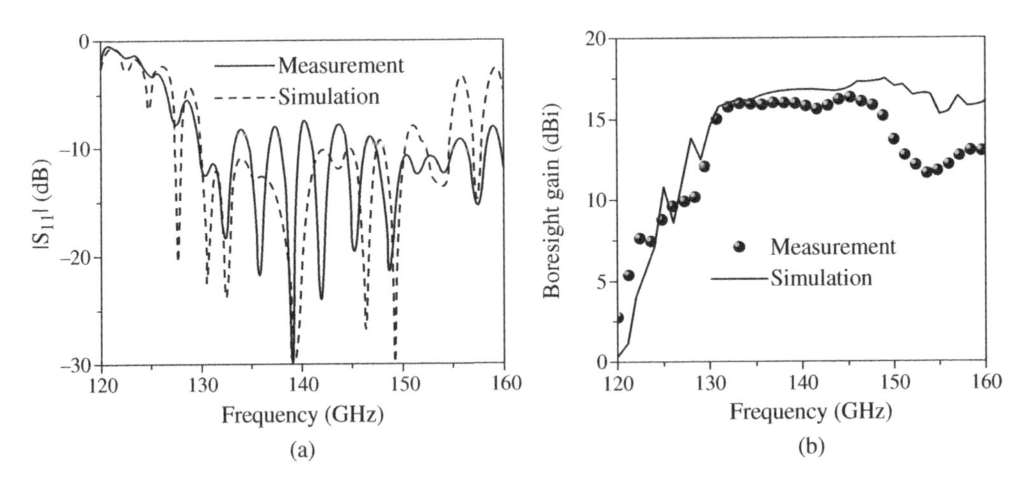

Figure 3.20 (a) Measured and simulated $|S_{11}|$ of the antenna array including the transition and (b) measured and simulated boresight gains of the antenna array including the transition.

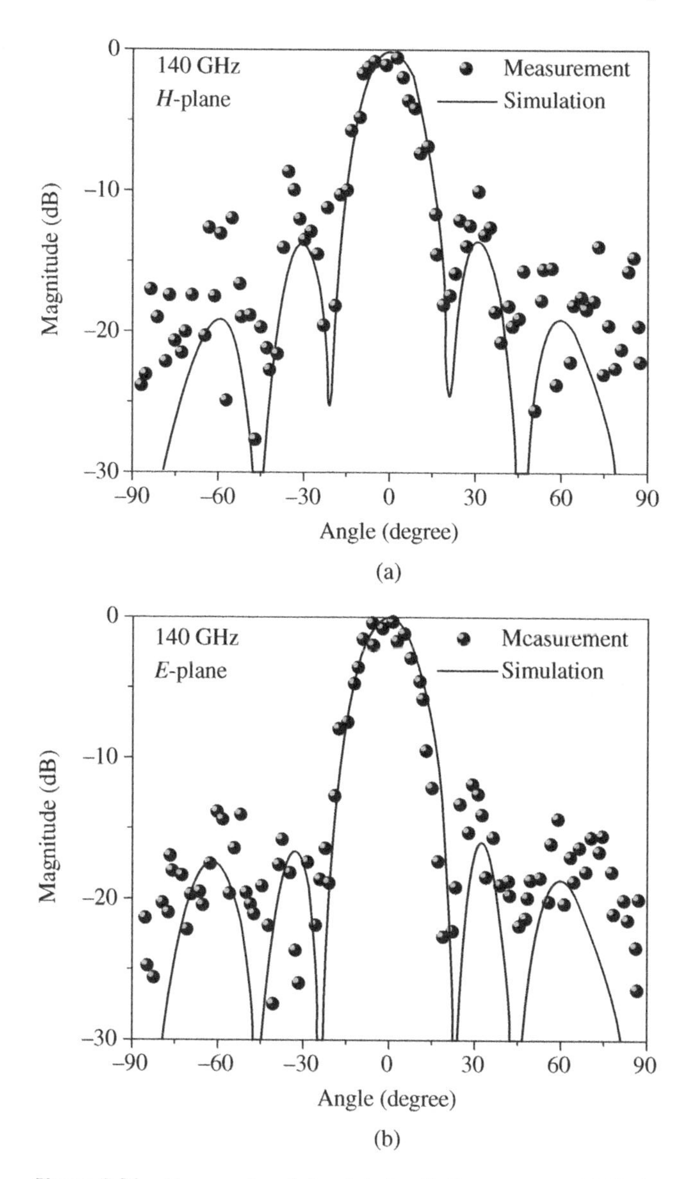

Figure 3.21 Measured and simulated radiation patterns. (a) *H*-plane and (b) *E*-plane radiation patterns.

3.2.3.2 140-GHz Slot Antenna Array with a Large-Via-Fence and Large-Slot Dielectric Loading

To further simplify the structure of the array in the previous sub-section, the antenna array operating at 140 GHz bands is designed with a large-via-fence and large-slot dielectric loading.

The array has a similar configuration to Figure 3.16a. To simplify the fabrication complexity and ease the fabrication limit, compared with the design of Figure 3.16b, the dielectric loading shown in Figure 3.22 is designed. The dielectric slab comprises of a large via-fenced structure in layers 1–2 and four large slots on top of layer 1 as shown in Figure 3.16a but the *y*-oriented narrow metal strips and most of the vias except the edge ones are removed. The dimensions of the dielectric loading are set to $s = 1.5$ mm, $w_{ac} = 1.1$ mm, $l_{ac} = 0.65$ mm, $c_x = 0.1$ mm, $c_y = 0.2$ mm. The dimensions of the subarrays, the power divider, and the transition are the same as the ones in Figure 3.16a.

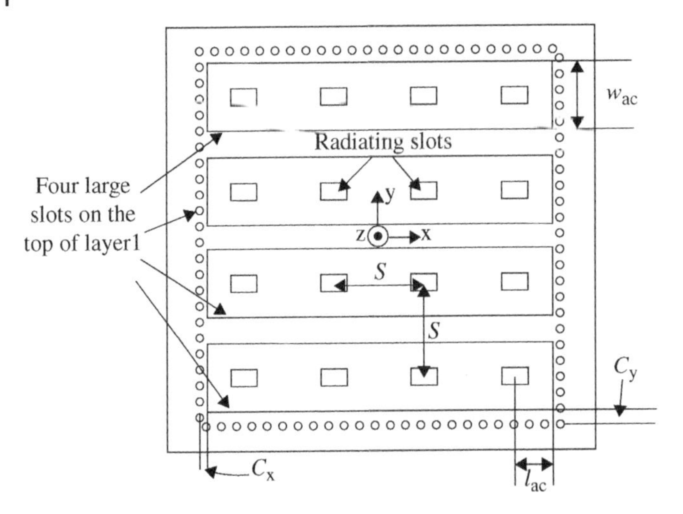

Figure 3.22 Large-via-fence and large-aperture dielectric loading of antenna array in LTCC.

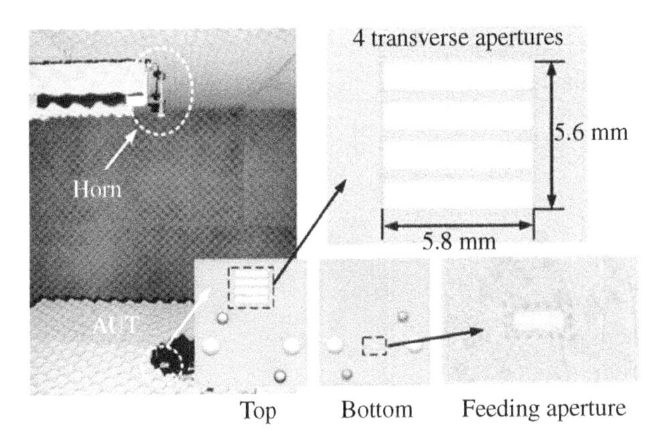

Figure 3.23 Photographs of the measurement setup with a standard horn and the antenna under test (AUT) in an anechoic chamber.

Figure 3.23 shows the prototype of a 4×4 antenna array with the large-via-fence and large-aperture dielectric loading. The LTCC substrate is Ferro A6-M with $\varepsilon_r = 5.9 \pm 0.20$ and $\tan\delta = 0.002$ at 100 GHz. Each co-fired layer is 0.095 mm thick. The conductor used for metallization is Au, with conductivity of 4.56×10^7 S m^{-1}. The antenna array was measured using an mmW antenna measurement system in an anechoic chamber as an inset shown in Figure 3.23.

Figure 3.24 shows the measured bandwidth of 126.2–158.2 GHz for $|S_{11}|$ less than −10 dB, agreeing with the simulation. Compared with the results in Figure 3.20a, the impedance performance in terms of $|S_{11}|$ is improved due to the added x-direction strips.

The measured and simulated gain of the antenna array at boresight including the transition is shown in Figure 3.25. The measured gain is 16.2 dBi at 140 GHz and above 14.1 dBi over the band of 130–152 GHz. The measured and simulated gain is in reasonable agreement.

Figure 3.26 shows the normalized radiation patterns at 140 GHz in both the H- and E-plane. Good agreement has been observed. The measured H-plane and E-plane SLLs are below −11 and −13 dB, respectively.

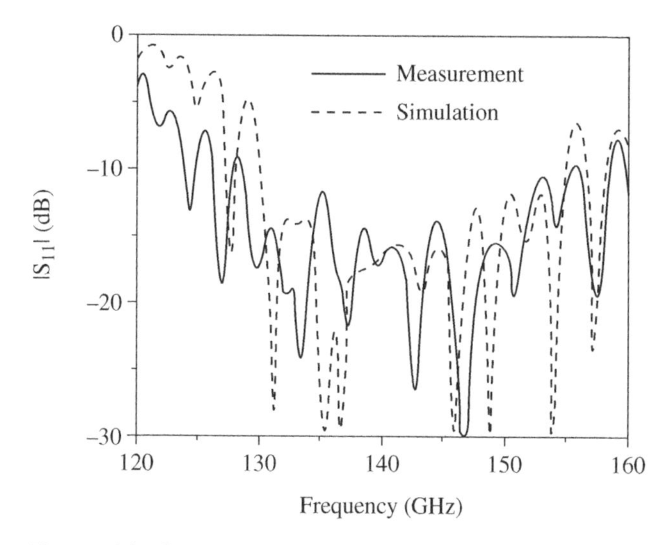

Figure 3.24 Comparison of measured and simulated $|S_{11}|$.

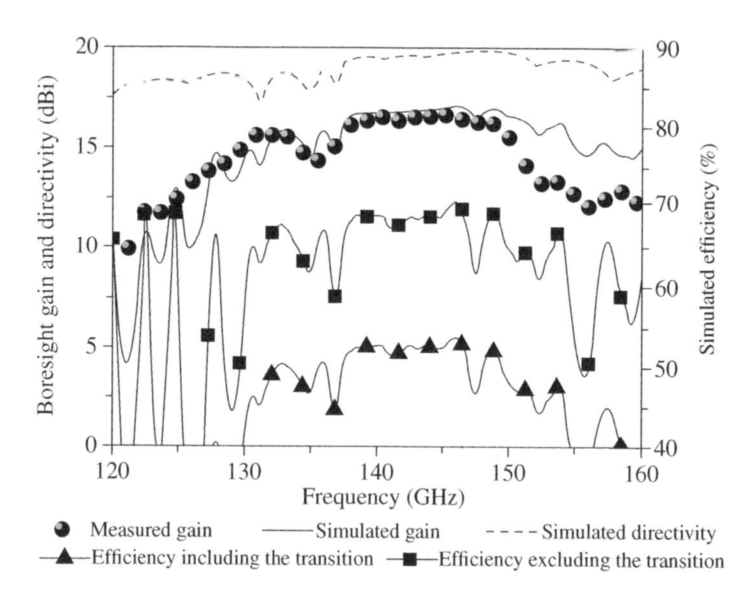

Figure 3.25 Comparison of measured and simulated gains as well as efficiency including and excluding the transition in the simulation.

In summary, using the four large aperture the fabrication of 140 GHz SIW-fed slot antenna array with a dielectric loading using a large via-fenced structure in LTCC has been further simplified. The design of the dielectric loading has removed the *y*-oriented narrow strips.

3.2.3.3 140-GHz Slot Antenna Array Operating at a Higher-Order Mode (TE_{20} Mode)

When a higher-order mode is used, the internal sidewalls are removed, so the width of waveguide is enlarged. Using a single-layer PCB, for instance, the higher-order modes in an SIW with an enlarged width were used to feed a linearly tapered slot antenna [30]. The method to use higher-order modes was also proposed in the design of hollow waveguides by reducing the sidewalls in order to reduce

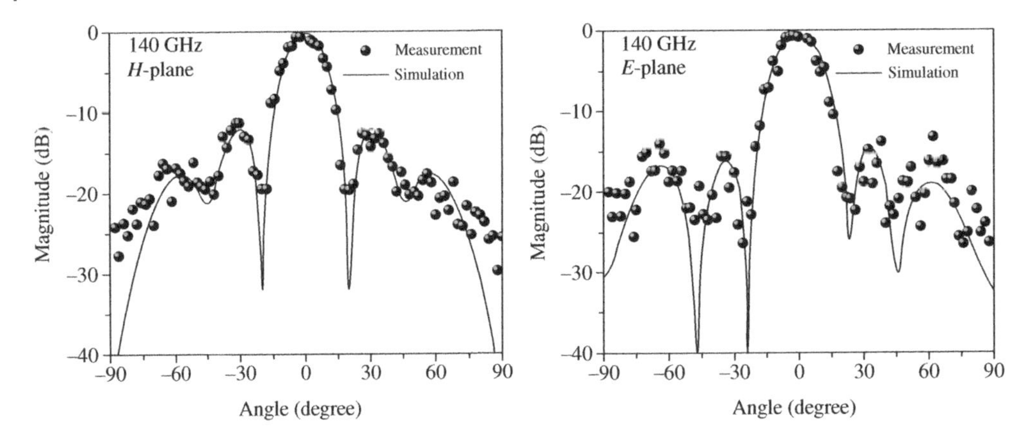

Figure 3.26 Measured and simulated radiation patterns at 140 GHz (a) in *H*-plane and (b) in *E*-plane.

the fabrication cost and ohmic loss of the antennas [31]. To further ease the fabrication and reduce the fabrication cost, a higher-order mode (TE_{20} mode) is used to simplify the structure of a multi-layered 8 × 8 array in LTCC.

The feeding network of the array is similar to the previous designs. To excite the higher order TE_{20} mode, the configuration is designed first. Figure 3.27 shows the top view of each of the functionality portions of the antenna array in LTCC. The portions include a power divider, *E*-plane couplers, four-element subarrays, and dielectric loadings.

The eight-way power divider in layers 8–9 consists of multiple cascaded T-junctions as shown in Figure 3.27a. The input of the power divider, plane *A*, is connected to an SIW-waveguide transition in layers 10 and 11. The incident power is distributed and coupled upward through the eight coupling slots on top of layer 8. Figure 3.27b shows the eight unique *E*-plane couplers in layers 6 and 7 to form the feeding network to support the TE_{20} mode. Each coupler divides the incident power upward into two subarrays through the coupling slots on top of layer 6. The TE_{20} mode is excited by the output of the coupler with a uniform amplitude and 180° phase shift.

Figure 3.27c shows the 16 subarrays formed in layers 3–5. Each subarray consists of two pairs of longitudinal radiating slots on top of layer 3. The slots are fed in parallel by the TE_{20} mode generated by the *E*-plane couplers. The longitudinal slots instead of transverse slots are used here because the longitudinal slots with opposite offsets generate the in-phase radiation. The spacing between the two slots along the *x*-direction is 1.9 mm when the spacing along the *y*-direction is 1.28 and 1.6 mm, respectively. The width of each subarray is doubled compared to the configuration using the dominant TE_{10} mode. Besides, the pair of the slots shares one dielectric loading with a doubled width via fence and aperture. Therefore, the number of vias is reduced and the structure is simplified.

Then the 8 × 4 dielectric loadings in layers 1 and 2 are formed as shown in Figure 3.27d. Each dielectric loading comprises a via fence and an aperture on top of layer 1. The dielectric loading positioned right above the pair of radiating slots is used to enhance bandwidth and gain of radiating slots. Finally, the antenna array was fabricated with 11 layers of LTCC substrate and the same materials as previous designs.

Design of Pair of Radiating Slots with a Dielectric Loading Figure 3.28a shows the three-dimensional schematic view of the pair of the radiating slots with a dielectric loading. The TE_{20} mode operation is simultaneously excited by ports 1 and 2 with the same amplitude and 180° out of phase. The two radiating slots are symmetrical in structure and location to avoid the excitation of other

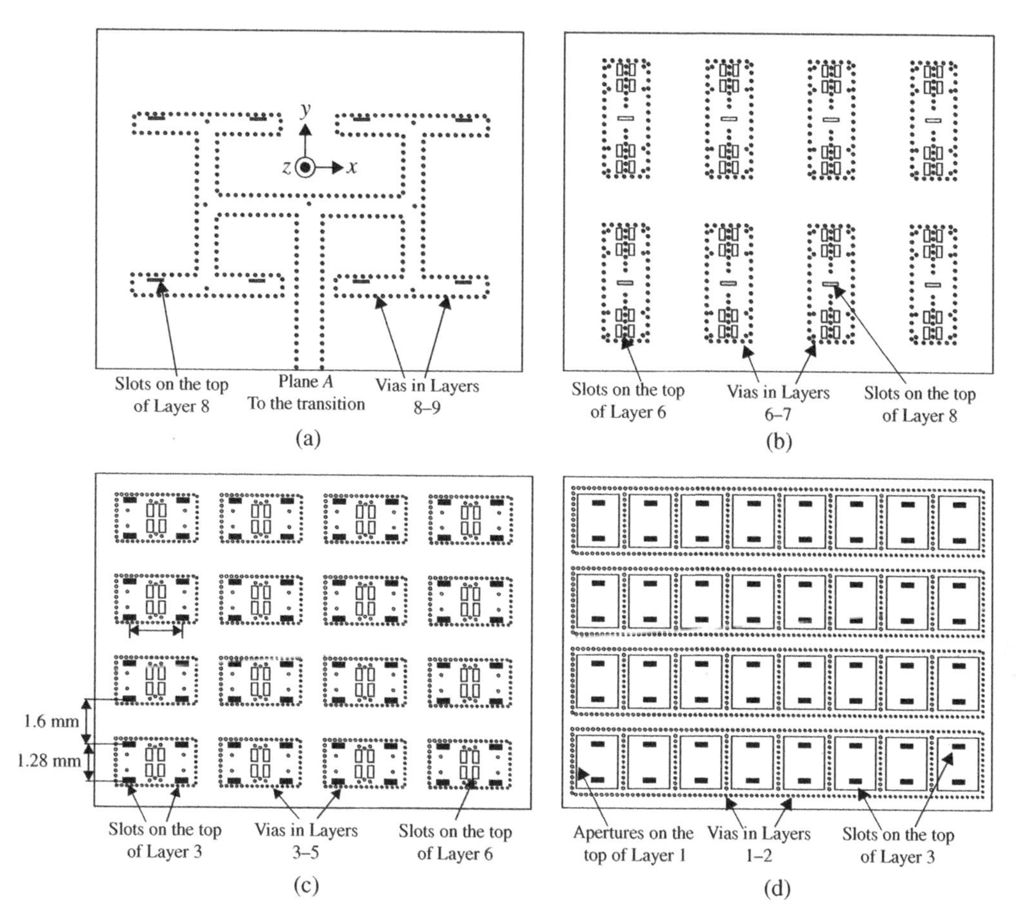

Figure 3.27 Top view of functionality portions of antenna array operating at higher order TE_{20} mode in LTCC. (a) Power divider in Layers 8–9, (b) *E*-plane couplers in Layers 6–7, (c) four-element subarrays in Layers 3–5, and (d) dielectric loadings Layers 1–2.

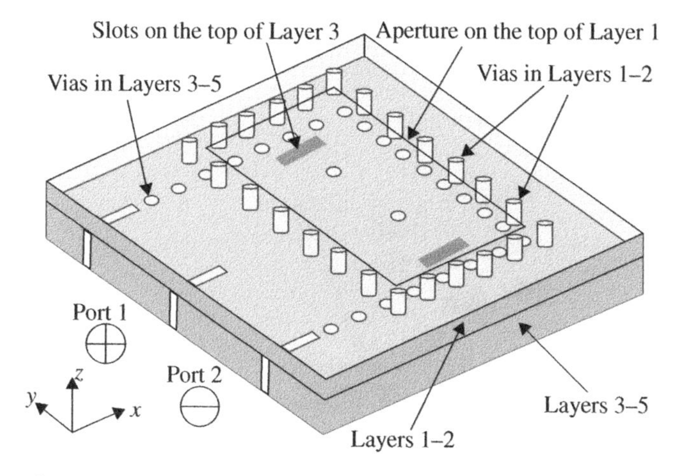

Figure 3.28 The three-dimensional schematic view of pair of the radiating slots with a dielectric loading.

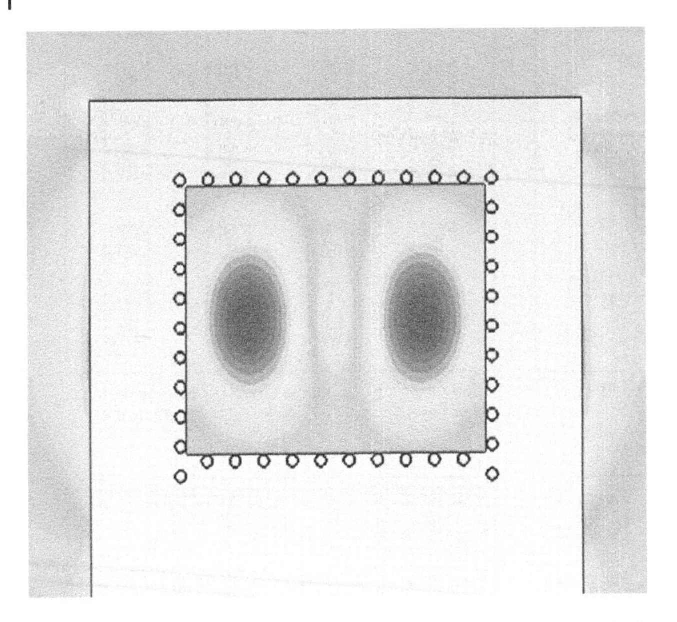

Figure 3.29 Simulated E_x component of pair of radiating slots with a dielectric loading on the top of Layer 1 at 140 GHz.

higher-order modes. The SIW with a width of w_1 can be considered as two branches of TE_{10} mode SIWs without a common sidewall. The dimensions and locations of the pair of the radiating slots are determined using a method similar to the TE_{10} mode longitudinal slot antenna design.

Figure 3.29 shows the simulated E_x component on top of layer 1. It is clear that a higher-order mode has been well excited inside the via fence. The simulated $|S_{11}|$ also shows the bandwidth for $|S_{11}|$ less than −10 dB is over 131–151 GHz.

Design of E-Plane Coupler Figure 3.30 shows the three-dimensional schematic view of the *E*-plane coupler. The incident power is excited by port 1 in layers 8–9. In the array, the incident power comes from the output of the pre-stage T-junction and is coupled into layers 6–7 through the slot

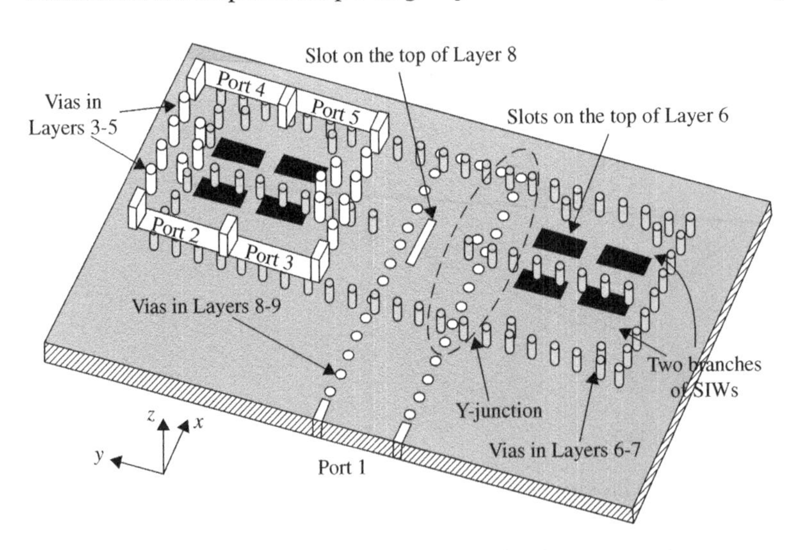

Figure 3.30 Three-dimensional schematic view of an *E*-plane coupler.

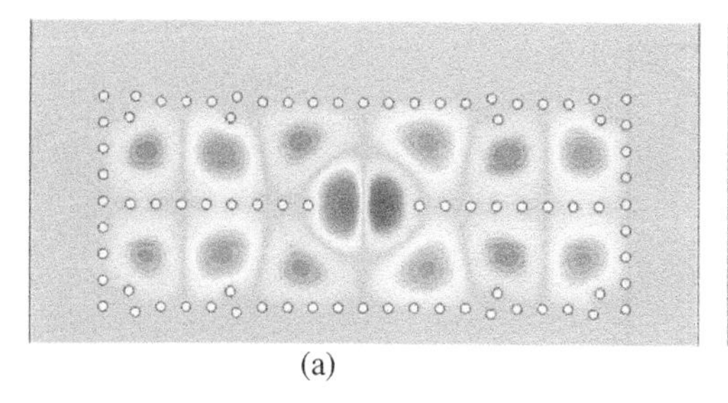
(a)

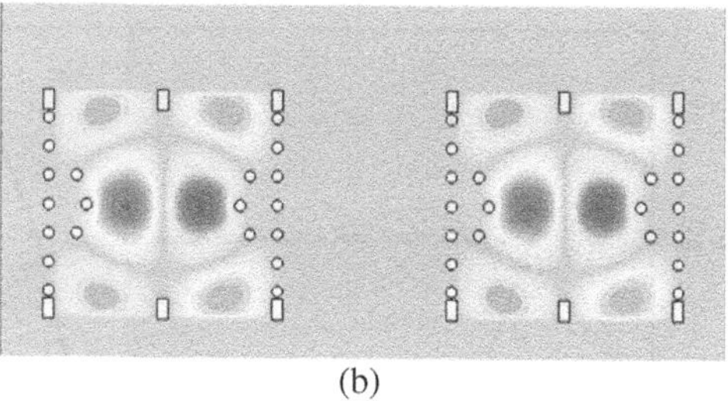
(b)

Figure 3.31 Simulated E_z component of the E-plane coupler at 140 GHz. (a) In Layer 6 and (b) in Layer 4.

on top of layer 8. The power transmitting along both $+y$ and $-y$ directions is further split by two "Y-junctions," namely, the dashed ellipse in Figure 3.30 and coupled into the four branches of SIWs. Each branch of the SIW has two coupling slots on top of layer 6. The two slots are located at a distance of a half guided wavelength of SIW so that the power coupled into layers 3–5 and transmitted into the two adjacent output ports, port 2 and port 3, is opposite in phase.

Figure 3.31 shows the simulated E_z component of the E-plane coupler at 140 GHz. In Figure 3.31b, the TE_{20} mode is well and symmetrically excited as expected.

The simulated S-parameters of the E-plane coupler are shown in Figure 3.32. Figure 3.32a shows the good impedance matching with the $|S_{11}|$ less than −20 dB over the bandwidth of 130–150 GHz. The mutual coupling between the ports $|S_{21}|$, $|S_{31}|$, $|S_{41}|$, and $|S_{51}|$ are −10.04, −9.94, −9.93, −9.83 dB at 140 GHz, respectively while Figure 3.32b shows the phases of S_{21}, S_{31}, S_{41}, and S_{51} are 129.4°, −50.4°, 129.5°, and −50.5°, respectively, very close to desired the 180° out of phase across the range of 130–150 GHz.

Design of an 8 × 8 Array An 8 × 8 array is designed and optimized by changing the dimension of v_2 to 0.41 mm because of the mutual couplings in the array environment. The 8 × 8 array is fabricated using the LTCC process. The photograph of the antenna prototype is shown in Figure 3.33a. The size of the effective radiating aperture is 15.1 mm × 10.8 mm. The overall size of the antenna prototype is 28 mm × 19 mm × 1.1 mm. The screw holes and positioning pin holes are used to connect the standard flange (UG387/U) of WR-6 waveguide to the bottom feeding aperture of the antenna array.

The photograph of the AUT in the measurement system is shown in Figure 3.33b. The accurate measurement of antenna gain is challenging at 140 GHz because of the difficulty of alignment between the AUT and the standard horn. For instance, the misalignment in the horizontal plane (xy-plane) causes the deviation of the measured gain from boresight. The rotation of the AUT around the z-axis causes additional polarization mismatch between the AUT and the horn, which may also lower the measured gain. Because of the small operating wavelength, for instance, 2.14 mm at 140 GHz, a careful calibration procedure is necessary for an accurate gain measurement.

Figure 3.34 compares the measured and simulated $|S_{11}|$ and achieved gain of the array with the transition. The simulated and measured bandwidths for $|S_{11}|$ less than −10 dB are 21.4 and 21.0 GHz or 128–149.4 GHz and 126.8–147.8 GHz as shown in Figure 3.34a. Figure 3.34b shows the simulated and measured maximum boresight gain of 21.6 dBi at 140 GHz and 21.3 dBi at 140.6 GHz, respectively, and the gain higher than 19.3 dBi over 130–150 GHz and 18.3 dBi over 129.2–146 GHz, respectively. The efficiency of the antenna array with the transition is calculated based on the simulated directivity and gain. As shown in Figure 3.34b, the simulated efficiency of the antenna array itself without the transition is about 46.6% at 140 GHz.

Figure 3.35 shows the measured and simulated normalized radiation patterns in both the H- and the E-planes at 140 GHz. All the main beams point at the boresight consistently without any

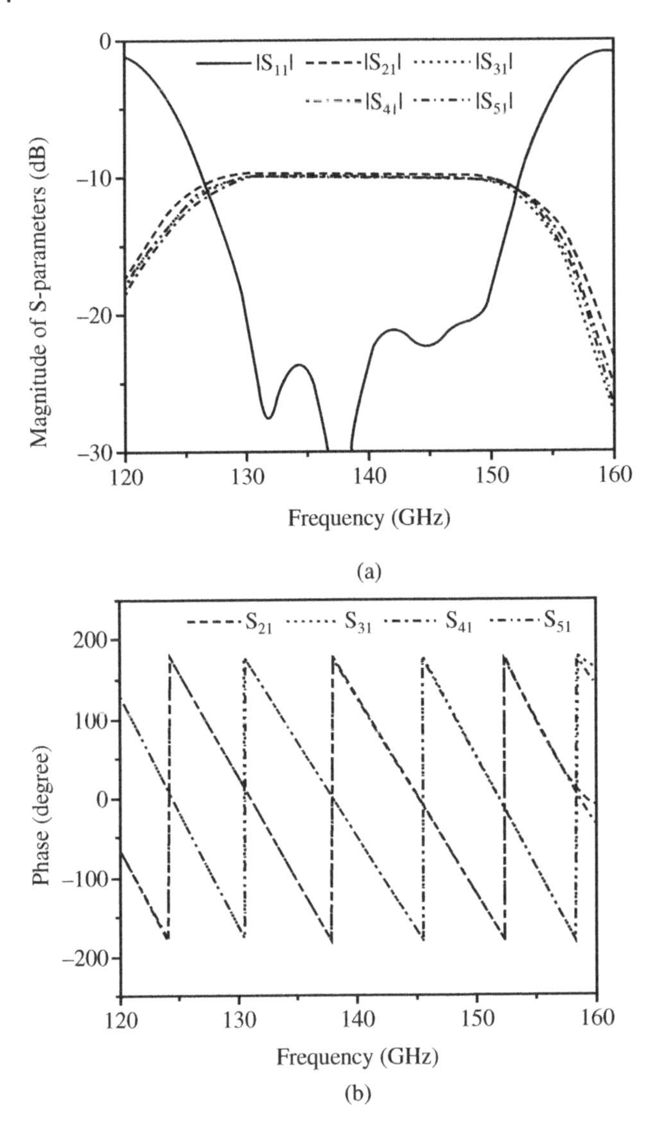

Figure 3.32 Simulated S-parameters of the *E*-plane coupler. (a) Amplitude and (b) phase.

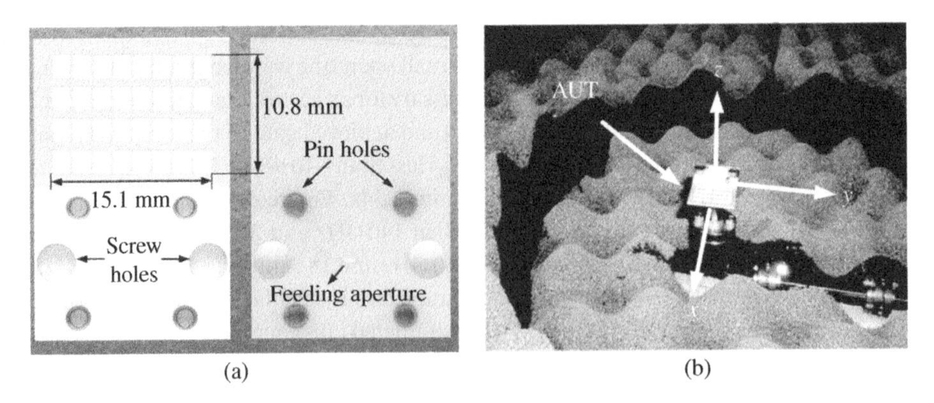

Figure 3.33 Photograph of the antenna array. (a) With transition (left: top view, right: bottom view) and (b) AUT in measurement setup.

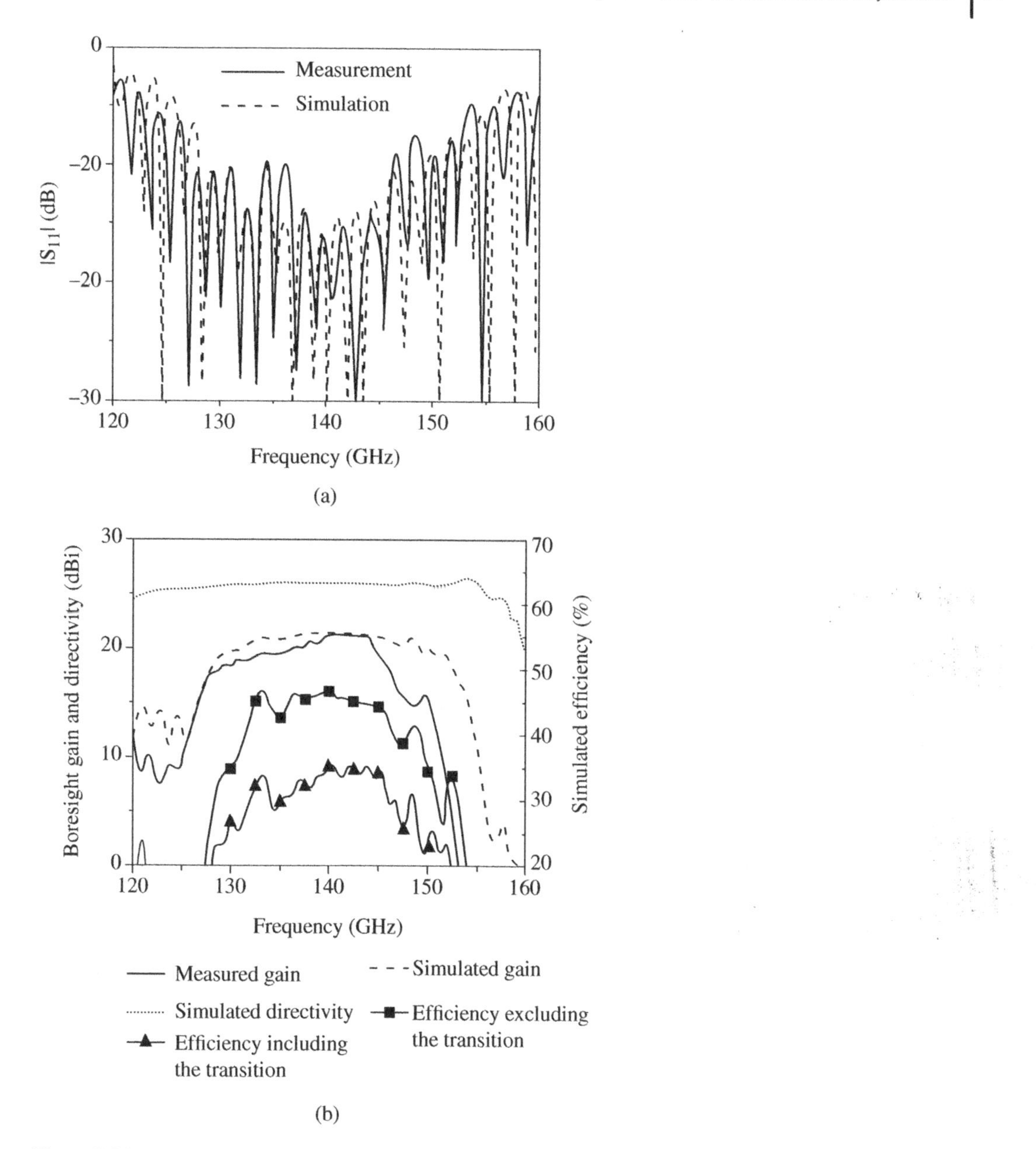

Figure 3.34 (a) Measured and simulated $|S_{11}|$ of the antenna array with the transition and (b) the maximum boresight gain, directivity and efficiency.

beam squinting even at edge frequencies. The simulated SLLs in the *H*- and *E*-plane are less than −13.2 and −11.8 dB, at 140 GHz, respectively. The measured ratio of co- to cross-polarization level is higher than 24 and 23 dB over 130–150 GHz in the *H*- and the *E*-plane, respectively.

In summary, to ease the fabrication and reduce the cost of fabrication, the higher-order TE_{20}-mode can be used to simplify the radiating and feeding structures by greatly reducing the number of vias. Such design challenges are unique in the design of antennas at higher frequency when the fabrication of antennas reaches the limit of conventional process, in particular in mass production.

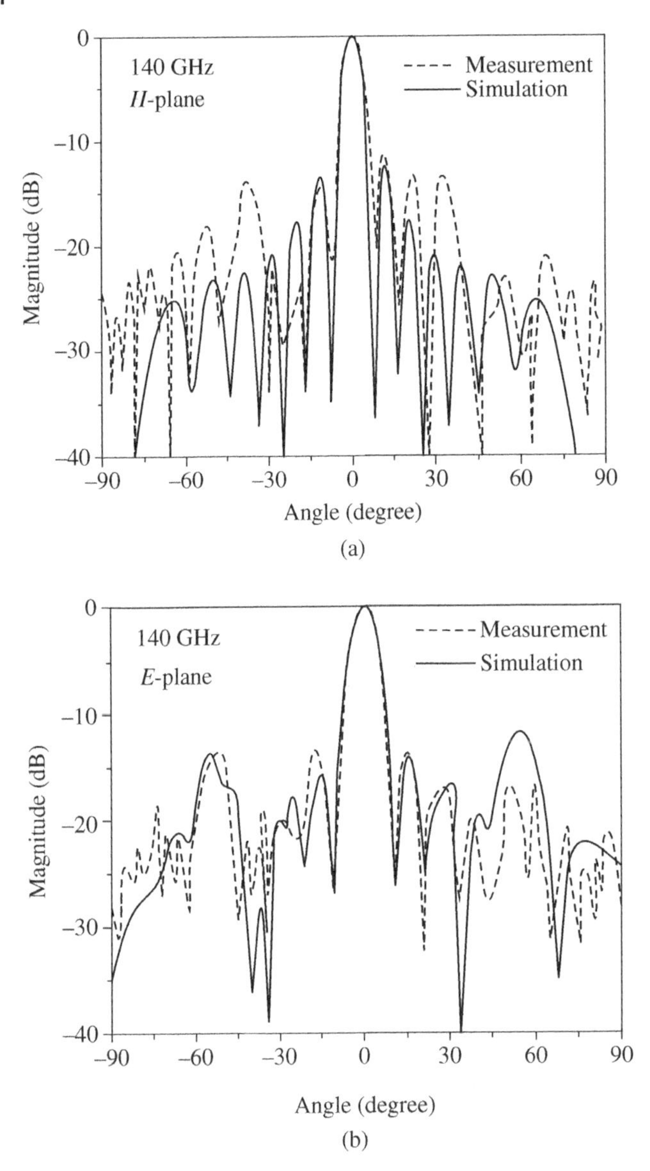

Figure 3.35 Measured and simulated radiation patterns in (a) *H*-plane and (b) *E*-plane.

3.2.4 Fully Substrate Integrated Antennas at 270 GHz

Beyond the wireless systems at 140–GHz bands, there have been some applications in radio astronomy, detection of weapons and contraband detection and wireless communication application operating at even higher frequencies around 300 GHz bands [32, 33]. The antenna design becomes a challenging issue again in the systems. There have been antennas operating at 270 GHz bands such as the lens antenna, frequency-selective surface-based polarization convertor, integrated corner-cube antenna, and slotted waveguide antenna [33–42]. Besides the conventional lens antennas operating at such high frequency bands, the planar and SIAs are still of interest because of the features such as the convenience in fabrication, installation, and integration with other circuits. Considering the issues related to cost and loss, the LTCC process is proposed and studied for planar high-gain antenna designs.

Previous design of antennas has clearly demonstrated the capability of the LTCC process at operating frequencies at the higher edge of mmW bands. To challenge the limit of the LTCC process in the design and fabrication of high-gain planar antennas at even higher frequency, wavelengths close to 1 mm, the antennas are designed, fabricated, and measured at 270 GHz. The experience to explore this design challenge shows that the design methodology works well for the high-gain planar LTCC array but at frequencies lower than 140 GHz, it encounters difficulties. One is that the losses caused by the large-scale feeding network are unacceptable for high-gain designs. The other one is that the pitch requirements in the implementation of LTCC-based vias are too close to its limit to meet the requirements to form SIWs.

3.2.4.1 Analysis of LTCC-Based Substrate Integrated Structures

An analysis is needed before the application of the LTCC-based substrate integrated structures at 270 GHz. No relevant detailed information is founded in the public literatures. Two of the most basic structures are examined. The LTCC substrate is still Ferro A6-M with $\varepsilon_r = 5.9 \pm 0.20$ and $\tan\delta = 0.002$ at 100 GHz. The thickness of each co-fired substrate layer is $t = 0.093$ mm. The conductor used for metallization is Au, whose conductivity is 4.56×10^7 S m^{-1}. The thickness of each metal layer is 0.005 mm. The same LTCC material is used in all designs through this chapter.

Case I: *SIW* As shown in Figure 3.36a, a pair of LTCC-based integrated waveguides are designed side by side in simulation. The performance of the waveguides operating at the 270 GHz band is examined and shown in Figure 3.36b while the performance of the waveguide operating at the 140 GHz band is used as reference as shown in Figure 3.36c.

From Table 3.2, it is clear that both structures at two frequency bands achieve good impedance matching with lower reflection at port 1. The transmission loss at the 270 GHz band is about 1 dB higher than that at the 140 GHz band. The coupling between two waveguides in the pair increases significantly by 24 dB between adjacent ports 1 and 3 and 12 dB between ports 1 and 4, respectively,

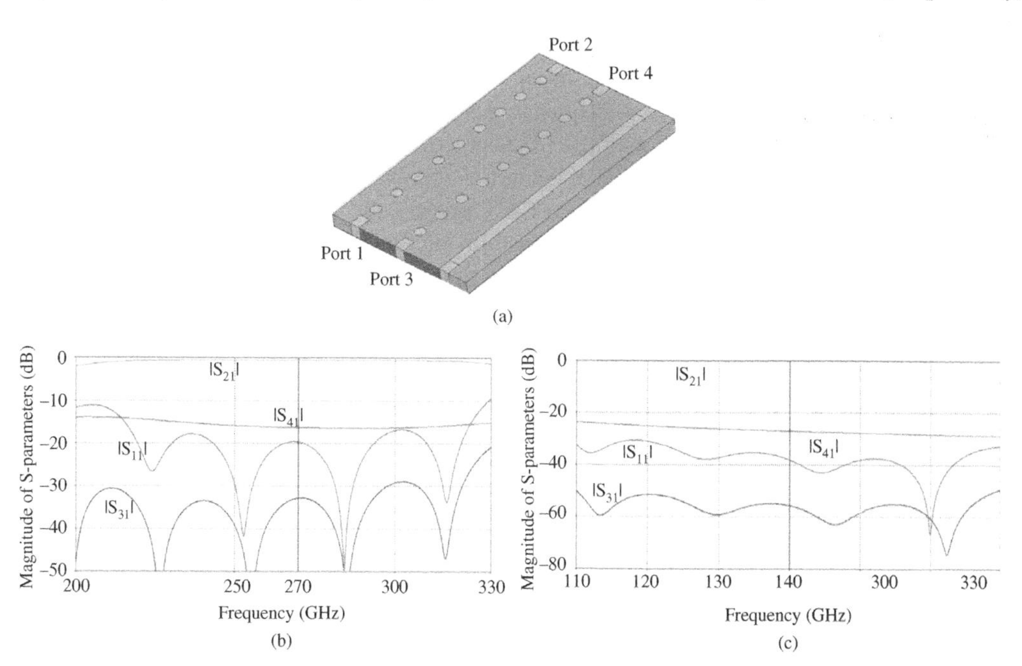

Figure 3.36 Comparison of simulated amplitudes of S-parameters of a pair of SIWs. (a) Geometry of waveguides, (b) performance at 270 GHz bands, and (c) reference at 140 GHz bands.

Table 3.2 Amplitudes of S-parameters of a pair of SIWs at 140 and 270 GHz bands.

Frequency	$\lvert S_{11}\rvert$	$\lvert S_{71}\rvert$	$\lvert S_{31}\rvert$	$\lvert S_{41}\rvert$
140 GHz	−38 dB	~0 dB	−57 dB	−28 dB
270 GHz	−20 dB	~−1 dB	−33 dB	−16 dB

when frequency increases from 140 to 270 GHz. Here, the loss tangent of the material at 270 GHz is still based on the data at 100 GHz. Therefore, it is concluded that the transmission and reflection performance at 140 GHz are much better than those at 270 GHz with the LTCC fabrication limits.

Case II: *Metal Strips at Sidewall* At lower frequencies like 30 GHz, the metal strips at the sidewalls of a cavity can be used to shield the transverse current. With small LTCC tape layer thickness of 0.1 mm, or about $\lambda/20$, the integration of metal strips with vias forms an electric-dense mesh structure to minimize the leakage at the sidewall of SIW or cavity structures.

The phenomenon is examined at 270 GHz. Figure 3.37 compares the distributions of simulated electric fields around the outer side of a circular SIC formed by LTCC with and without strips. The circles positioned closely to the edge of an LTCC medium are metalized through vias to enclose the electric fields. Figure 3.37a shows no observable leaked electric fields out of the via sidewall while Figure 3.37b shows more leakage of the electric field out of the sidewall when the strip is introduced. The vias connect the top and bottom strips. The leakage of electric field occurs because the tape layer thickness is electrically large, $\sim\lambda/4$, so that any two adjacent parallel strips serve as a parallel plate waveguide to guide the wave to leak from the LTCC substrate. The mechanism is similar to the parallel plate used in a leaky wave antenna for enlarged leakage rate.

Therefore, two analyses show the unique challenges of substrate integrated structures at 270 GHz. With such constraints, the antenna operating at 270 GHz can't be designed by simply and directly scaling down the low-frequency design because some of the critical physical dimensions are too electrically large with the LTCC process limits. This fact raises unique design considerations such as the rectangular SIW being not suitable for a 270 GHz design due to the relatively large via diameters and pitches in the LTCC process, the special transition between SIW and guided wave structure, easy fabrication by simplifying the design with desired performance, no metal strips at sidewalls, and shortening the power transmission path. In short, all the antenna design considerations at 270 GHz are caused by the LTCC limits of available minimum dimensions and potential higher losses.

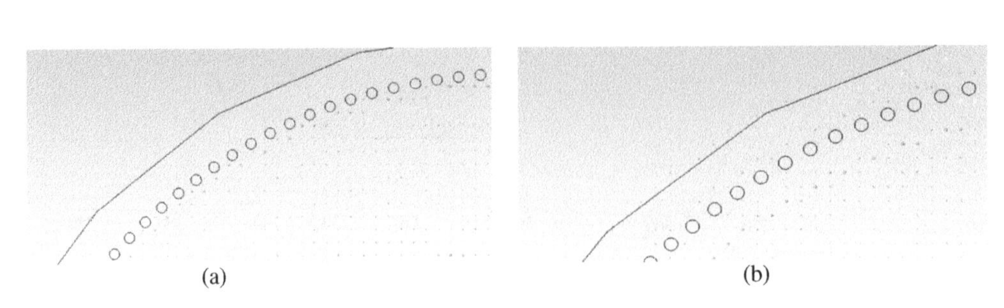

Figure 3.37 Comparison of distributions of simulated electric fields in LTCC substrate integrated cavity. (a) No strips in the sidewall and (b) with the multilayered strips in the sidewall.

3.2.4.2 Fresnel Zone Plate Antenna in LTCC at 270 GHz

Fresnel zone plates (FZPs) is a type of lens antennas constructed in a planar form with a zone plate and a far separated feed. They have been used in mmW bands such as quarter-wave subzones of the lens at 60 GHz, a high gain FZP reflector at 94 GHz, and a folded FZP and a resonant strip dipole as the feed at 230 GHz [43–45]. The FZP antennas feature the merits of planar surface, reduced weight, and ease of construction compared with the conventional dielectric lenses. However, the FZP antennas also suffer from the losses, in particular the spillover loss caused by the leakage of the feed energy from the edge of the lens, and the reflection loss caused at the higher-permittivity substrate-air interface. Therefore, a fully integrated FZP antenna in LTCC at 270 GHz is presented [35].

Figure 3.38 shows the configuration of the substrate integrated FZP antenna into LTCC. Figure 3.38a shows the top view of the design, a typical pattern of FZP and sidewall. The black portions are the metal (M1) whereas the white rings are the 13 concentric annular slots. The radii of the alternative rings are r_i (i = 1–25). For easy fabrication, the diameters (d_1) and pitches (p_1) of vias forming the annular sidewall should be kept as large as possible. Two rows of vias are positioned along the edge of a disc with a radius of v_1 and v_2, respectively, to suppress the leakage radiation from the edge. The spacing between the vias on the inner and outer circles is $p_1 = v_2 - v_1$. It should be noted that the sidewall suppresses the spillover loss but degrades the impedance matching due to a higher reflection of the antenna. With the sidewall, the $|S_{11}|$ decreases over the bandwidth of 257–264 GHz because of the out-of-phase cancelation of the reflection from the FZP and the sidewall whereas the $|S_{11}|$ increases over the bandwidth of 264–280 GHz because of the in-phase superposition of the reflection.

Figure 3.38b shows the cross-sectional sketch of the antenna. The antenna consists of 20 substrate layers and three metal layers. Metal layer M1 is the zone plate with a focal length equal to the thickness of the substrate layers Sub1–19, $fl = t \times 19 = 1.767\ \text{mm} = 3.8\lambda$, where λ is the wavelength in the substrate at 270 GHz. Benefiting from the shorter wavelength, the fully integrated structure can be achieved without any aired layer. There is a sidewall formed by metallic vias on the edge of Sub1–20. The FZP is fed through the slot at the center of metal layer M2, located at the focal point as shown in Figure 3.38c. A feeding transition between an external WR-03 waveguide and the substrate is designed accordingly in the substrate layer Sub20. The metal layers M2 and M3 are designed for impedance matching. A via fence is positioned in Sub20 and the feeding aperture is cut on M3.

Based on a parametric study, the dimensions are optimized as given in Table 3.3. Then the fully substrate integrated planar FZP antennas were prototyped using multilayer LTCC as shown in Figure 3.39. The radius of radiating aperture is 7.3 mm, and the overall size of the square prototype is 32 mm × 32 mm × 1.87 mm. The back view shows the feeding aperture wherein the center portion is zoomed in for a clear view. The four pin holes and four screw holes are used for positioning and connecting the antenna to a self-built fixture in measurement.

Figure 3.40a shows the comparison of measured and simulated $|S_{11}|$ of the FZP antenna. The simulated $|S_{11}|$ is −10.7 dB at 270 GHz and less than −10 dB over 253.7–272.6 GHz. The measured $|S_{11}|$ is −8.6 dB at 270 GHz and less than −8 dB in 250–272 GHz. Figure 3.40b shows the measured gain, simulated gain, and directivity at boresight of the FZP antenna. The maximum simulated gain is 23.0 dBi at 263.5 GHz. The simulated gain is greater than in 20 dBi in 262.3–270.5 GHz except a drop at 266.2 GHz. The measured gain is greater than in 17.9 dBi in 263.4–276.9 GHz, and the maximum measured gain is 20.9 dBi at 270 GHz.

Figure 3.41 shows the simulated and measured normalized radiation patterns in both *H*- and *E*- plane. The main lobe is always pointing at boresight without any beam squinting with the symmetrical feeding structure. Figure 3.41a shows the simulated SLLs in *H*-planes are below −19.1, −18.9, and −21.6 dB at 262.25, 266.25, and 270.25 GHz, respectively, while the simulated SLLs in

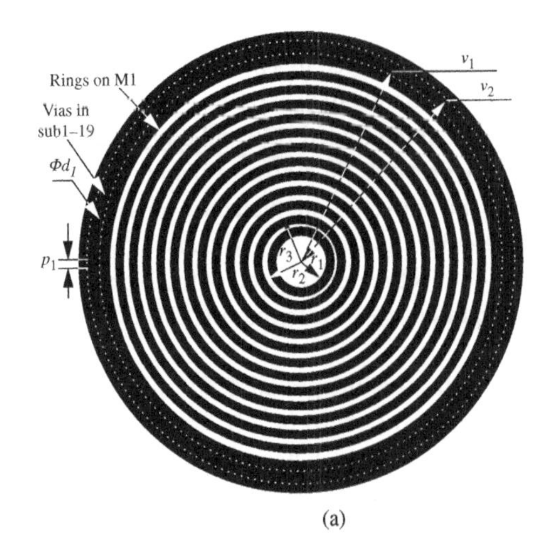

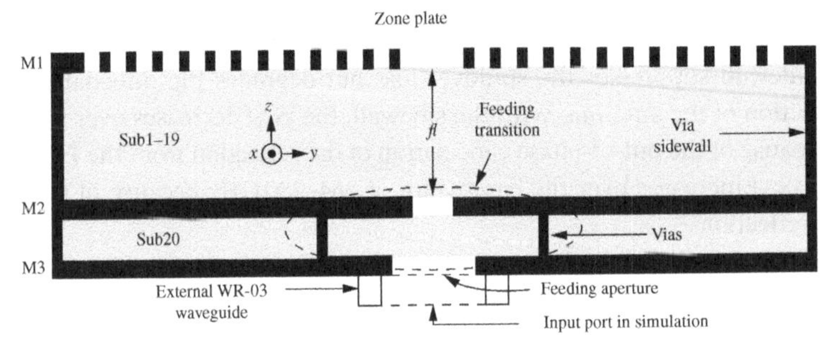

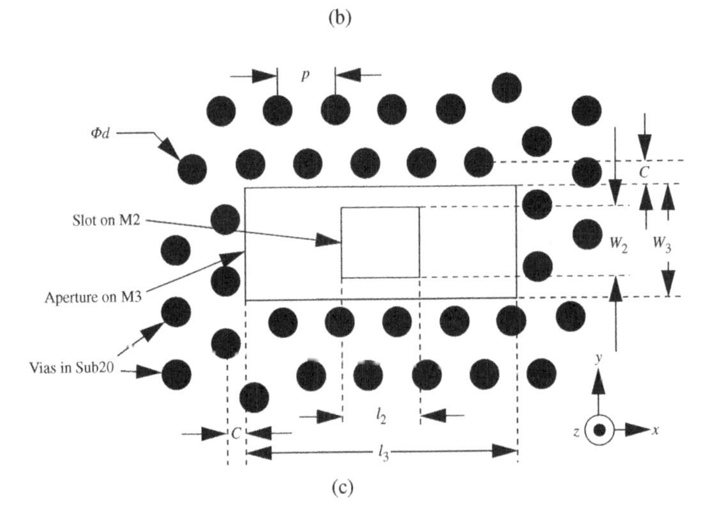

Figure 3.38 Configuration of the FZP antenna in LTCC. (a) Top view of the FZP and the sidewall, (b) cross-sectional view, and (c) top view of feeding transition.

Table 3.3 Dimensions of the FZP antenna (unit: mm).

v_1	7.94	v_2	8.21	d_1	0.15	p_1	0.32	r_1	0.93
r_2	1.35	r_3	1.71	r_4	2.02	r_5	2.32	r_6	2.60
r_7	2.88	r_8	3.14	r_9	3.40	r_{10}	3.66	r_{11}	3.91
r_{12}	4.16	r_{13}	4.41	r_{14}	4.66	r_{15}	4.91	r_{16}	5.15
r_{17}	5.39	r_{18}	5.63	r_{19}	5.87	r_{20}	6.11	r_{21}	6.35
r_{22}	6.59	r_{23}	6.83	r_{24}	7.07	r_{25}	7.30	d	0.08
p	0.16	c	0.06	l_2	0.21	w_2	0.18	l_3	0.75
w_3	0.29								

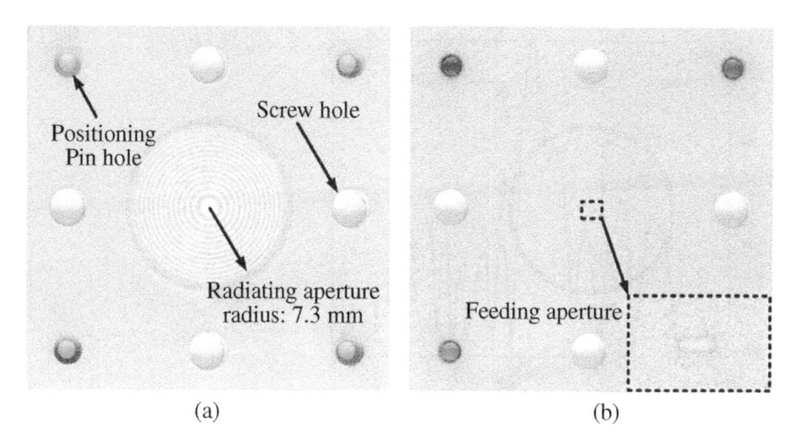

Figure 3.39 Photograph of the LTCC-FZP antenna. (a) Front view and (b) back view.

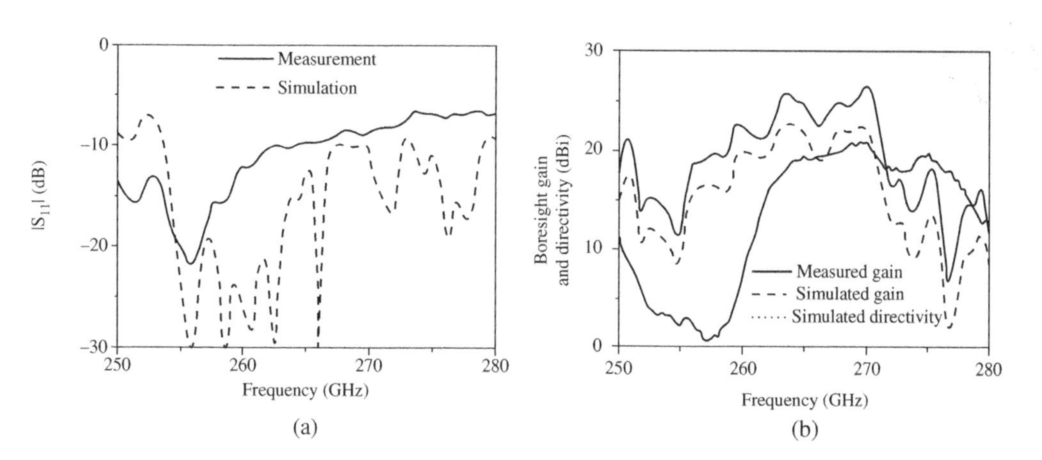

Figure 3.40 Measured and simulated results. (a) $|S_{11}|$ and (b) boresight gain and directivity.

E-planes are below −5, −11.8, and −15.3 dB at 262.25, 266.25, and 270.25 GHz, respectively, as shown in Figure 3.40b. Due to the tolerance of fabrication and installation with the feeding structure, the measurement challenging. Figure 3.41c compares the measured and simulated radiation patterns at 270 GHz after modifying the location of the feeding slot in the simulation. The good agreement verifies the design and fabrication as well as the installation tolerance.

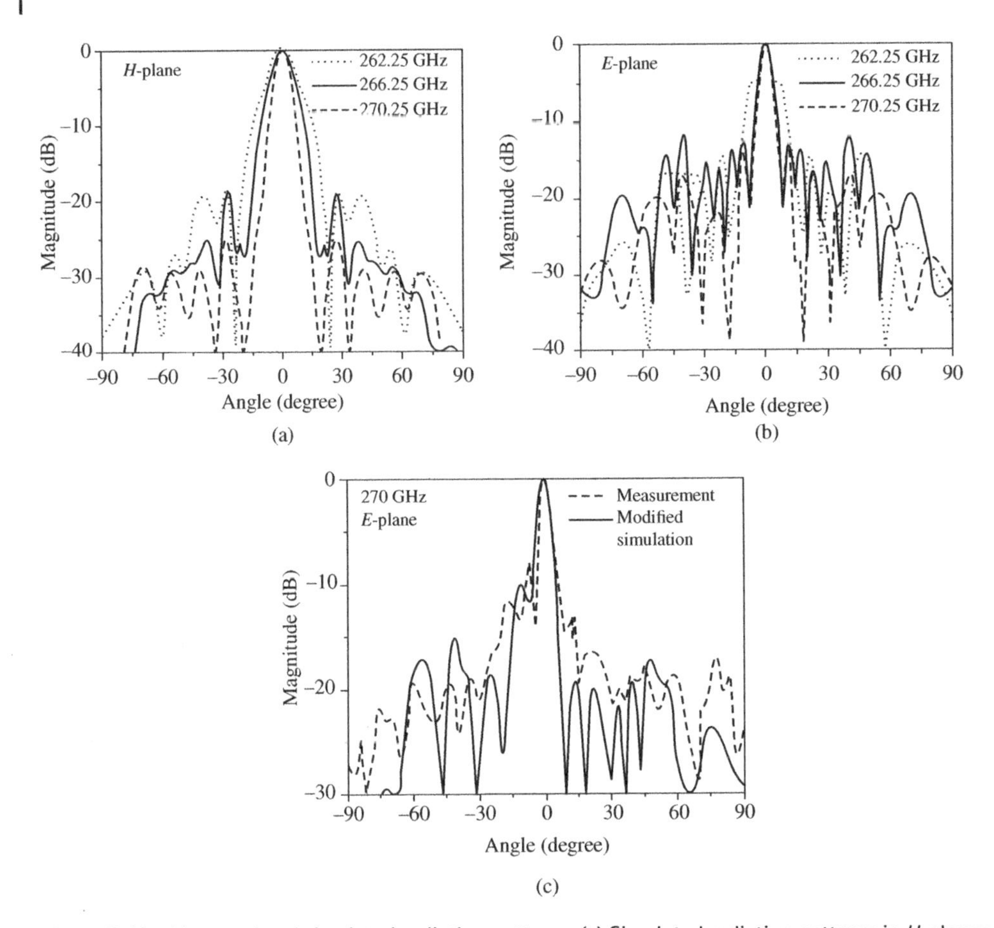

Figure 3.41 Measured and simulated radiation patterns. (a) Simulated radiation patterns in *H*-planes, (b) simulated radiation patterns in *E*-planes, and (c) measured radiation pattern versus simulated pattern in *E*-plane at 270 GHz.

3.3 Summary

Planar substrate integrated design of arrays with complicated feeding structures is a good option for high-gain mmW antennas. By addressing the unique design challenges such as losses and fabrication tolerance, the design of the arrays at mmW bands needs the development of new radiating elements, array configuration, and feeding structures. As a result, designs of large-scale arrays have been achieved using LTCC-based SIW slot antennas or lens antennas. In particular, the three-dimensional feeding structures have been designed for consistent radiation performance of large-scale high-gain array over a wide bandwidth at mmW bands.

References

1 Hirokawa, J., Arai, H., and Goto, N. (1989). Cavity-backed wide slot antenna. *IEE Proc. H Microwave Antennas Propag.* 136 (1): 29–33.

2 Yan, L., Hong, W., Hua, G. et al. (2004). Simulation and experiment on SIW slot array antennas. *IEEE Microwave Wireless Compon. Lett.* 14 (9): 446–448.

3 Seki, T., Honma, N., Nishikawa, K., and Tsunekawa, K. (2005). A 60-GHz multilayer parasitic microstrip array antenna on LTCC substrate for system-on-package. *IEEE Microwave Wireless Compon. Lett.* 15: 339–341.
4 Wi, S., Sun, Y., Song, I. et al. (2006). Package-level integrated antennas based on LTCC technology. *IEEE Trans. Antennas Propag.* 54 (8): 2190–2197.
5 Korisch, I.A., Manzione, L.T., Tsai, M.J., and Wong, Y.H. (1999). Antenna package for a wireless communications device. US09/396,948, 15 September 1999.
6 Chen, Z.N., Liu, D., Pfeiffer, U.R., and Zwick, T.M. (2006). Apparatus and methods for packaging antennas with integrated circuit chips for millimeter wave applications. EP1085597A3, European Patent Office, 5 June 2006.
7 Chen, Z.N., Liu, D., Pfeiffer, U.R., and Zwick, T.M. (2005). Apparatus and methods for packaging antennas with integrated circuit chips for millimeter wave applications. US20060276157A1, United States, 3 June 2005.
8 Chen, Z.N., Qing, X., Chia, M.Y.W. et al. (2011). Microwave, millimeter wave, and Terahertz technologies in Singapore. In: *Proceedings of the 41st European Microwave Conference*, 1–4.
9 Chen, Z.N. et al. (2012). Research and Development of Microwave and Millimeter-wave Technology in Singapore. In: *Proceedings of the 42nd European Microwave Conference*, 1–4.
10 Xu, F., Hong, W., Chen, P., and Wu, K. (2009). Design and implementation of low sidelobe substrate integrated waveguide longitudinal slot array antennas. *IET Microwave Antennas Propag.* 3 (5): 790–797.
11 Hosseininejad, S.E. and Komjani, N. (2013). Optimum design of traveling-wave SIW slot array antennas. *IEEE Trans Antennas Propag.* 61 (4): 1971–1975.
12 Hamadallah, M. (1989). Frequency limitations on broad-band performance of shunt slot arrays. *IEEE Trans. Antennas Propag.* 37 (7): 817–823.
13 Xu, J., Chen, Z.N., Qing, X., and Hong, W. (2011). Bandwidth enhancement for a 60 GHz substrate integrated waveguide fed cavity array antenna on LTCC. *IEEE Trans. Antennas Propag.* 59 (3): 826–832.
14 A6-M datasheet. Ferro Materials Company. http://www.ferro.com/non-cms/ems/EPM/LTCC/A6-M-LTCC-System.pdf. https://www.ferro.com/-/media/files/resources/electronic-materials/ferro-electronic-materials-a6m-e-ltcc-tape-system.pdf
15 FCC online table of frequency allocations. Federal Communications Commission. http://transition.fcc.gov/oet/spectrum/table/fcctable.pdf (accessed 19 December 2020).
16 Goldsmith, P.F., Hsieh, C.T., Huguenin, G.R. et al. (1993). Focal plane imaging systems for millimeter wavelengths. *IEEE Trans. Microwave Theory Tech.* 41 (10): 1664–1675.
17 Herrero, P. and Schoebel, J. (2008). Planar antenna array at D-band fed by rectangular waveguide for future automotive radar systems. In: *Proceedings of the 38th European Microwave Conference*, 1030–1033.
18 Digby, J.W., McIntosh, C.E., Parkhurst, G.M. et al. (2000). Fabrication and characterization of micromachined rectangular waveguide components for use at millimeter-wave and terahertz frequencies. *IEEE Trans. Microwave Theory Tech.* 48 (8): 1293–1302.
19 Hirata, A., Kosugi, T., Takahashi, H. et al. (2006). 120-GHz-band millimeter-wave photonic wireless link for 10-Gb/s data transmission. *IEEE Trans. Microwave Theory Tech.* 54 (5): 1937–1944.
20 Dongjin, K., Hirokawa, J., Ando, M., and Nagatsuma, T. (2011). Design and fabrication of a corporate-feed plate-laminated waveguide slot array antenna for 120 GHz-band. In: *IEEE International Symposium on Antennas and Propagation*, 3044–3047.
21 Jakoby, R. (1996). A novel quasi-optical monopulse-tracking system for millimeter-wave application. *IEEE Trans. Antennas Propag.* 44 (4): 466–477.

22 Yeap, S.B., Chen, Z.N., Qing, X. et al. (2012). 135GHz antenna array on BCB membrane backed by polymer-filled cavity. In: *6th European Conference on Antennas and Propagation (EUCAP)*, 1. 4.

23 Wang, R., Sun, Y., Kaynak, M. et al. (2012). A micromachined double-dipole antenna for 122 – 140 GHz applications based on a SiGe BiCMOS technology. In: *IEEE/MTT-S International Microwave Symposium Digest*, 1–4.

24 Jong-Hoon, L., Kidera, N., DeJean, G. et al. (2006). A V-band front-end with 3-D integrated cavity filters/duplexers and antenna in LTCC technologies. *IEEE Trans. Microwave Theory Tech.* 54 (7): 2925–2936.

25 Xu, J., Chen, Z.N., Qing, X., and Hong, W. (2012). 140-GHz planar broadband LTCC SIW slot antenna array. *IEEE Trans. Antennas Propag.* 60 (6): 3025–3028.

26 Xu, J., Chen, Z.N., Qing, X., and Hong, W. (2012). 140-GHz planar SIW slot antenna array with a large-via-fence dielectric loading in LTCC. In: *The 6th European Conference on Antennas and Propagation (EUCAP)*, 1–4.

27 Xu, J., Chen, Z.N., Qing, X., and Hong, W. (2012). Dielectric loading effect on 140-GHz LTCC SIW slot array antenna. In: *IEEE-APS Topical Conference on Antennas and Propagation in Wireless Communications (APWC)*, 1–4.

28 Xu, J., Chen, Z.N., Qing, X., and Hong, W. (2013). 140-GHz TE_{20}-mode dielectric-loaded SIW slot antenna array in LTCC. *IEEE Trans. Antennas Propag.* 61 (4): 1784–1793.

29 Jin, H., Che, W., Chin, K. et al. (2018). Millimeter-wave TE20-mode SIW dual-slot-fed patch antenna array with a compact differential feeding network. *IEEE Trans. Antennas Propag.* 66 (1): 456–461.

30 Cheng, Y.J., Hong, W., and Wu, K. (2008). Design of a monopulse antenna using a dual V-type linearly tapered slot antenna (DVLTSA). *IEEE Trans. Antennas Propag.* 56 (9): 2903–2909.

31 Kimura, Y., Hirokawa, J., Ando, M., and Goto, N. (1997). Frequency characteristics of alternating-phase single-layer slotted waveguide array with reduced narrow walls. In: *IEEE International Symposium on Antennas and Propagation*, 1450–1453.

32 Appleby, R. and Wallace, H.B. (2007). Standoff detection of weapons and contraband in the 100 GHz to 1 THz region. *IEEE Trans. Antennas Propag.* 55 (11): 2944–2956.

33 Song, H.J., Ajito, K., Hirata, A. et al. (2009). 8 Gbit/s wireless data transmission at 250 GHz. *Electron. Lett* 45 (22): 1121–1122.

34 Abbasi, M., Gunnarsson, S.E., Wadefalk, N. et al. (2011). Single-chip 220-GHz active heterodyne receiver and transmitter MMICs with on-chip integrated antenna. *IEEE Trans. Microwave Theory Tech.* 59 (2): 466–478.

35 Xu, J., Chen, Z.N., and Qing, X. (2013). 270-GHz LTCC-integrated high gain cavity-backed fresnel zone plate lens antenna. *IEEE Trans. Antennas Propag.* 61 (4): 1679–1687.

36 Euler, M., Fusco, V., Cahill, R., and Dickie, R. (2010). 325 GHz single layer sub-millimeter wave FSS based split slot ring linear to circular polarization convertor. *IEEE Trans. Antennas Propag.* 58 (7): 2457–2459.

37 Gearhart, S.S., Ling, C.C., and Rebeiz, G.M. (1991). Integrated millimeter-wave corner-cube antennas. *IEEE Trans. Antennas Propag.* 39 (7): 1000–1006.

38 Yi, W., Maolong, K., Lancaster, M.J., and Jian, C. (2011). Micromachined 300-GHz SU-8-based slotted waveguide antenna. *IEEE Antennas Wireless Propag. Lett.* 10: 573–576.

39 Xu, J., Chen, Z.N., and Qing, X. (2013). 270-GHz LTCC-integrated strip-loaded linearly polarized radial line slot array antenna. *IEEE Trans. Antennas Propag.* 61 (4): 1794–1801.

40 Chen, Z.N., Qing, X., Yeap, S.B., and Xu, J. (2014). Design and measurement of substrate integrated planar millimeter wave antenna arrays at 60–325 GHz. In: *IEEE Radio and Wireless Symposium (RWS)*, 1–4.

41 Qing, X. and Chen, Z.N. (2014). Measurement setups for millimeter-wave antennas at 60/140/270 GHz bands. In: *Internt'l Workshop Antenna Techno.: Small Antennas, Novel EM Structures and Materials, and Applications (iWAT)*, 1–4.

42 Chen, Z.N., Qing, X., Sun, M. et al. (2015). Substrate integrated antennas at 60/77/140/270 GHz. In: *IEEE 4th Asia-Pacific Conference on Antennas and Propagation (APCAP)*, 1–2.

43 Hristov, H.D. and Herben, M.H.A.J. (1995). Millimeter-wave Fresnel-zone plate lens and antenna. *IEEE Trans. Microwave Theory Tech.* 43 (12): 2779–2785.

44 Nguyen, B.D., Migliaccio, C., Pichot, C. et al. (2007). W-band Fresnel zone plate reflector for helicopter collision avoidance radar. *IEEE Trans. Antennas Propag.* 55 (5): 1452–1456.

45 Gouker, M.A. and Smith, G.S. (1992). A millimeter-wave integrated-circuit antenna based on the Fresnel zone plate. *IEEE Trans. Microwave Theory Tech.* 40 (5): 968–977.

4

Broadband Metamaterial-Mushroom Antenna Array at 60 GHz Bands

Wei Liu and Zhi Ning Chen

Department of Electrical and Computer Engineering, National University of Singapore, Singapore 117583, Republic of Singapore

4.1 Introduction

With the increasing demands on wireless data traffic, the mmW communication technology covering the frequency range of 30–300 GHz has been widely accepted as a promising candidate for more and more wireless systems, in particular, 5G systems, due to its wide bandwidth and unique features [1–6]. As a key component, mmW antennas have been studied for various applications as mentioned in Chapter 1. For example, the unlicensed 60 GHz band has showed its potential in indoor short-range wireless communications for uncompressed high-definition streaming and outdoor point-to-point systems. Broadband high-gain antenna arrays are required to compensate for high path loss of mmW links. It is challenging to design such antenna arrays because of the severe losses at mmW bands caused by dielectric, conductors, and surface waves from their radiators and feeding networks [7].

The microstrip patch antenna has found extensive applications in military, industrial, and commercial wireless systems due to its advantages of low profile, conformability, ease of fabrication, and compatibility with integrated circuit technology [8–12]. However, it suffers from an inherent narrow bandwidth. A number of techniques have been developed to increase the bandwidths of patch antennas. One of the basic solutions is the utilization of a thick substrate with low permittivity for a low quality factor [13]. When a U-shaped slot is cut onto the patch on a thick air substrate to introduce a capacitance to compensate for the inductance caused by the long probe feed, a bandwidth of 30% is easily obtained [14, 15]. The L-probe feed provides an alternative way to compensate for the inductance, and a bandwidth larger than 30% can be achieved with an air substrate with a thickness of $0.1\lambda_0$ (λ_0 is the free-space wavelength at the operating frequency) [16, 17]. One drawback in both the U-slotted and L-probe cases is that, at the upper edges of the operating bands, the cross-polarization levels are high in the H-planes and the radiation patterns are distorted in the E-planes. Incorporated with a separate aperture coupling structure, a bandwidth of 25% can be obtained but with high backward radiation levels due to the aperture resonance [18–20]. Two stacked patches coupled by an aperture can further broaden the bandwidth, but the backward radiation is still high [21, 22]. A proximity-coupled cavity-backed patch antenna of more than $0.1\lambda_0$ thickness can achieve a bandwidth of 40% but with deteriorated gain and beam squinting at the higher operating frequencies [23]. The above-mentioned patch antenna designs achieve the broadened operating bandwidth but suffer from increased substrate thickness.

Substrate-Integrated Millimeter-Wave Antennas for Next-Generation Communication and Radar Systems, First Edition.
Edited by Zhi Ning Chen and Xianming Qing.

Multilayer LTCC technology stands out in the mmW antenna design because of its attractive features of low fabrication tolerance, flexible metallization, convenient perpendicular interconnection, and easy implementation of blind, buried, through vias, and air cavity compared with conventional PCB process. Some LTCC antennas and arrays at 60 GHz bands have been developed [24–28]. It is observed that severe surface waves are launched in the electrically thick LTCC substrate with high permittivity, and thus the antennas suffer from reduced efficiency and deteriorated radiation patterns. A number of techniques have been reported to suppress the surface waves as well as the mutual coupling between antennas, such as adding metal-topped via-fences [24–26], lowering effective substrate permittivity with an embedded air cavity [27], and utilizing open air cavities around the radiating edges of a patch antenna [28]. The reported techniques are usually with much high manufacturing complexity and fabrication cost caused by the extra specific LTCC processing. Thus, a broadband mmW antenna on a low-profile LTCC substrate is preferred to suppress the surface waves for good radiation performance.

During the last two decades, metamaterials and metasurfaces have attracted great interest because of the exhibited exotic electromagnetic properties [29–32]. Metasurface comprises a two-dimensional array of electrically small inclusions to control the electromagnetic fields for unique characteristics. Due to the simplified configuration with less loss and ease of fabrication compared with the typical three-dimensional metamaterials, metasurfaces have been widely applied in various electromagnetic designs [33–36].

Mushroom structures can be used to form a kind of metasurface possessing a composite right/left-handed (CRLH) dispersion of guided waves [32]. Figure 4.1 shows the simulation model and equivalent circuit of a mushroom unit cell. A typical unbalanced CRLH dispersion of a mushroom structure is obtained using the full-wave simulation of the unit cell. Compared with a conventional microstrip patch, the CRLH-mushroom structure can generate much more resonances, including the positive-, zeroth-, and negative-order resonances. Usually, the negative and zeroth order resonance CRLH-mushroom antennas offer an alternative way for size reduction or multiband operation but suffer from very narrow operating bandwidth [37–41]. By combining the adjacent resonances with low quality factor, the CRLH mushroom structure can be used to design broadband low-profile antennas.

Furthermore, the mushroom structure can be periodically arranged to realize the electromagnetic band-gap (EBG) surface or high impedance surface [42–45]. Figure 4.2 shows a typical dispersion diagram of the surface waves in the EBG mushroom structure, exhibiting a frequency band gap through which the surface wave cannot propagate for any incident angles and polarization states. The EBG mushroom structures integrated into antenna designs have been used to reduce the mutual coupling between the antenna elements.

The metamaterial-mushroom structure can be engineered for a broadband low-profile antenna with a self-decoupling capability by simultaneously controlling the CRLH dispersion of guided waves and the EBG characteristic of surface waves. The mutual couplings between metamaterial-mushroom antenna elements, even closely deployed in a large-scale array, would be low without any extra decoupling structure.

This chapter introduces the design and working principle of the CRLH-mushroom antenna. Then, a unique metamaterial-mushroom antenna array in LTCC is exemplified at 60 GHz band with the advantages of low profile, broad bandwidth, high gain, high aperture efficiency, low cross-polarization level, and low mutual coupling.

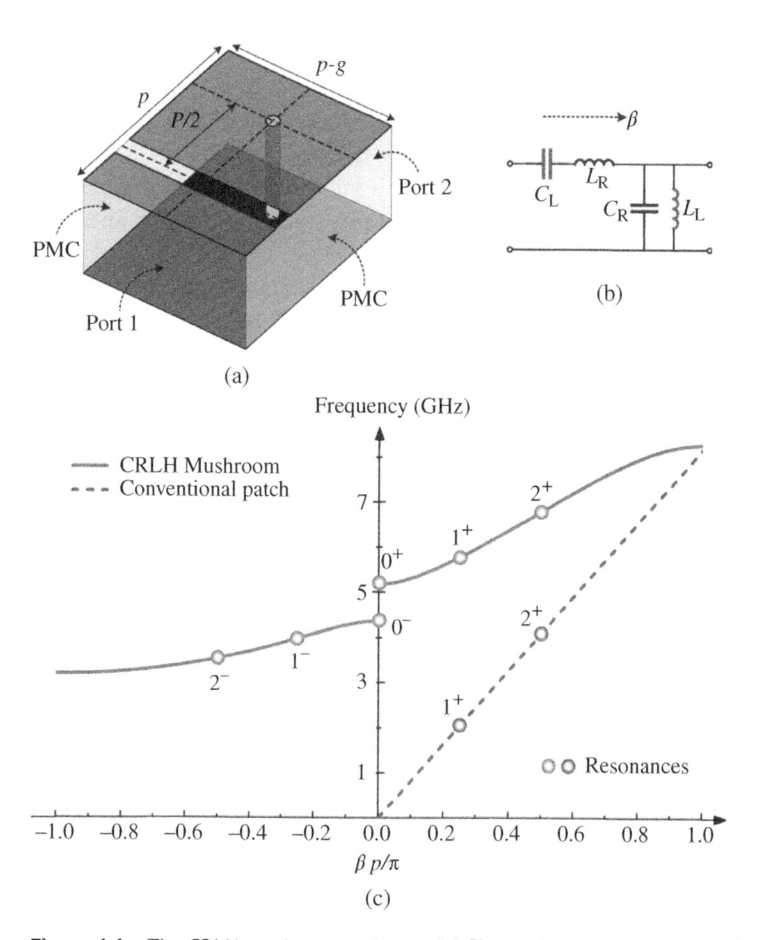

Figure 4.1 The CRLH mushroom unit cell. (a) Simulation model, (b) equivalent circuit, and (c) dispersion diagram of the CRLH mushroom cell in comparison with the conventional microstrip patch cell.

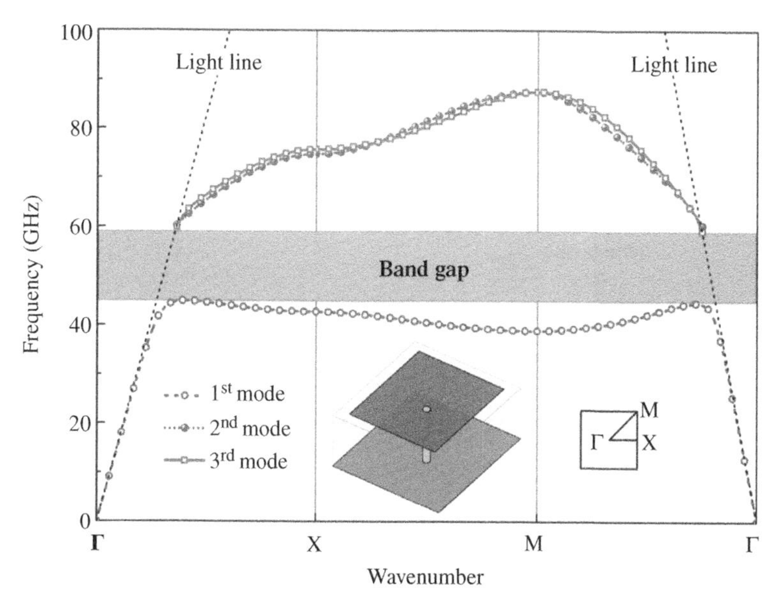

Figure 4.2 Dispersion diagram of surface waves of EBG mushroom unit cell.

4.2 Broadband Low-Profile CRLH-Mushroom Antenna

A CRLH-mushroom antenna operating at a 5 GHz band is studied in this section. The two-dimensional periodic mushroom structure, acting as the radiator, exhibits a CRLH dispersion relation of guided waves. The antenna is center-fed by a microstrip-line-slot to generate two adjacent modes for broadband operation. The normal PCB process is used for the antenna implementation.

As shown in Figure 4.3, the CRLH-mushroom antenna is fed at the center by a microstrip line through a slot cut on a finite ground plane with a size of $G_L \times G_W$. The 50-Ω microstrip line has a width of W_m. The mushroom structure and microstrip line are printed onto a piece of Rogers RO4003C ($\varepsilon_r = 3.38$ and $\tan\delta = 0.0027$) substrate with a thickness of h and h_m, respectively. The mushroom unit cell consists of a square metallic patch with a side width of w and a shorting via with a diameter of D_{via}, where the shorting via connects the center of the square patch and the ground plane. The mushroom cells are two-dimensionally distributed on the center of the dielectric substrate with a periodicity of p and a gap width of g in between, and $w = p - g$. The number of the mushroom cells along the x- and y-axis directions is N_x and N_y, respectively. The design procedure and considerations are discussed here, and the related values of the parameters can be found in [46].

4.2.1 Working Principle

The full-wave simulation of the unit-cell model shown in Figure 4.1a is conducted by exciting the two waveguide ports, and the propagation constant β_{mr} of the CRLH mushroom unit cell can be extracted from the simulated S-parameters. The dispersion diagram of the CRLH mushroom unit cell is provided in Figure 4.4 for the following analysis of the operating modes of the antenna.

When the mushroom substrate thickness h is very small compared with the free-space wavelength λ_0 and the mushroom array width W_p ($h \ll \lambda_0$, $h \ll W_p$), the transmission line model is utilized for the mode analysis by integrating the dispersion relation of the mushroom unit cell. Due to the fringing fields at the open edges of the mushroom array, an additional extended length ΔL at each end is approximated to that of the corresponding entire rectangular patch with the same width of W_p, and it is given by

$$\frac{\Delta L}{h} = 0.412\frac{(\varepsilon_{re} + 0.3)(W_p/h + 0.262)}{(\varepsilon_{re} - 0.258)(W_p/h + 0.813)} \tag{4.1}$$

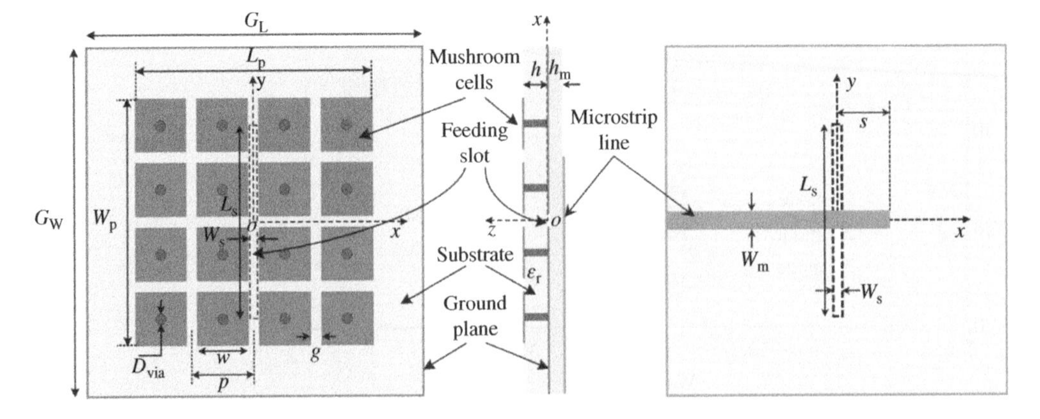

Figure 4.3 Configuration of the microstrip-line-slot fed CRLH-mushroom antenna.

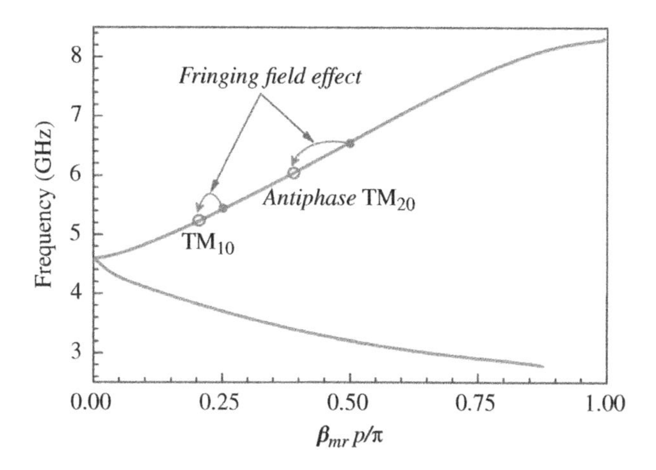

Figure 4.4 Dispersion diagram of the CRLH-mushroom unit cell.

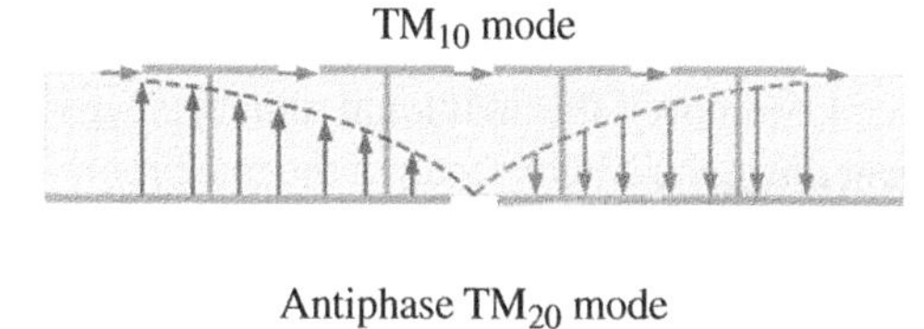

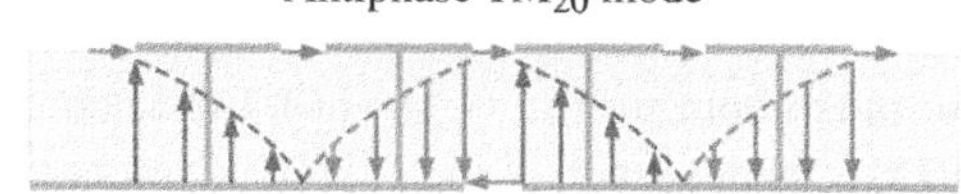

Figure 4.5 Sketch of the electric field distributions at TM_{10} and antiphase TM_{20} modes.

$$\varepsilon_{re} = \frac{\varepsilon_r + 1}{2} + \frac{\varepsilon_r - 1}{2}\left(1 + 12\frac{h}{W_p}\right)^{-1/2} \tag{4.2}$$

$$W_p = pN_y - g \tag{4.3}$$

The propagation constant β_e in the extended region is given by

$$\beta_e = k_0\sqrt{\varepsilon_{re}} = \frac{2\pi f}{c}\sqrt{\varepsilon_{re}} \tag{4.4}$$

where k_0 is the wavenumber in free space, f is the operating frequency, and c is the velocity of light in free space.

Similar to the conventional rectangular patch antenna, the TM_{10} mode can be generated in the mushroom antenna as illustrated in Figure 4.5, and its resonant frequency can be approximately determined by

$$\beta_{mr} pN_x + 2\beta_e \Delta L = \pi \tag{4.5}$$

The mushroom antenna operating at a TM_{10} mode exhibits an in-phase electric field distribution along the mushroom gaps and radiating edges at two ends for the boresight directional radiation.

Meanwhile, an antiphase TM_{20} mode is generated due to the slot excitation on the ground plane and the central mushroom gap, as depicted in Figure 4.5. When operating at the antiphase TM_{20} mode, the electric fields at the central region are out of phase such that an in-phase electric field distribution is also excited along the mushroom gaps and open edges. Based on the field distribution, the resonant frequency of the antiphase TM_{20} mode is estimated by

$$\beta_{mr} pN_x/2 + 2\beta_e \Delta L = \pi \tag{4.6}$$

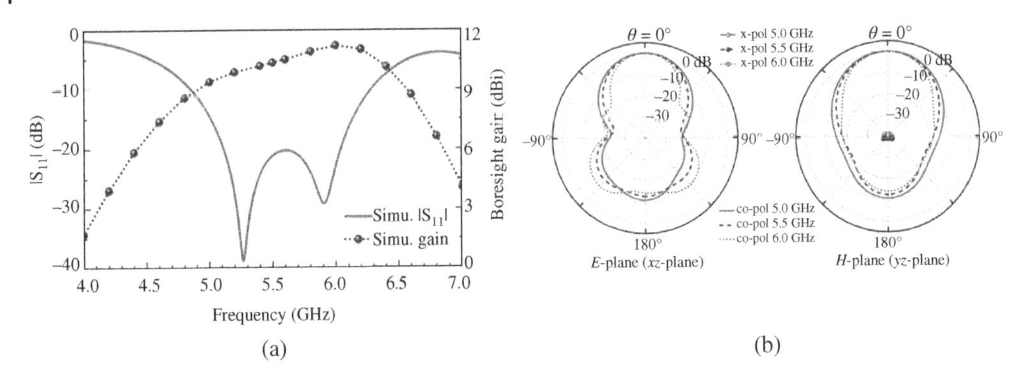

Figure 4.6 Simulated results of the microstrip-line-slot fed CRLH-mushroom antenna. (a) $|S_{11}|$ and realized boresight gain and (b) normalized radiation patterns.

For a conventional rectangular patch antenna, the resonant frequency of the TM_{20} mode is about twice that of the TM_{10} mode. In contrast, the frequency ratio of the antiphase TM_{20} mode to the TM_{10} mode of the CRLH-mushroom antenna is less than two and can be freely controlled by engineering the dispersion relation of the mushroom unit cell. Moreover, the additional radiating gaps at both the TM_{10} and antiphase TM_{20} modes help to lower down the quality factor of the mushroom antenna, which benefits the bandwidth enhancement. By combining the two adjacent operating modes, a low-profile CRLH-mushroom antenna is capable of achieving the consistent radiation at boresight over a broad impedance bandwidth.

By applying the extracted dispersion relation of the mushroom unit in (4.1)–(4.6), the calculated resonant frequencies of the TM_{10} and antiphase TM_{20} modes are 5.23 and 6.05 GHz, respectively. Figure 4.6a shows the simulated reflection coefficient and realized boresight gain of the microstrip-line-slot fed CRLH-mushroom antenna. The mushroom antenna with a thickness of $0.06\lambda_0$ can achieve an impedance bandwidth of 26% from 4.85 to 6.28 GHz, wherein the realized gain varies from 8.8 to 11.2 dBi. Figure 4.6b shows the simulated radiation patterns at 5.0, 5.5, and 6.0 GHz, demonstrating the consistent radiation at boresight across the operating band. The simulated cross-polarization levels are less than −130 and −37 dB in the E-plane (xz-plane) and H-plane (yz-plane), respectively. The low cross-polarization levels are mainly attributed to the mushroom structure and the slot feed compared with the probe fed patch antenna.

As shown in Figure 4.6a, the first two frequencies with the local minimum reflection are 5.28 and 5.94 GHz, very close to the calculated resonant frequencies of the two operating modes. Figure 4.7 shows the simulated electric field distributions of the antenna on the planes $y = 5$ mm

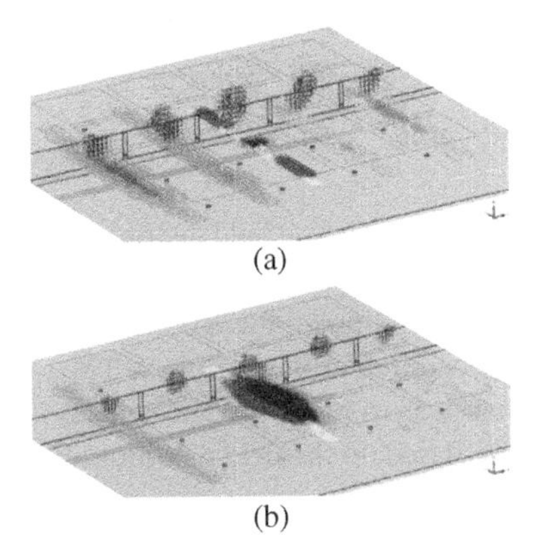

Figure 4.7 Simulated electric field distributions on the planes of $y = 5$ mm and $z = 0.5$ mm at (a) 5.28 GHz and (b) 5.94 GHz.

and $z = 0.5$ mm at 5.28 and 5.94 GHz, further validating the mode analysis of the mushroom antenna.

4.2.2 Impedance Matching

The radiation performance of the CRLH-mushroom antenna is mainly determined by the mushroom structure, while the feeding slot does not affect the radiation performance but the impedance matching. As shown in Figure 4.8, the effects of the parameters of the feeding slot (L_s, W_s, s) on the input impedance are examined to provide a guideline for the optimization of the CRLH-mushroom antenna.

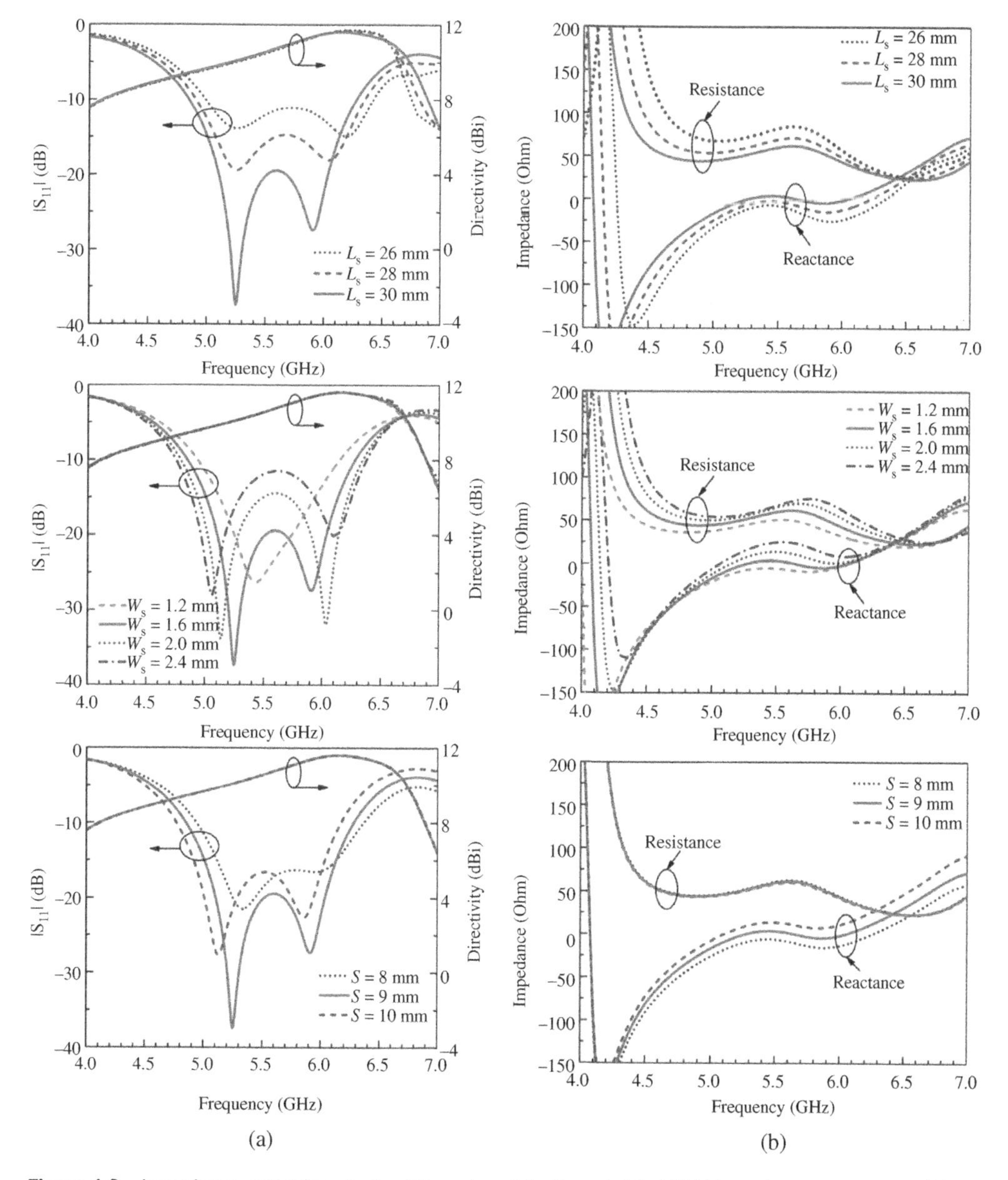

Figure 4.8 Impedance matching study of the microstrip-line-slot fed CRLH-mushroom antenna with different slot length L_s, width W_s, and distance s. (a) Simulated $|S_{11}|$ and directivity and (b) input impedance.

Over the operating frequency range of the antenna, increasing the slot length L_s brings about a lower input resistance and a larger input reactance, while increasing the slot width W_s leads to a higher input resistance and a more severe variation in the input reactance. With the increase of the distance s from its initial value about one quarter guided wavelength, the input resistance decreases slightly, and the input reactance increases significantly.

Therefore, a proper slot length L_s can be found to make the input resistance close to the required value of 50 Ω, and then the distance s could be tuned to shift the input reactance to zero over the desired operating band. Last, a larger slot width W_s causes a wider impedance bandwidth but a worse in-band reflection coefficient. The final chosen slot width would be a compromise between the required bandwidth and acceptable in-band reflection coefficient.

4.3 Broadband LTCC Metamaterial-Mushroom Antenna Array at 60 GHz

The CRLH-mushroom antenna is a promising candidate for the antenna array design at mmW bands due to its merits of low profile, broad bandwidth, high gain, high efficiency, and low cross-polarization levels. As an example, the design process and unique features of a CRLH-mushroom antenna array at 60 GHz are introduced in detail [47].

The LTCC multilayer technology is used in designing a 60-GHz antenna array because of its fabrication accuracy and design freedom. To reduce the losses caused by the feeding network and suppress the backward radiation, a SIW is used for the feeding network [48, 49]. A new transmission-line-based simulation model is proposed to predict the mode resonance frequencies directly. Besides the known CRLH dispersion of the guided waves, the mushroom antenna itself also serves as an EBG structure for the surface wave suppression, and thus the mushroom antenna elements possess the self-decoupling capability, which is convenient for the antenna array design. Metamaterial-mushroom antenna is capable of simultaneously exhibiting the CRLH dispersion of guided waves and EBG characteristic of surface waves. A broadband SIW slot fed 8×8 metamaterial-mushroom antenna array in LTCC is demonstrated at the 60 GHz band to ease the most critical design issue of high gain, and the detailed dimensions can be found in [47].

4.3.1 SIW Fed CRLH-Mushroom Antenna Element

The SIW fed CRLH-mushroom antenna element is implemented in a seven-layer LTCC substrate, Ferro A6-M tape ($\varepsilon_r = 5.9 \pm 0.2$ and $\tan\delta = 0.002$ at 100 GHz). Each substrate and metal layer is 0.095 mm and 0.017 mm thick, respectively. The conductor used is Au, whose conductivity is 4.56×10^7 S m^{-1}. Figure 4.9 shows the 4×4 mushroom cells in the substrate layers Sub1–2 form the radiating element. The mushroom unit cell comprises a square patch on the metal layer M1 and a thin via connecting the center of the patch and a ground plane on the metal layer M2. The longitudinal feeding slot is cut onto the broadwall (on the metal layer M2) of the SIW in Sub3–7.

Figure 4.10a shows the transmission-line model for directly extracting the resonant frequency of the TM_{10} mode of the mushroom antenna element. The two opposite edges of the 4×4 mushroom cells are connected to the microstrip line with the same overall width of $W_p = 4 \times p - g$. Waveguide ports are set at the two ends with the reference planes located at ΔL away from the opposite edges of the mushroom structure to take into account the effect of the fringing field. The extended length ΔL is calculated using (4.1, 4.2) and can be integrated into the full-wave simulation to obtain the phase

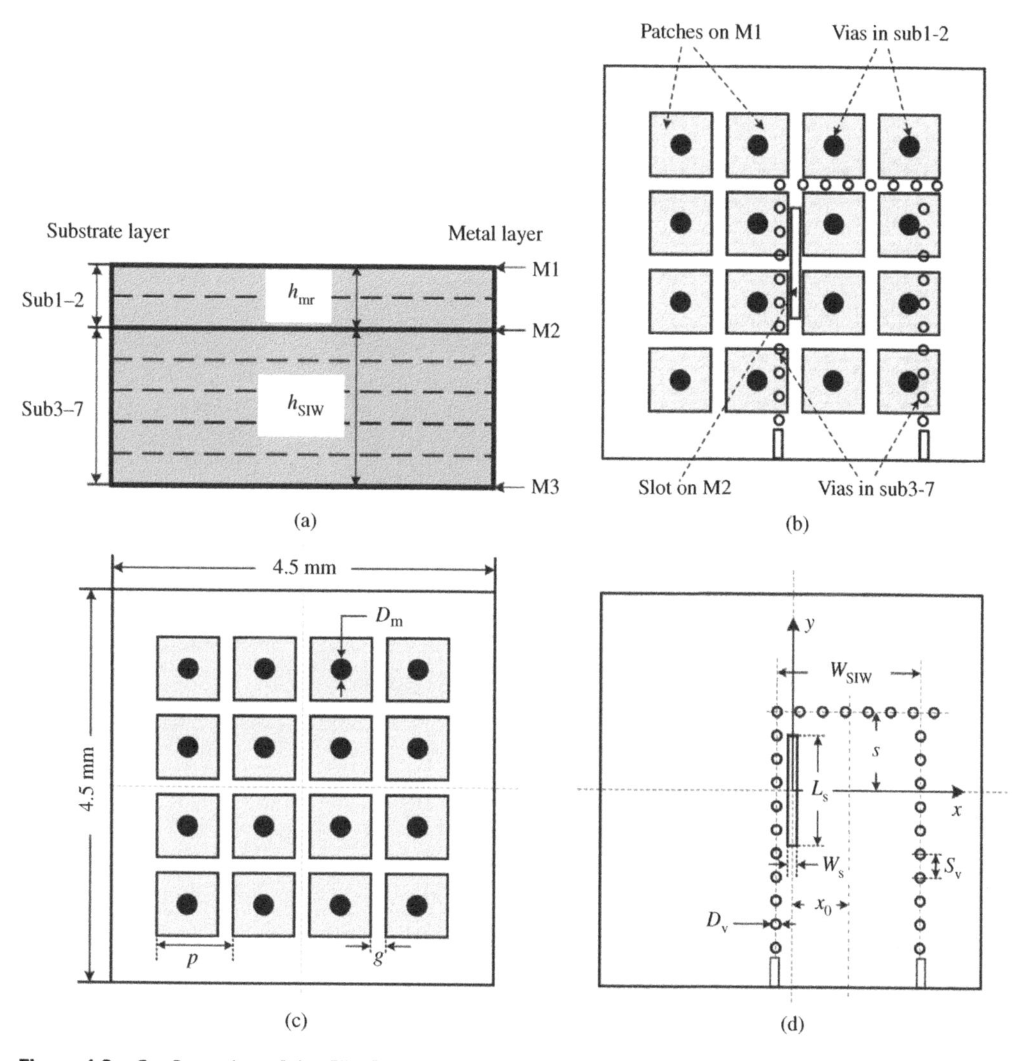

Figure 4.9 Configuration of the SIW fed metamaterial-mushroom antenna element on LTCC. (a) Side view, (b) top view, (c) mushroom structure, and (d) SIW single longitudinal slot feed.

shift between the reference planes A and B directly. The TM_{10} mode is located at the frequency where the phase shift equals −180°. Similarly, the simulation model for estimating the resonant frequency of the antiphase TM_{20} mode is presented in Figure 4.10b, where the 2×4 mushroom cells are considered instead. The antiphase TM_{20} mode occurs at the frequency where the phase shift becomes −180°.

From the simulated phase-shift responses of the proposed transmission-line models shown in Figure 4.11, it is found that the resonant frequencies of the TM_{10} and the antiphase TM_{20} modes of the mushroom antenna element are 58.0 and 64.0 GHz, respectively. As shown in Figure 4.12, the SIW-fed mushroom antenna element has a simulated impedance bandwidth of 12.2% from 57.0 to 64.4 GHz, wherein the realized boresight gain is higher than 8.0 dBi. The mushroom structure has a small thickness of $0.038\lambda_0$ and an aperture size of $0.7\lambda_0 \times 0.7\lambda_0$, ($\lambda_0$ is the free-space wavelength

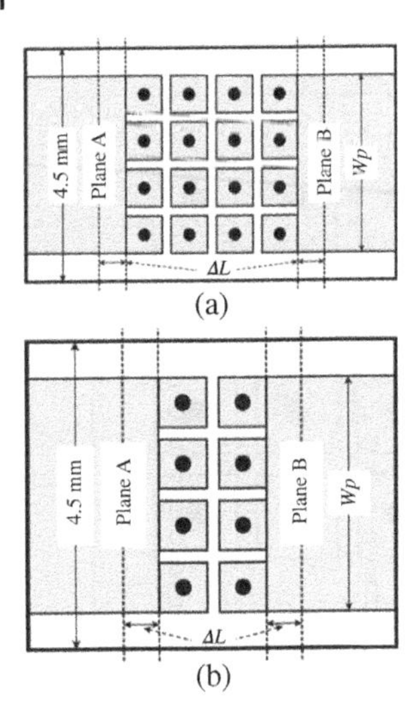

Figure 4.10 Transmission-line models for estimating the resonance frequencies of operating modes of the metamaterial-mushroom antenna element. (a) TM_{10} mode and (b) antiphase TM_{20} mode.

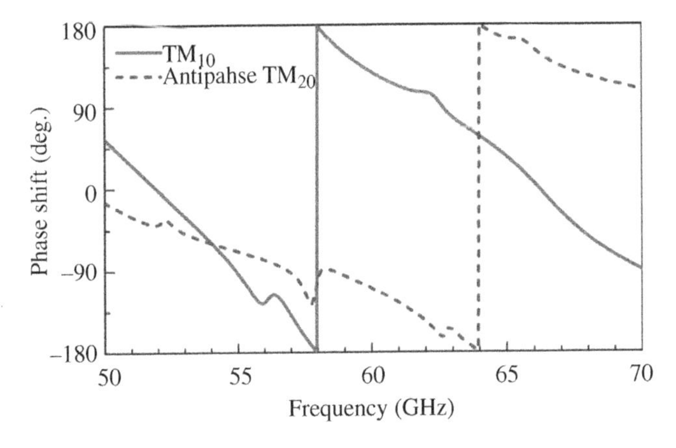

Figure 4.11 Simulated phase shift between Planes A and B.

at the center operating frequency of 60.7 GHz). The first two frequencies with the local minimum reflection are 58.7 and 63.0 GHz, close to the calculated resonant frequencies of 58.0 and 64.0 GHz, respectively. Using the SIW slot feed, the simulated backlobe levels in both *E*- and *H*-plane are suppressed to less than −23 dB across the operating band.

Figure 4.13 shows the SIW longitudinal slot fed patch antennas for comparison. The radiating rectangular patch is printed on the top layer. The dimensions of the SIW and the substrate area are the same as the mushroom antenna element design. An additional metallic via is positioned in the SIW for reflection cancelation. The simulated reflection coefficient and realized gain of the antennas are compared in Figure 4.14. The patch antenna with the same two-layer LTCC substrate as the mushroom antenna element has a much narrower impedance bandwidth of 59.2–61.8 GHz (4.3%), the patch antenna with four-layer LTCC substrate can achieve a comparable impedance bandwidth of 57.0–64.7 GHz (12.7%). However, both patch antennas have much lower gain compared with the mushroom antenna element. In short, the mushroom antenna element shows superior performance in terms of low profile, broad bandwidth, and high gain.

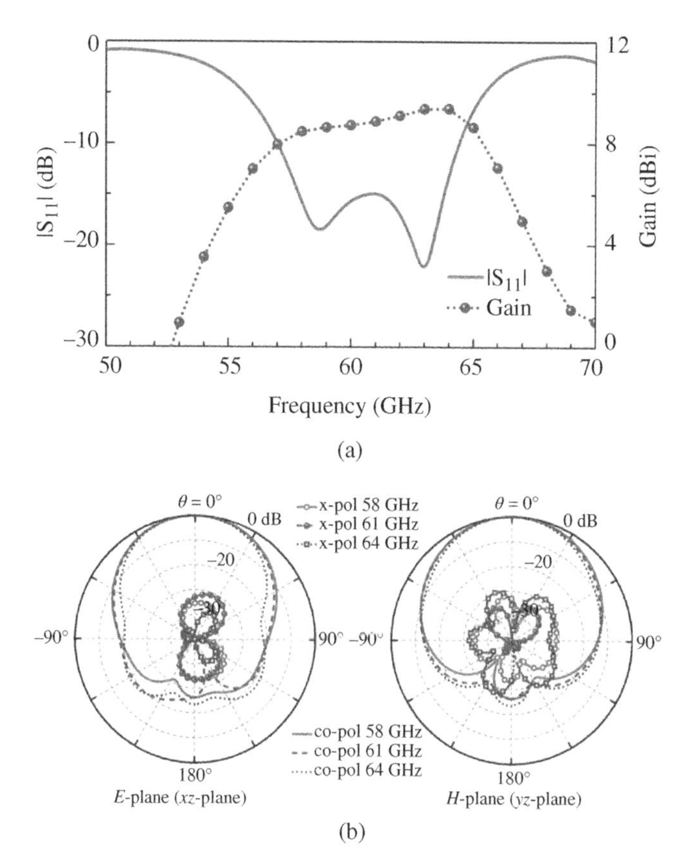

Figure 4.12 Simulated results of the metamaterial-mushroom antenna element. (a) $|S_{11}|$ and realized gain and (b) normalized radiation patterns.

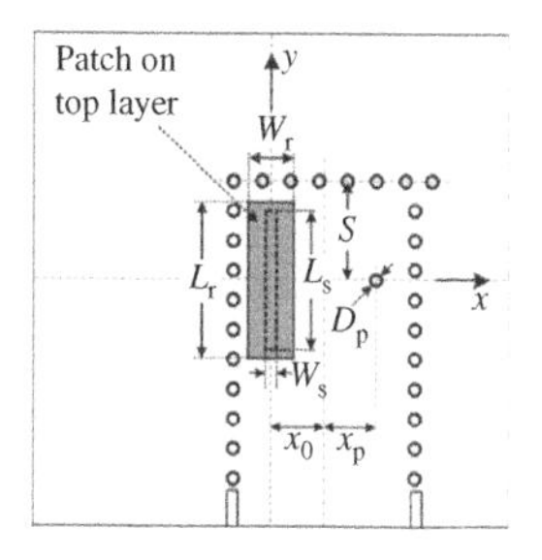

Figure 4.13 Configurations of the SIW-slot fed patch antennas.

4.3.2 Self-Decoupling Functionality

The 4 × 4 mushroom cells not only form the radiating structure of the CRLH-mushroom antenna element but also serve as the EBG structure for suppressing surface waves. The self-decoupling functionality of the mushroom antenna without any extra decoupling structures is demonstrated.

In the mushroom antenna and the patch antennas, transverse magnetic (TM) surface waves are excited in the *E*-plane, whereas transverse electric (TE) surface waves propagate in the *H*-plane [50, 51]. In the grounded two/four-layer LTCC substrate, only the fundamental TM_0 mode surface wave exists in the frequency range below 70 GHz [51]. The surface wave band gap obtained from the mushroom EBG unit cell simulation is shown in Figure 4.15. A surface wave forbidden

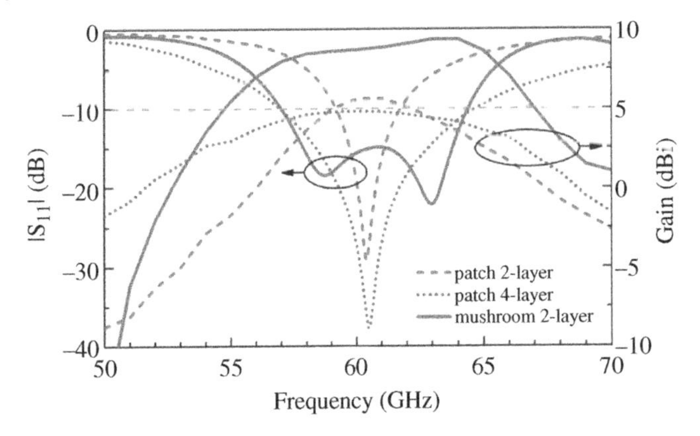

Figure 4.14 Simulated $|S_{11}|$ and realized gain of the SIW-slot fed patch antenna elements in comparison with the SIW-slot fed mushroom antenna element.

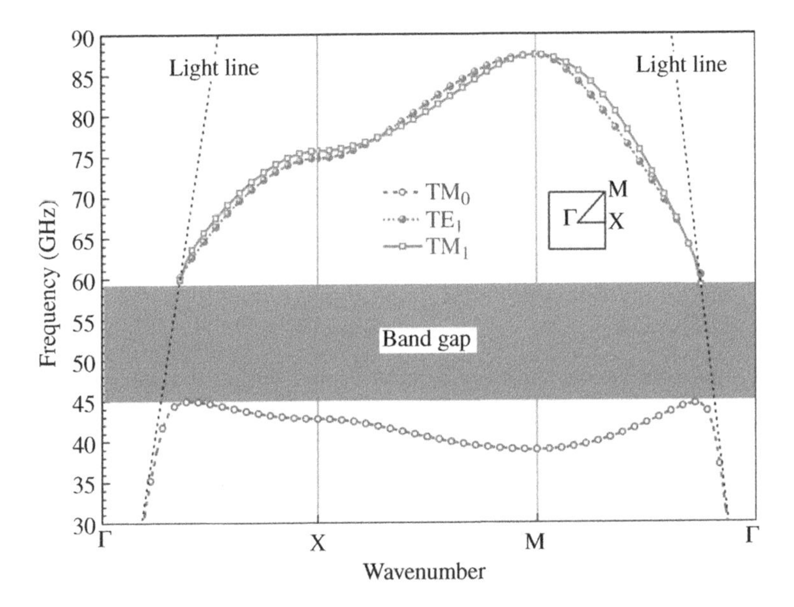

Figure 4.15 Surface wave band gap of the mushroom structure served as an EBG surface.

band of 45–59 GHz emerges in the EBG-mushroom structure. As shown in Figure 4.16, the mushroom/patch antenna elements are positioned collinearly along the *E*- and *H*-plane with a center-to-center separation of s_e and s_h, respectively. As the antenna elements are used for the array design, the center-to-center separation rather than the edge-to-edge separation is chosen to examine the mutual coupling.

Figure 4.17a plots the mutual couplings of the *E*-plane arrayed mushroom and patch antenna elements with the same center-to-center separation of 4.2 mm. The two-layer patch antenna elements exhibit less mutual coupling than the four-layer patch antenna elements in the *E*-plane, due to the weaker TM surface wave propagating along the grounded substrate area. Because of the additional EBG characteristic of the mushroom structure, the two-layer mushroom antenna elements enjoy significantly suppressed *E*-plane inter-element mutual coupling of −33 dB at 60 GHz, 15 dB lower than that of the two-layer patch antenna elements, although there is a much smaller edge-to-edge separation of 0.72 mm between the mushroom antenna elements in the *E*-plane. Over the operating

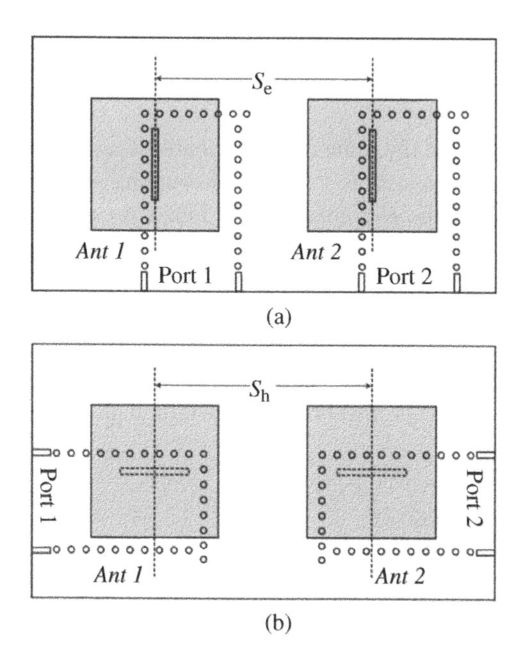

Figure 4.16 Arrangement of the general antenna elements for the mutual coupling study in (a) *E*-plane and (b) *H*-plane.

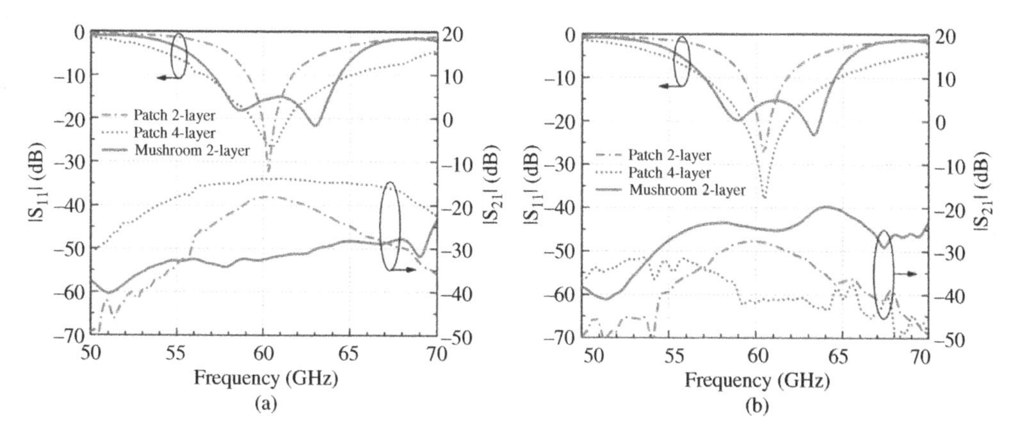

Figure 4.17 Mutual couplings of the antenna elements arrayed (a) in the *E*-plane with $s_e = 4.2$ mm and (b) in the *H*-plane with $s_h = 4.2$ mm.

frequency range of 57.0–64.4 GHz, the *E*-plane mutual coupling of the mushroom antenna elements is less than −28.6 dB.

Figure 4.17b presents the *H*-plane mutual couplings of the mushroom and patch antenna elements with the same center-to-center separation of 4.2 mm. Because the surface waves are not excited in the grounded two/four-layer substrate along the *H*-plane of the antenna elements, the fringing field coupling dominates in the *H*-plane coupled antenna elements [50]. The four-layer patch antenna elements show less *H*-plane inter-element mutual coupling than the two-layer patch antenna elements because of the decreased size of the radiating patch. The two-layer mushroom antenna elements arrayed in the *H*-plane have an edge-to-edge separation of 0.72 mm, much smaller than the corresponding separation of 3 mm in the four-layer patch antenna elements, leading to a higher *H*-plane mutual coupling but still less than −19.8 dB over the operating band.

The *H*-plane mutual coupling of the mushroom antenna elements increases from −24.8 dB at 60 GHz to −19.8 dB at 64 GHz, due to the TE surface wave propagating in the mushroom structure starting from 60 GHz.

With the same operating bandwidth of 57.0–64.4 GHz and the same center-to-center separation of 4.2 mm, the two-layer mushroom antenna elements demonstrate low mutual coupling of less than −28.6 dB in the *E*-plane and less than −19.8 dB in the *H*-plane, and the four-layer patch antenna elements exhibit a mutual coupling of less than −13.9 dB in the *E*-plane and less than −34.3 dB in the *H*-plane. The accompanying EBG characteristic contributes to the low mutual coupling of the CRLH-mushroom antenna elements, especially in the *E*-plane.

Therefore, by combining the CRLH dispersion of guided waves and EBG characteristic of surface waves, the so-called metamaterial-mushroom antenna element demonstrates its superiority over the conventional patch antenna because of the merits of low profile, broad bandwidth, high gain, and self-decoupling capability.

4.3.3 Self-Decoupled Metamaterial-Mushroom Subarray

In order to reduce the number of SIW T-junctions for a simplified feeding network, a two-element self-decoupled metamaterial-mushroom subarray is designed using double longitudinal feeding slots in one SIW section, as shown in Figure 4.18. By positioning the SIW longitudinal slots with either the same offsets and even times of half guided wavelength distance or the opposite offsets and odd times of half guided wavelength distance, an in-phase radiation can be generated. In the designed subarray, the distance between the two longitudinal slots (s_p) is 3.9 mm, which is three times of the half guided wavelength in the SIW at the operating frequency of 61 GHz. The two slots are positioned at the opposite sides at an offset of 0.4 mm with respect to the center line of the SIW. The two mushroom antenna elements are aligned to the two feeding slots in the center, respectively. As shown in Figure 4.19, the subarray has a simulated −10 dB impedance bandwidth of 15.9% from 55.5 to 65.1 GHz, wherein the realized boresight gain is higher than 9.5 dBi with a peak gain of 11.8 dBi at 61 GHz.

As shown in Figure 4.20, the *E*/*H*-plane coupled subarrays are positioned along the *x*/*y*-axis direction with an inter-element spacing of s_e/s_h. With an inter-element spacing of 4.2 mm, the *E*/*H*-plane mutual coupling of the closely placed mushroom subarrays is less than −26.6/−22.5 dB over the operating bandwidth of 55.5–65.1 GHz, as shown in Figure 4.21. The simulated electric field distributions at 56.5 GHz on the top surface of the *E*/*H*-plane coupled subarrays are illustrated in Figure 4.22. When one mushroom subarray is activated as the radiating elements, the

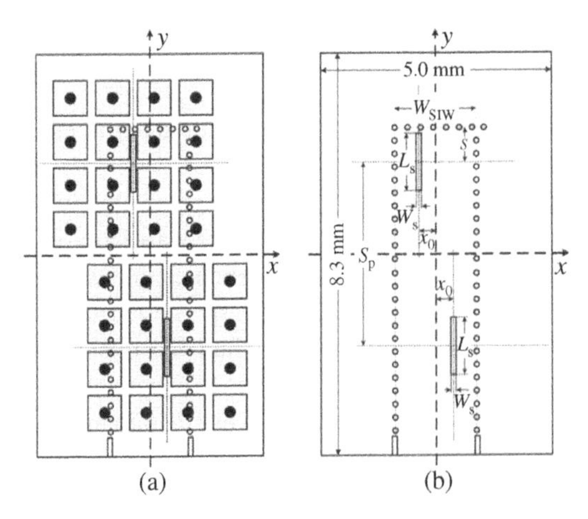

Figure 4.18 Configuration of the two-element metamaterial-mushroom antenna subarray. (a) Top view and (b) SIW double longitudinal slots feed.

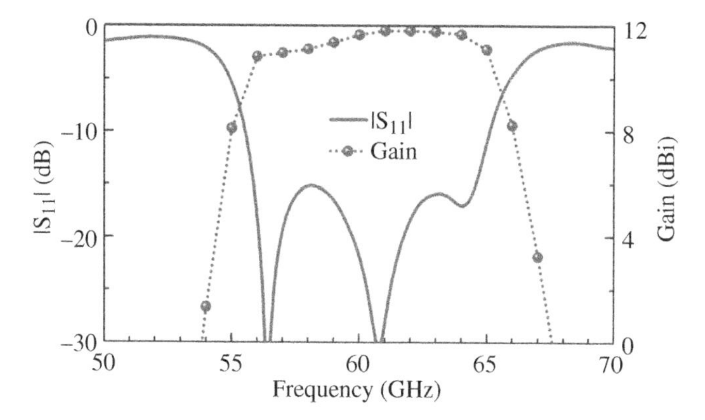

Figure 4.19 Simulated $|S_{11}|$ and realized boresight gain of the metamaterial-mushroom antenna subarray.

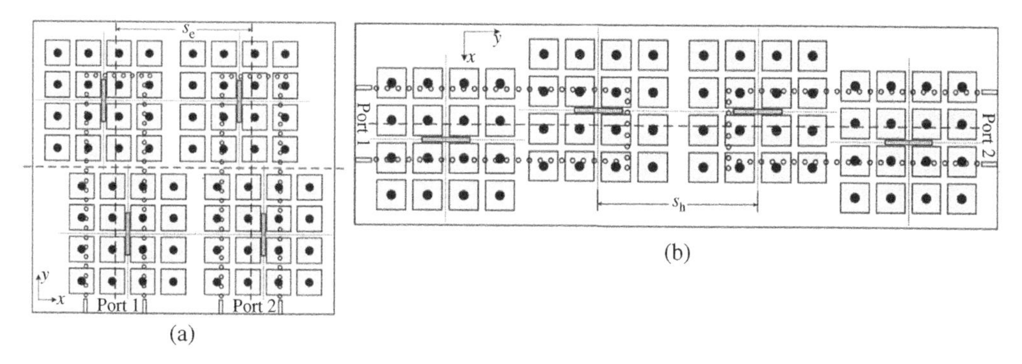

Figure 4.20 Geometry of the metamaterial-mushroom antenna subarrays configured to study the mutual coupling in (a) *E*-plane and (b) *H*-plane.

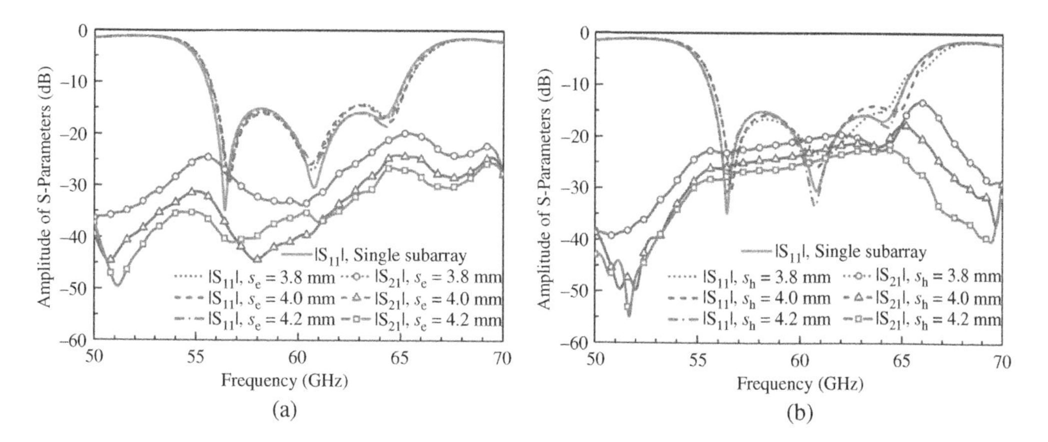

Figure 4.21 Mutual couplings of the subarrays arrayed in (a) *E*-plane, and (b) *H*-plane.

other mushroom subarray serves as the EBG decoupling structure. Therefore, the coupled electric fields in the subarrays are greatly suppressed at 56.5 GHz.

4.3.4 Metamaterial-Mushroom Antenna Array

The self-decoupled metamaterial-mushroom subarrays are employed in the antenna array design. As shown in Figure 4.23, the 8 × 8 antenna array is formed by 8 × 4 mushroom subarrays with an inter-element spacing of 4.2 mm ($0.84\lambda_0$ at 60 GHz) horizontally and 3.9 mm ($0.78\lambda_0$ at 60 GHz)

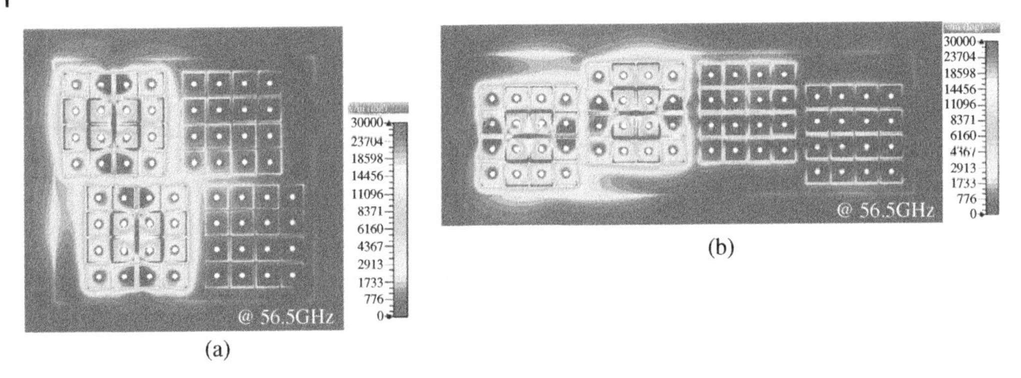

Figure 4.22 Simulated electric filed amplitude distributions at 56.5 GHz on the top surface of (a) E-plane coupled subarrays with $s_e = 4.2$ mm and (b) H-plane coupled subarrays with $s_h = 4.0$ mm.

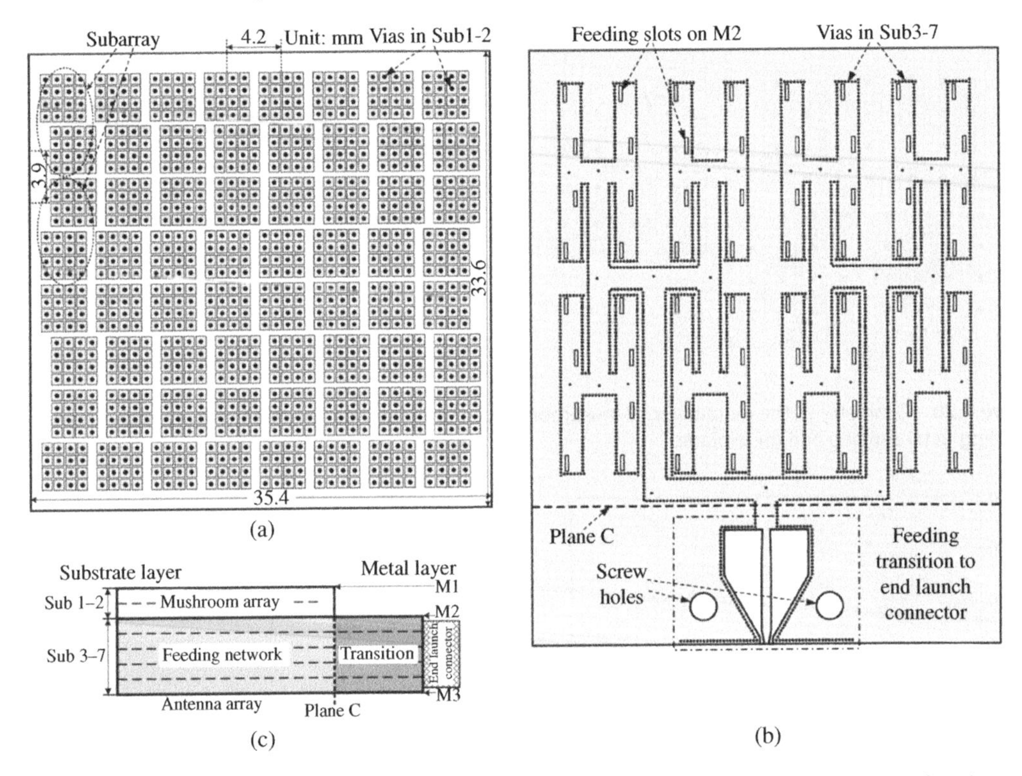

Figure 4.23 Configuration of the antenna array in LTCC. (a) Top view of the mushroom array (M1 and Sub1–2), (b) top view of the feeding network with a transition to the End Launch Connector (M2 and Sub3–7), and (c) side view.

vertically. The feeding network comprising 31 T-junctions and two H-plane bends is simplified for less transmission loss. The antenna array with the feeding network is on the left side of plane C, while the transition between SIW and the 50-Ω End Launch Connector (ELC) is designed on the right side of plane C for measurement. The transition consists of a 50-Ω microstrip line with a tapered section in each end, a surrounding via fence, and grounded vias in the interface. The surrounding via fence is constructed to suppress the server surface waves in the thick high-permittivity LTCC substrate. Over the bandwidth of 55–64 GHz, the transition achieves the return loss of less than 20 dB and the insertion loss below 1 dB. The detailed dimensions of the array can be found in [47].

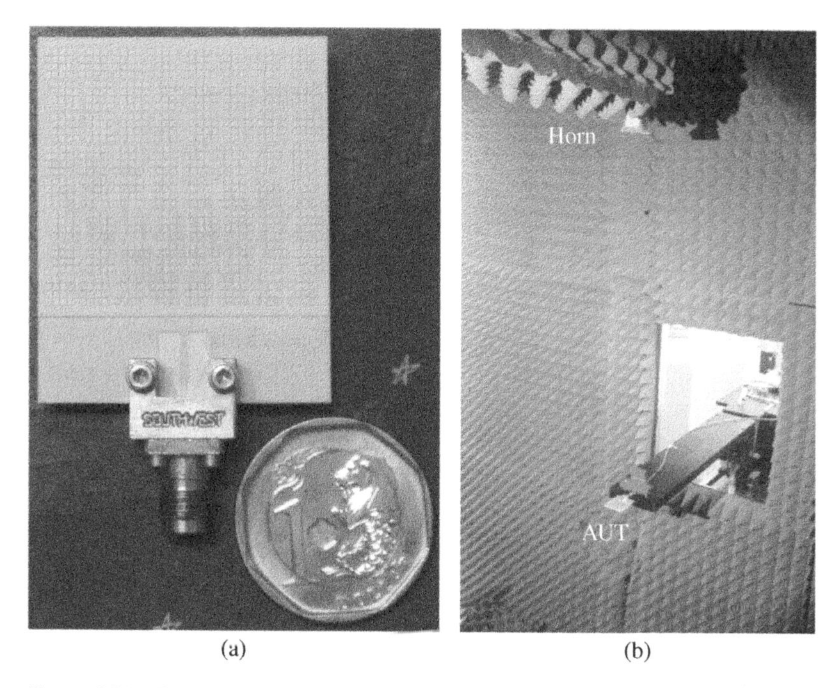

(a) (b)

Figure 4.24 Photographs of (a) the LTCC antenna array and (b) measurement setup.

The fabricated LTCC array is connected to the ELC as shown in Figure 4.24a with an overall size of $35.4 \times 43.8 \times 0.7\,\text{mm}^3$ and a radiating aperture area of $35.4 \times 33.6\,\text{mm}^2$. The antenna was measured in an anechoic chamber using a self-built mmW antenna measurement system at Institute for Infocomm Research, Singapore, as shown in Figure 4.24b [52].

The measured and simulated $|S_{11}|$ of the antenna array with the transition are presented in Figure 4.25a. The simulated and measured −10 dB impedance bandwidths are 55.1–65.3 GHz and 56.2–67 GHz, respectively. The slight upward shift of the measured operating frequencies is mainly attributed to the fabrication tolerance and material properties.

The aperture efficiency (η_{ap}) of an aperture antenna is defined by

$$\eta_{ap} = \frac{D}{D_i} = \frac{D\lambda_0^2}{4\pi A_p} \tag{4.7}$$

where A_p is the physical aperture size of $35.4 \times 33.6\,\text{mm}^2$, λ_0 is the free-space wavelength at the operating frequency, D is the directivity of the antenna, and D_i is the ideal directivity corresponding to a uniform field distribution over the physical aperture.

The radiation performance of the antenna array itself without the transition is examined first. As compared in Figure 4.25b, the simulated directivity is close to the corresponding ideal directivity, and the simulated gain of the array is higher than 22.8 dBi over 55.5–65 GHz with the peak gain of 24.9 dBi at 62 GHz. As shown in Figure 4.25c, the simulated antenna efficiency of the array is higher than 41% over 55.5–65 GHz with a maximum efficiency of 55.1% at 60.5 GHz. The reduced antenna efficiency of the large-scale array is mainly due to the losses in the feeding network. The high aperture efficiency of 90.6% at 61.5 GHz demonstrates an excellent uniform aperture field distribution due to the closely placed self-decoupled high-gain mushroom antenna elements.

As shown in Figure 4.25b, the simulated gain of the array with the transition is higher than 21.5 dBi over 55.5–65.0 GHz and approaches the peak value of 24.1 dBi at 61 GHz. The measured

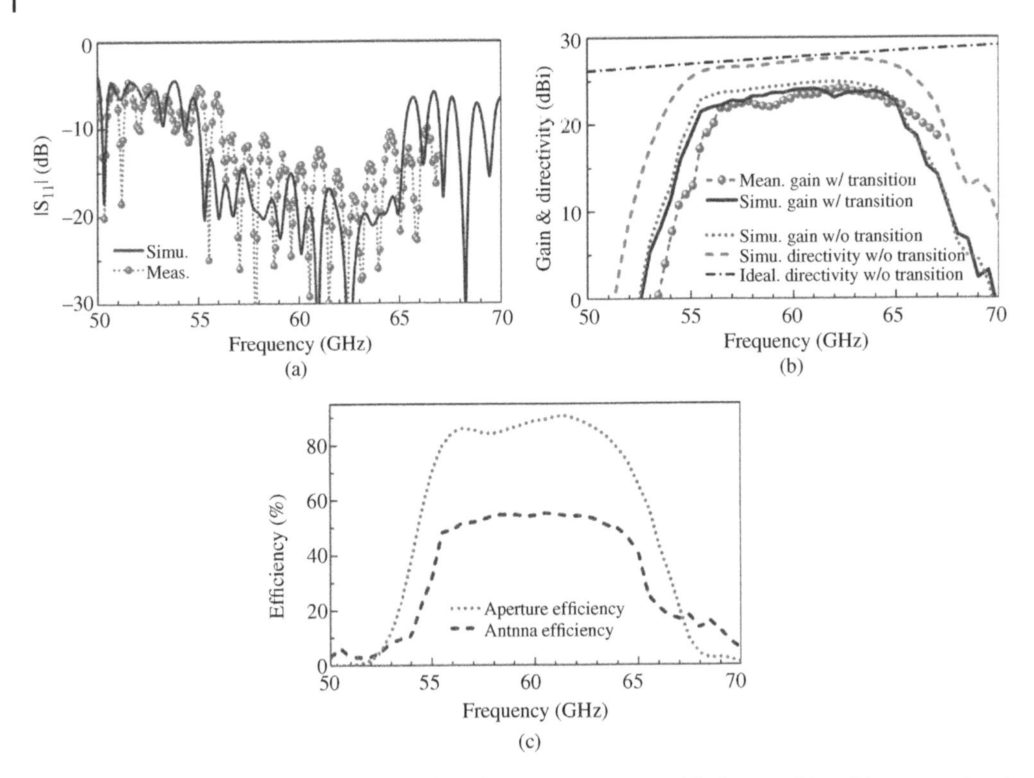

Figure 4.25 (a) Measured and simulated $|S_{11}|$ of the antenna array with the transition, (b) measured and simulated gains/directivity at boresight of the antenna array with/without the transition, and (c) simulated antenna efficiency and aperture efficiency of the array without the transition.

gain of the array with the transition shows a maximum value of 24.2 dBi at 62.3 GHz and a 3-dB gain bandwidth of 56.3–65.7 GHz.

Figure 4.26 shows measured normalized radiation patterns in the *E*/*H*-plane of the array with the transition and ELC in comparison with the simulated ones at 60 GHz. The simulated sidelobe levels (SLLs) are lower than −14.4 dB in the *E*-plane and lower than −12.1 dB in the *H*-plane, while the measured SLLs are less than −13.6 dB in the *E*-plane and less than −11.0 dB in the *H*-plane. The measured backlobe levels are suppressed to less than −34 dB. The measured cross-polarization levels are lower than −24.1 and −21.9 dB in the *E*-plane and *H*-plane, respectively. It is found from the simulation that the worsened cross-polarization levels are due to the presence of the transition part and the ELC. The array itself without the transition and ELC exhibits much lower simulated cross-polarization levels of less than −48.4 and − 37.3 dB in the *E*-plane and *H*-plane, respectively.

In general, the measure results are in good agreement with the simulated ones. The SIW fed metamaterial-mushroom array in LTCC proves to be a promising candidate in the mmW antenna array design with high gain and broad bandwidth.

4.4 Summary

In this chapter, a metamaterial-mushroom antenna array in LTCC has been introduced for broadband and high-gain operation at mmW bands. The unique operating modes and design guidelines of a CRLH-mushroom antenna element at the 5 GHz band have been reviewed to design

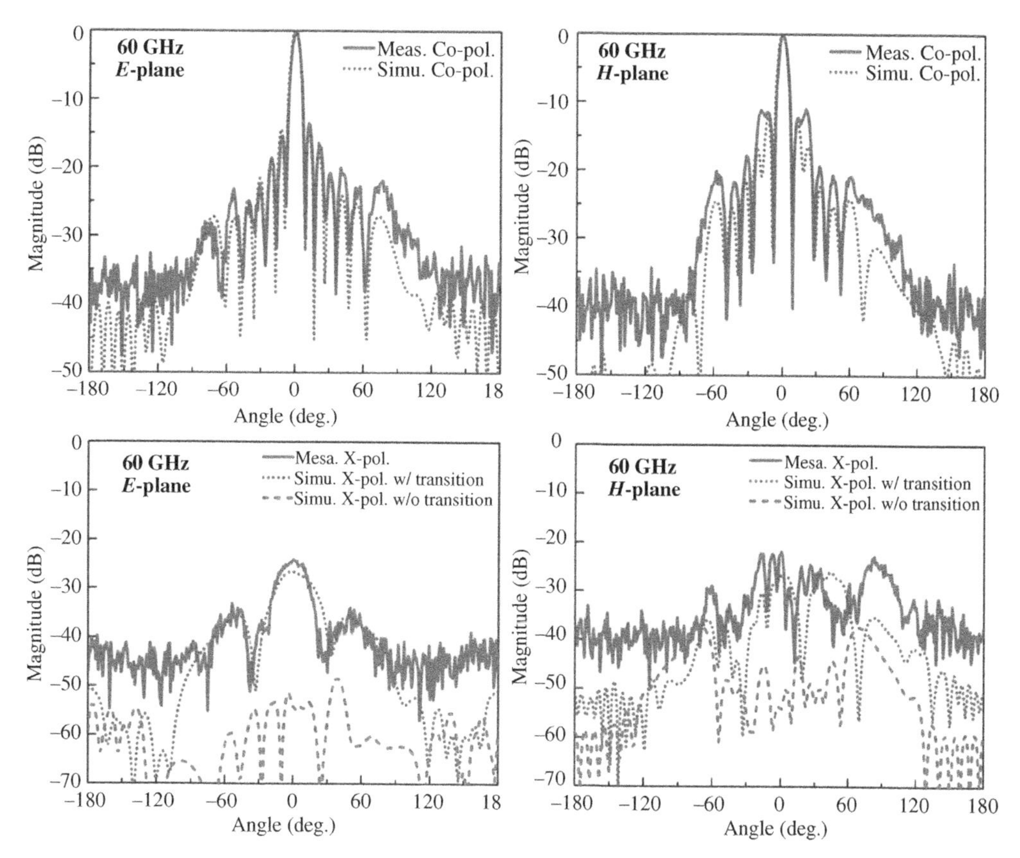

Figure 4.26 Measured and simulated normalized co/cross-polarized radiation patterns in the *E*/*H*-plane of the antenna array operating at 60 GHz.

a broadband antenna with low profile. By combining the CRLH dispersion of guided waves and EBG characteristic of surface waves, the 60 GHz SIW slot fed metamaterial-mushroom antenna demonstrates its superiority with the merits of low profile, broad bandwidth, high gain, high efficiency, low cross-polarization levels, and particularly the self-decoupling capability. The LTCC metamaterial-mushroom antenna array has provided a promising solution to broadband high-gain array design at mmW bands.

References

1 Smulders, P. (2002). Exploiting the 60 GHz band for local wireless multimedia access: prospects and future directions. *IEEE Commun. Mag.* 40 (1): 140–147.

2 Rappaport, T.S., Murdock, J.N., and Gutierrez, F. (2011). State of the art in 60-GHz integrated circuits and systems for wireless communications. *Proc. IEEE* 99 (8): 1390–1436.

3 Pi, Z. and Khan, F. (2011). An introduction to millimeter-wave mobile broadband systems. *IEEE Commun. Mag.* 49 (6): 101–107.

4 Rappaport, T.S. et al. (2013). Millimeter wave mobile communications for 5G cellular: it will work! *IEEE Access* 1: 335–449.

5 Rangan, S., Rappaport, T., and Erkip, E. (2014). Millimeter wave cellular wireless networks: potentials and challenges. *Proc. IEEE* 102 (3): 366–385.

6 Roh, W. et al. (2014). Millimeter-wave beamforming as an enabling technology for 5G cellular communications: theoretical feasibility and prototype results. *IEEE Commun. Mag.* 52 (2): 106–113.
7 Pozar, D.M. (1983). Considerations for millimeter wave printed antennas. *IEEE Trans. Antennas Propag.* 31 (5): 740–747.
8 Deschamps, G.A. (1953). Microstrip microwave antennas. *Proc. 3rd USAF symposium on Antennas*, 22–26.
9 James, J.R., Hall, P.S., and Wood, C. (1981). *Microstrip Antenna Theory and Design*. Stevenage, U.K.: Peter Peregrinus.
10 Chen, Z.N. and Chia, M.Y.W. (2005). *Broadband Planar Antennas*. New York, NY, USA: Wiley.
11 Lee, K.F. and Luk, K.M. (2010). *Microstrip Patch Antennas*. London, U.K.: Imperial College Press.
12 Lee, K.F. and Tong, K.F. (2012). Microstrip patch antennas—basic characteristics and some recent advances. *Proc. IEEE* 100 (7): 2169–2180.
13 Chang, E., Long, S., and Richards, W. (1986). An experimental investigation of electrically thick rectangular microstrip antennas. *IEEE Trans. Antennas Propag.* 34 (6): 767–772.
14 Lee, K.F., Luk, K.M., Tong, K.F. et al. (1997). Experimental and simulation studies of the coaxially fed U-slot rectangular patch antenna. *IEE Proc. Microwaves Antennas Propag.* 144 (5): 354–358.
15 Ansari, J.A. and Ram, R.B. (2008). Analysis of broad band U-slot microstrip patch antenna. *Microwave Opt. Technol. Lett.* 50 (4): 1069–1073.
16 Nakano, H., Yamazaki, M., and Yamauchi, J. (1997). Electromagnetically coupled curl antenna. *Electron. Lett.* 33 (12): 1003–1004.
17 Luk, K.M., Mak, C.L., Chow, Y.L., and Lee, K.F. (1998). Broadband microstrip patch antenna. *Electron. Lett.* 34 (15): 1442–1443.
18 Croq, F., Papiernik, A., and Brachat, P. (1990). Wideband aperture coupled microstrip subarray. In: *Proceedings of IEEE Antennas and Propagation Society International Symposium*, Dallas, TX, USA, 1128–1131.
19 Croq, F. and Papiernik, A. (1990). Large bandwidth aperture-coupled microstrip antenna. *Electron. Lett.* 26 (16): 1293–1294.
20 Targonski, S.D. and Pozar, D.M. (1993). Design of wideband circularly polarized aperture-coupled microstrip antennas. *IEEE Trans. Antennas Propag.* 41 (2): 214–220.
21 Croq, F. and Papiernik, A. (1991). Stacked slot-coupled printed antenna. *IEEE Microwave Guided Wave Lett.* 1 (10): 288–290.
22 Targonski, S.D., Waterhouse, R.B., and Pozar, D.M. (1998). Design of wide-band aperture-stacked patch microstrip antennas. *IEEE Trans. Antennas Propag.* 46 (9): 1245–1251.
23 Sun, D. and You, L. (2010). A broadband impedance matching method for proximity-coupled microstrip antenna. *IEEE Trans. Antennas Propag.* 58 (4): 1392–1397.
24 Xu, J.F., Chen, Z.N., Qing, X., and Hong, W. (2011). Bandwidth enhancement for a 60 GHz substrate integrated waveguide fed cavity array antenna on LTCC. *IEEE Trans. Antennas Propag.* 59 (3): 826–832.
25 Xu, J.F., Chen, Z.N., Qing, X., and Hong, W. (2013). 140-GHz TE20-mode dielectric-loaded SIW slot antenna array in LTCC. *IEEE Trans. Antennas Propag.* 61 (4): 1784–1793.
26 Li, Y., Chen, Z.N., Qing, X. et al. (2012). Axial ratio bandwidth enhancement of 60-GHz substrate integrated waveguide-fed circularly polarized LTCC antenna array. *IEEE Trans. Antennas Propag.* 60 (10): 4619–4626.
27 Lamminen, A.E.I., Säily, J., and Vimpari, A.R. (2008). 60-GHz patch antennas and arrays on LTCC with embedded-cavity substrates. *IEEE Trans. Antennas Propag.* 56 (9): 2865–2874.

28 Yeap, S.B., Chen, Z.N., and Qing, X. (2011). Gain-enhanced 60-GHz LTCC antenna array with open air cavities. *IEEE Trans. Antennas Propag.* 59 (9): 3470–3473.
29 Smith, D.R., Padilla, W.J., Vier, D.C. et al. (2000). Composite medium with simultaneously negative permeability and permittivity. *Phys. Rev. Lett.* 84 (18): 4184–4187.
30 Shelby, R.A., Smith, D.R., and Schultz, S. (2001). Experimental verification of a negative index of refraction. *Science* 292 (5514): 77–79.
31 Eleftheriades, G.V. and Balmain, K.G. (2005). *Negative-Refraction Metamaterials: Fundamental Principle and Applications*, 62–82. New York, NY, USA: Wiley.
32 Caloz, C. and Itoh, T. (2005). *Electromagnetic Metamaterials: Transmission Line Theory and Microwave Applications*. New York, NY, USA: Wiley.
33 Holloway, C.L., Kuester, E.F., Gordon, J.A. et al. (2012). An overview of the theory and applications of metasurfaces: the two-dimensional equivalents of metamaterials. *IEEE Antennas Propag. Mag.* 54 (2): 10–35.
34 Glybovskia, S.B., Tretyakovb, S.A., Belova, P.A. et al. (2016). Metasurfaces: from microwaves to visible. *Phys. Rep.* 634: 1–72.
35 Chen, H.-T., Taylor, A.J., and Yu, N. (2016). A review of metasurfaces: physics and applications. *Rep. Prog. Phys.* 79: 076401-1–076401-40.
36 Li, A.B., Singh, S., and Sievenpiper, D. (2018). Metasurfaces and their applications. *Nanophotonics* 7 (6): 989–1011.
37 Lee, C.J., Leong, K.M.K.H., and Itoh, T. (2006). Composite right/left-handed transmission line based compact resonant antennas for RF module integration. *IEEE Trans. Antennas Propag.* 54 (8): 2283–2291.
38 Dong, Y. and Itoh, T. (2010). Miniaturized substrate integrated waveguide slot antennas based on negative order resonance. *IEEE Trans. Antennas Propag.* 58 (12): 3856–3864.
39 Lai, A., Leong, K.M.K.H., and Itoh, T. (2007). Infinite wavelength resonant antennas with monopolar radiation pattern based on periodic structures. *IEEE Trans. Antennas Propag.* 55 (3): 868–876.
40 Park, J.H., Ryu, Y.H., Lee, J.G., and Lee, J.H. (2007). Epsilon negative zeroth-order resonator antenna. *IEEE Trans. Antennas Propag.* 55 (12): 3710–3712.
41 Pyo, S., Han, S.M., Baik, J.W., and Kim, Y.S. (2009). A slot-loaded composite right/left-handed transmission line for a zeroth-order resonant antenna with improved efficiency. *IEEE Trans. Microwave Theory Tech.* 57 (11): 2775–2782.
42 Sievenpiper, D., Zhang, L., Broas, R.F.J. et al. (1999). High-impedance electromagnetic surfaces with a forbidden frequency band. *IEEE Trans. Antennas Propag.* 47 (11): 2059–2074.
43 Sievenpiper, D.F. (1999). High-impedance electromagnetic surfaces. Ph.D. dissertation. Univercity of California at Los Angeles.
44 Yang, F. and Rahmat-Samii, Y. (2003). Microstrip antennas integrated with electromagnetic band-gap (EBG) structures: a low mutual coupling design for array applications. *IEEE Trans. Antennas Propag.* 51 (10): 2936–2946.
45 Yang, F. and Rahmat-Samii, Y. (2008). *Electromagnetic Band Gap Structures in Antenna Engineering*. Cambridge, U.K.: Cambridge University Press.
46 Liu, W., Chen, Z.N., and Qing, X. (2014). Metamaterial-based low-profile broadband mushroom antenna. *IEEE Trans. Antennas Propag.* 62 (3): 1165–1172.
47 Liu, W., Chen, Z.N., and Qing, X. (2014). 60-GHz thin broadband high-gain LTCC metamaterial-mushroom antenna array. *IEEE Trans. Antennas Propag.* 62 (9): 4592–4601.
48 Uchimura, H., Takenoshita, T., and Fujii, M. (1998). Development of a 'laminated waveguide'. *IEEE Trans. Microwave Theory Tech.* 46 (12): 2438–2443.

49 Huang, Y., Wu, K.L., Fang, D.G., and Ehlert, M. (2005). An integrated LTCC millimeter-wave planar array antenna with low-loss feeding network. *IEEE Trans. Antennas Propag.* 53 (3): 1232–1234.
50 Balanis, C.A. (1997, ch. 14). *Antenna Theory: Analysis and Design*, 2e, 764–767. New York, NY, USA: Wiley.
51 Colin, R. (1991, ch. 11). *Field Theory of Guided Waves*, 2e, 697–708. New York, NY, USA: IEEE Press.
52 Qing, X. and Chen, Z.N. (2014). Measurement setups for millimeter-wave antennas at 60/140/270 GHz bands. In: *IEEE International Workshop on Antenna Technology (iWAT)*, Sydney, NSW, Australia, 281–284.

5

Narrow-Wall-Fed Substrate Integrated Cavity Antenna at 60 GHz

Yan Zhang

School of Information Science and Engineering, Southeast University, Nanjing 211111, People's Republic of China

5.1 Introduction

Due to the rapid development and commercialization of the 5G wireless communications, mmW antennas have been widely explored and implemented in recent years by both academia and industries [1–3]. On one hand, the mmW frequency bands can offer a relatively wide bandwidth to support ultra-high date rate and ultra-high throughput wireless transmission [4]. On the other hand, the small antenna size at the mmW bands enables the high-density integration of large-scale arrays [5], for instance, 64-, 128-, and 256-element arrays, which can provide highly dense coverage and multiple diversity to further enhance the performance of mmW communication systems. At present, the most concerning mmW operating bands include *Ka*-band (24.25–29.5 GHz), *Q*-band (37–43.5 GHz, 45.5–47 GHz, 47.2–48.2 GHz), *V*-band (57–64 GHz), *E*-band (71–86 GHz), and *W*-band (91–97 GHz), as mentioned in Chapter 1. However, from the antenna engineering point of view, the recorded research activities on mmW antennas can be tracked back to around 1950s according to the IEEE Xplore database [6], and it has become a worldwide research hot spot since around 2008 for two reasons.

First, the *V*-band usually known as 60 GHz band has been released from its original satellite communication applications to the commercial high-rate wireless personal area networks (WPANs), which was promoted by the IEEE 802.15.3 Task Group 3c (TG3c) during 2005–2009 and an IEEE 802.15.3 standard was finally delivered [7]. Stimulated by the re-raised potential applications, antennas operating in *V*-band have been widely investigated to meet the low-cost, low-profile, and high-integration requirements. Later on, the mmW communications entered into an unprecedented period of full bloom, among which the 5G wireless communication has played a predominant role.

Second, the emergence and development of a batch of innovative mmW transmission-line technology, including SIW, half-mode SIW, folded SIW, and substrate integrated coaxial line (SICL) technologies [8–10], provide both novel concepts and high-integrated co-planar implementations for designing high-integrated mmW components and antennas. The SIW is a widely explored and exploited technology for mmW applications, and it can be implemented on the low-profile substrate with the standard PCB process. It has been widely demonstrated that the SIW transmission line and SIW-based components exhibit better performance compared with those using conventional transmission lines which have open boundaries, such as microstrip line, co-planar waveguide, in terms of both insertion loss and radiation loss. Further, the SIW can be regarded as a planar PCB case of conventional metallic rectangular waveguide, which

Substrate-Integrated Millimeter-Wave Antennas for Next-Generation Communication and Radar Systems, First Edition.
Edited by Zhi Ning Chen and Xianming Qing.

supports transverse electric (TE) modes only and is dominated with the TE_{10} mode [11]. Hence, most of conventional rectangular waveguide-based component and antenna designs can be easily implemented to SIW based designs. As a result, the development of SIW technology exhibits a rapid development during 2003–2008 due to the conceptual design based on conventional waveguide technology [12–15]. As is well known, the waveguide slot antenna or slot array is the most popular form of waveguide antennas, and it has widely been used in radar systems. However, the waveguide slot antennas suffer from inherent limited bandwidth because of the sharply varied input impedance of the slot element and the dispersive properties of electromagnetic waves in the waveguide. Similarly, the bandwidth constraint applies to SIW-based slot antennas as well.

An SIW slot antenna has been the most popular type of antenna including longitudinal slot, inclined slot, transverse slot, and so on, for decades [13, 16, 17]. However, these slot antennas suffer from narrow bandwidth, typically ~3%–5%. To widen the bandwidth, an SIW long-slot antenna has been proposed as a leakage-wave antenna, and the bandwidth enhancement is at the expense of the frequency-dependent beam squinting [18]. Then, the SIW has been used as the feeding line of other types of antennas, such as dipole antenna [19], Yagi-Uda antenna [20], log-periodic antenna [21], and patch antenna [22], as well as dielectric resonator antenna [23]. In these antennas, the SIW feeding network usually occupies a large area, which either causes a difficulty in forming large-scale arrays or greatly enlarges the profile in the case of stacked structures. In recent years, advanced solutions with low-profile radiators embedded into or stacked on the SIW feeding lines have been widely proposed with enhanced performance.

In this chapter, a class of SIAs, operating at the aforementioned potential mmW frequency bands, is summarized to reveal the schemes that successfully broaden the bandwidth as well as enhance the radiation performance. Only the low-cost standard PCB-based SIAs are highlighted here; the other SIAs using a LTCC process and a silicon-micro machine process are not addressed.

5.2 Broadband Techniques for Substrate Integrated Antennas

The mmW communication systems are commonly broadband and even multi-band for specific applications. For mmW antennas, the broadband or wideband impedance matching characteristic is not only a key design specification but also an essential specification. In addition, the other specifications such as gain, efficiency, beam steering, sidelobe level, and cross-polarization level must be considered simultaneously. In other words, the bandwidth of the antenna refers not only to the impedance bandwidth but also to the antenna radiation performance.

SIW antennas can be classified in terms of radiation direction as the broadside and end-fire SIW antennas. Considering the PCB process, as shown in Figure 5.1, the SIW antennas with beams along the normal direction of the PCB board are categorized as broadside SIW antennas, and the others with beams lying in the plane of PCB boards are categorized as end-fire SIW antennas. For the category of broadside SIW antenna, it can have either single layer or multi-layered PCB to implement the antenna structure, and RF circuits can be integrated with the stacked layers as shown in Figure 5.1a. Besides a relatively low-profile and high-density integration, this type of SIW antenna can support a two-dimensional beam sweeping by controlling the feeding phase of array elements. Although the end-fire SIW antenna can also be integrated with RF circuits as illustrated in Figure 5.1b, the RF circuits placed nearby the antenna would deteriorate the radiation patterns, and only one-dimensional beam sweeping is allowed in this case. Only broadside SIW antennas are reviewed in this chapter, and the end-fire SIW antennas are not included, both of which can find their applications in specific scenarios.

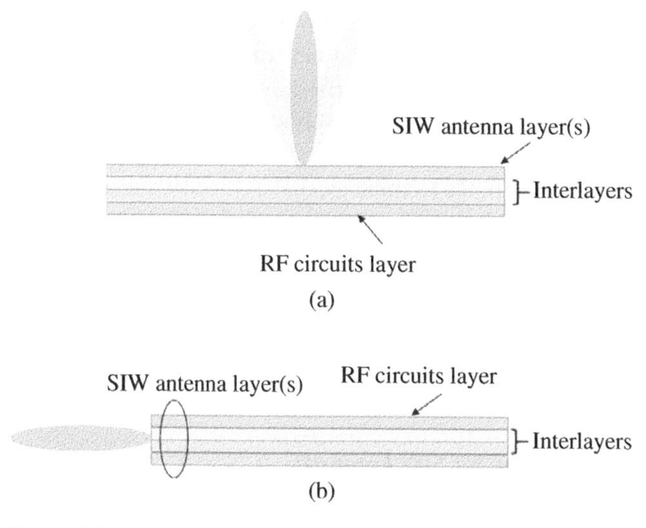

Figure 5.1 Illustration of the SIW antennas integrated with RF circuits. (a) Broadside antennas and (b) end-fire antennas.

At the beginning, the SIW slot antennas as shown in Figure 5.2 are the most used antenna forms that are derived from the conventional rectangular waveguide slot antennas. Their design approaches are very mature, and they are particularly suitable for designing large-scale slot arrays. However, the SIW slot antenna usually suffers from narrow bandwidth, typically ~5%–7%. Considering the SIW slot array, the series feeding scheme is commonly used to achieve a high-density large-scale array, so that the bandwidth is further reduced to only 3% [13, 16]. The narrow bandwidth severely restricts the applications of SIW slot antennas in mmW bands, except for some mmW radar systems, for instance, 77 GHz automobile collision-avoiding radar systems.

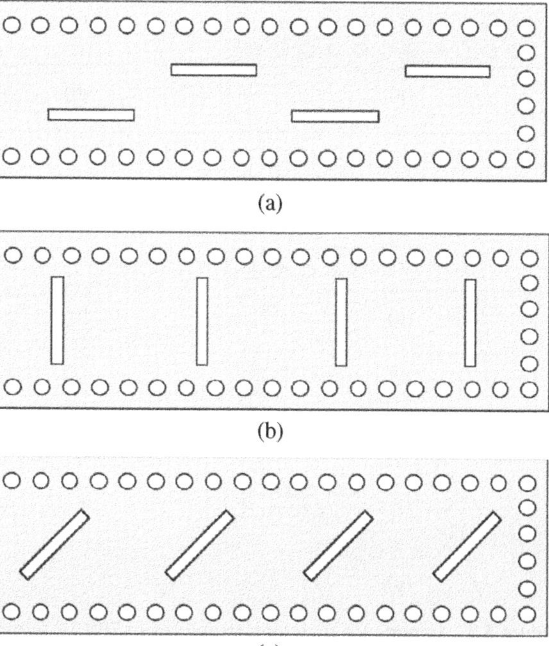

Figure 5.2 Typical SIW slot antenna and array. (a) SIW longitudinal slot antenna, (b) SIW transvers slot array, and (c) SIW inclined slot array.

For this reason, in the last decade, the research into broadband SIW antennas has been widely carried out, and much effort has been put into both SIW feeding networks and SIW radiators. The techniques developed for broadband SIW antennas are categorized into the three main types. The first one is the impedance matching approach, which can effectively alleviate the narrow bandwidth problem caused by the sharp variation of the radiation impedance of the SIW slots or other types of SIW radiators. The second one is the multi-mode technology, that is, the SIW radiator generates multiple resonant modes simultaneously to jointly widen the operating bandwidth. The last one is utilization of a compound of multiple types of radiators, such as patches, slots, spiral, and so on, which are coupled, or loaded by SIWs in either planar form or stacked form to achieve multiple resonances for bandwidth enhancement. Although we are focusing on SIW antenna designs at mmW bands, it should be noted that all these techniques are also applicable for the antenna design at microwave bands.

5.2.1 Enhancement of the Impedance Matching for SIW Antennas

Impedance matching approaches are widely used in microwave engineering, which can successfully transfer the loaded impedance into any desired system standard impedance. The widely used impedance matching structures are microstrip-line-based impedance transformers, such as the quarter-wavelength impedance transformer, stepped impedance transformer, and open- and short-circuited stubs. In designs of SIW antennas and circuits, there are also quite similar impedance transformers widely used or adopted for a broadband purpose. As shown in Figure 5.3a, the iris window induced by two symmetrically distributed vias in an SIW is the basic building block for either the SIW filter design or the impedance matching design of SIW antennas. Its equivalent circuit is depicted in Figure 5.3b, and the value of the introduced inductances are

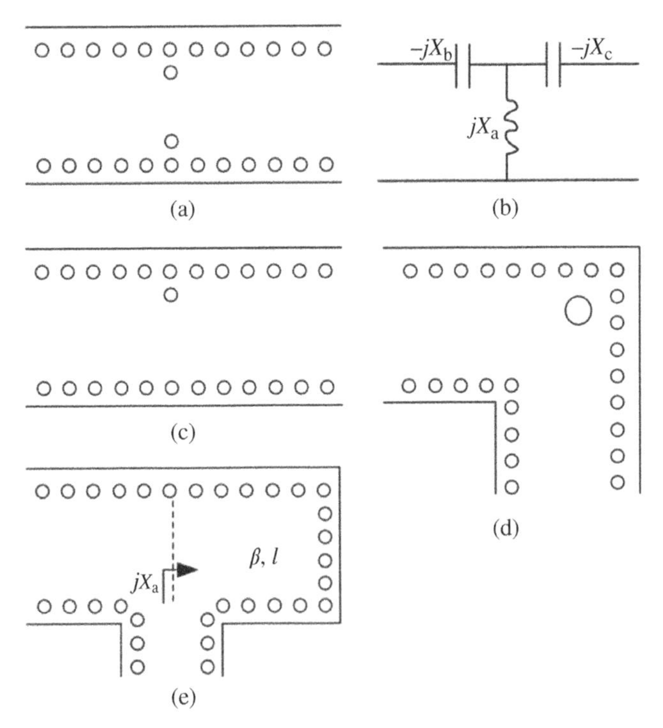

Figure 5.3 Impedance matching structures in SIW. (a) Symmetry iris window, (b) equivalent circuit of (a), (c) asymmetry iris window, (d) inductive corner, and (e) short-circuited SIW stub.

functions of the geometry, including the width of SIW, the width of the window, and the diameter of vias. Using commercial full-wave simulation software packages, the values of susceptance can be easily extracted from the scattering parameters, as [14]:

$$j\frac{X_a}{Z_0} = \frac{2S_{11}}{(1-S_{11})^2 - S_{21}^2} \quad (5.1)$$

$$-j\frac{X_b}{Z_0} = \frac{1+S_{11}-S_{21}}{1-S_{11}+S_{21}} \quad (5.2)$$

These susceptances rely on the width of the window; thus, a relationship can be mapped between the physical dimensions and the susceptances. Note that the value of X_b is usually very small and can be omitted in some cases; instead, the inductance X_a dominates the impedance matching. Thus, this kind of iris window is also called inductance window. To achieve a wideband impedance matching, such iris windows can be cascaded with SIW sections inserted in between. With the presence of extracted values, a cascades SIW iris windows structure can be optimized in the view of an equivalent circuit that could produce a desired impedance matching in a rapid way.

To further illustrate the impedance matching method of using cascaded SIW iris windows, an SIW-fed circularly polarized patch antenna as developed in [22] is presented as an example, as shown in Figure 5.4. The substrate used is Rogers 5880 ($\varepsilon_r = 2.2$, $\tan\delta = 0.0009$) with a thickness of 0.508 mm. Figure 5.4a shows the whole structure of the circularly polarized patch antenna that is orthogonally fed through an SIW 90° coupler. Two iris windows are inserted in the SIW feeding section, as shown in Figure 5.4b, to form a second-order impedance matching network. Its equivalent circuit is shown in Figure 5.4c, and the tuning procedure with the circuit model is depicted in a

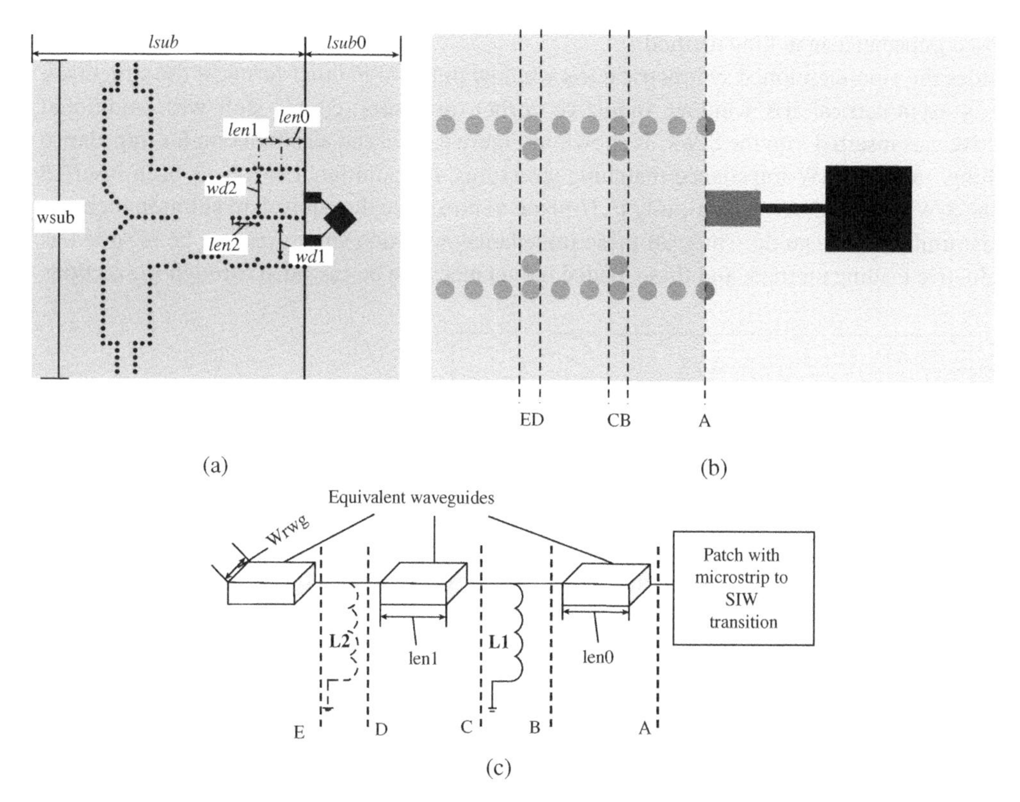

Figure 5.4 An SIW fed patch antenna at Q-band [20]. (a) Circularly polarized patch antenna model, (b) the full-wave model of single fed patch antenna, and (c) equivalent circuit model.

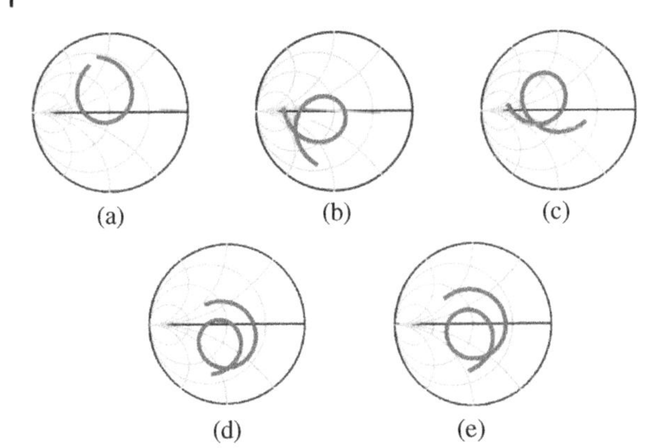

Figure 5.5 Input impedance (simulated by the circuit model) seen from Plane A to Plane E. The simulated frequency ranges from 39 to 48 GHz. (a) The original reflect coefficient seen from Plane A, (b) seen from Plane B after tuning *len*0, (c) seen from Plane C after tuning *L*1, (d) seen from Plane D after tuning *len*1, and (e) seen from Plane E after tuning *L*2.

Smith Chart as shown in Figure 5.5. Advanced Design System (ADS) is used to simulate the model. The patch with a microstrip-to-SIW transition is simulated and then de-embedded to plan A by ANSYS HFSS, which is taken as a loaded impedance during the tuning of the impedance matching network. The optimized parameters are as follows: $len0 = 2$ mm, $len1 = 2.08$ mm, $L1 = 0.347$ nH, and $L2 = 1.027$ nH. The width of the equivalent waveguides is fixed as 3.75 mm during the tuning process. Then the value of the two inductors are mapped to the real size of inductive windows. The sizes are as follows: $wd1 = 2.65$ mm and $wd2 = 3.05$ mm. The results in Figure 5.6 show that the power transmission is higher than 80% over the frequency range from 39 to 48 GHz. The final designed circularly polarized patch antenna achieved an impedance bandwidth of 38.5–48 GHz in terms of $|S_{11}| < -10$ dB, and a 3-dB axial ratio bandwidth of 38.5–47 GHz, which verifies the proposed impedance matching method.

Besides the aforementioned symmetrical iris window in the SIW, other forms of the structures, such as asymmetrical iris window, inductive corner, and short-circuit stub with additional inductive vias inserted into the SIWs, as shown in Figure 5.3c–e, can also function for impedance matching. In those SIW impedance matching structures, the additional vias have been inserted into the SIW to generate a discontinuity in *H*-plane to introduce the equivalent shunt inductance into the uniform waveguide. Thus, all these impedance-matching structures can be regarded as an inductive loading method, and these loaded inductances can be cascaded through the sections

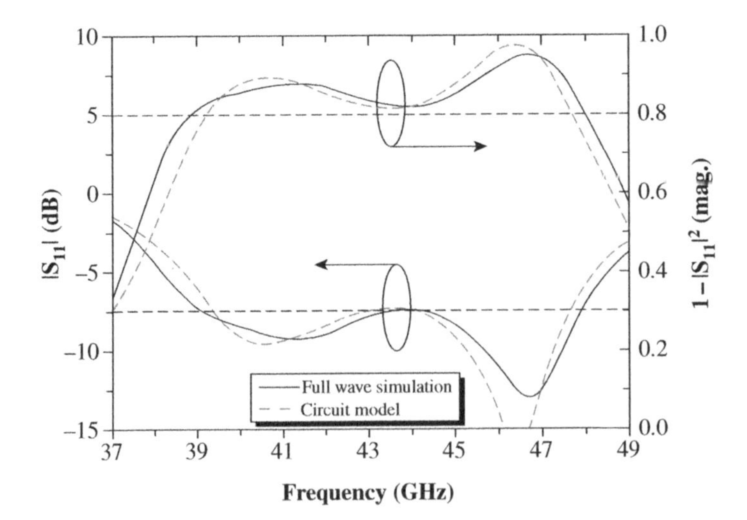

Figure 5.6 Reflection coefficients and power transmissions of the matched patch antenna.

of SIW to achieve any desired impedance matching, quite similar to the design of inductance-only filters. These impedance-matching structures in SIWs have the advantages: (i) the inserted inductive vias will not change the outline or geometry of SIWs and thus there is no extra space required in antenna design; (ii) the SIW impedance-matching structure doesn't cause any spurious radiations; and (iii) the inductive vias can be inserted in the SIW with much freedom, for any type of radiators. Actually, the aforementioned impedance-matching methods have been widely used for bandwidth enhancement in the SIW based antenna designs, [24, 25].

5.2.2 Multi-Mode Substrate Integrated Cavity Antennas

The substrate integrated cavity (SIC) antennas have been proposed and demonstrated for superior performance over conventional SIW slot antennas. The so-called SIC is built with enclosed vias in the laminate substrate, quite similar to the SIW, and the geometry of the cavity could be rectangular, circular, or even some special shapes, as shown in Figure 5.7. The resonant behaviors of such SICs can be analyzed as conventional waveguide cavities. The substrate usually has a very thin thickness, so there is no E-field variation along the thickness direction. A rectangular SIC as shown in Figure 5.7a is exemplified to illustrate the analysis of the resonant behaviors. First, the SIC is realized by four rows of vias with the constraints on the diameter (d) and periodicity (p) of vias, as:

$$d/p \geq 0.5, d/\lambda_0 \leq 0.1 \tag{5.3}$$

where d, p are the diameter and the periodicity of the vias, respectively, and λ_0 denotes the wavelength in free space, [11, 12, 26]. The SIC is equivalent into a conventional waveguide cavity in terms of dimensions, and then the resonant frequencies can be calculated following the approximation formulas:

$$L_{\text{eff}} = L_{\text{c}} - \frac{d^2}{0.95p} \tag{5.4}$$

$$W_{\text{eff}} = W_{\text{c}} - \frac{d^2}{0.95p} \tag{5.5}$$

$$f_{mnp} = \frac{1}{2\sqrt{\varepsilon\mu}}\sqrt{\left(\frac{m}{L_{eff}}\right)^2 + \left(\frac{n}{W_{eff}}\right)^2 + \left(\frac{p}{h}\right)^2} \tag{5.6}$$

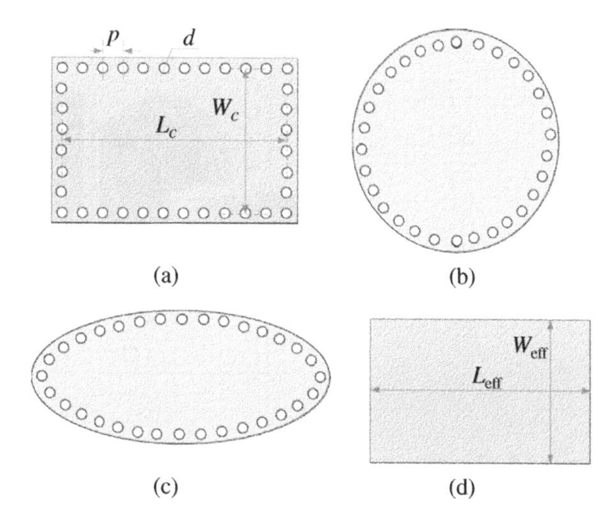

Figure 5.7 Typical Substrate integrated cavities (SICs). (a) Rectangular SIC, (b) circular SIC, (c) elliptical SIC, and (d) equivalent waveguide cavity of (a).

where ε and μ are the permittivity and permeability of the substrate, respectively; W_c and L_c denote the width and length of the SIC, respectively; L_{eff} and W_{eff} are the width and length of the equivalent cavity, respectively; and h denotes the thickness of the substrate.

The SIC can be used as the radiator directly or a carrier as well as a supporter which can further have additional radiators embedded in/on it, such as slots, apertures, slot coupled patches, aperture coupled patches, or spirals. Herein, the SIC antennas are used to indicate all these types of SIC-based antennas. Generally there are two types of methods to widen the bandwidth of SIC antennas. First, an SIC can be designed with multi modes, and these modes can operate jointly to achieve a wide bandwidth. Second, the coupled modes from both the cavity and additional resonators by loading additional resonators in/on the SIC can operate jointly, covering the desired wide bandwidth. Both of the methods mentioned above are based on the combination of multiple modes generated simultaneously. Therefore, "multi-mode" design is another key technique for the design of broadband SIC antenna.

5.2.3 Substrate Integrated Cavity Backed Slot Antenna

The SIC backed slot antenna has been originally proposed by Luo in 2008 [26], as shown in Figure 5.8a. The cavity is fed through a microstrip line printed on the top side of the substrate,

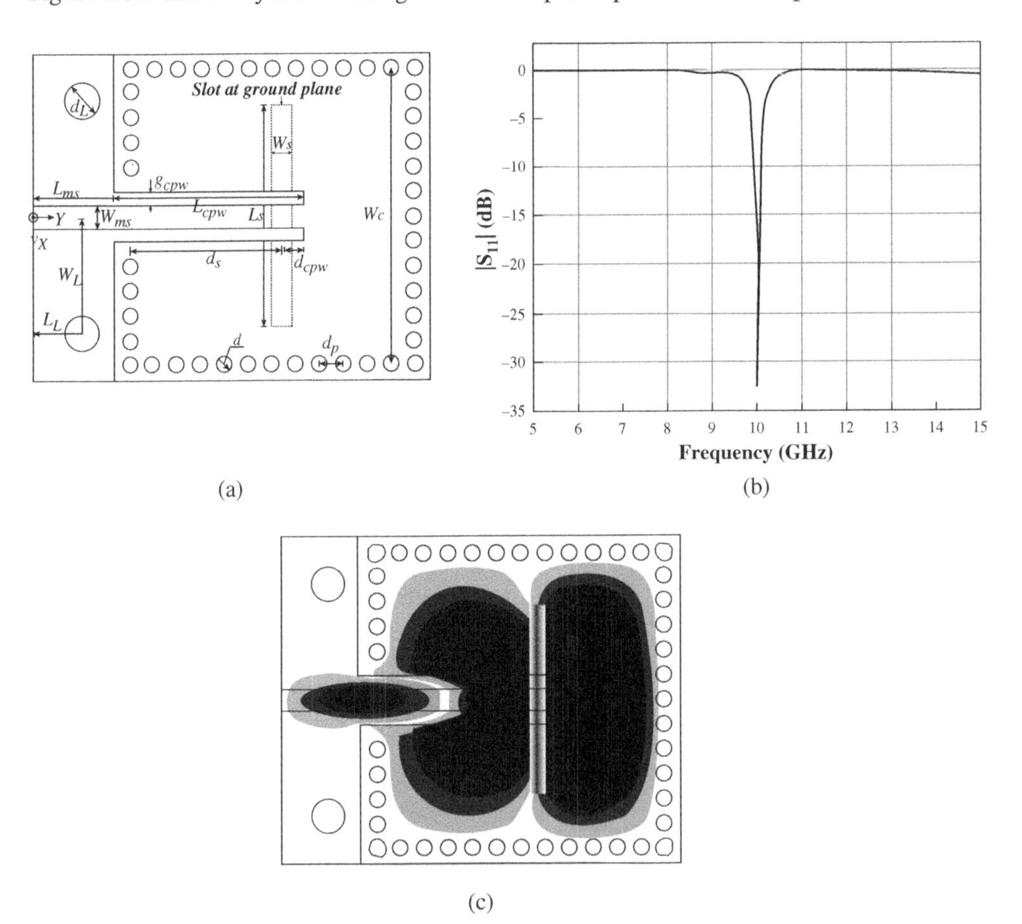

Figure 5.8 Typical substrate integrated cavity backed single slot antenna [26]. (a) Geometry, (b) $|S_{11}|$, and (c) E-field distribution of TE_{120} mode in the cavity.

which is stretched into the center of the cavity, and the radiating slot is etched onto the back side of the substrate. The SIC is designed to operate at its TE_{120} mode at 10 GHz, and the corresponding $|S_{11}|$ is shown in Figure 5.7b. As can be seen in Figure 5.7c, the *E*-field at the two sides of the slot is out of phase, so the slot penetrates the surface currents and radiate into space. The achieved impedance bandwidth is only 1.4% centered at 10 GHz, because the input impedance of the slot varies sharply with the frequency and only one cavity mode is used to excite the slot. Even there is no additional impedance matching applied, the development of such a cavity backed slot antenna provides a new opportunity to apply the SIC in the planar antenna design.

In 2009, an SIC backed crossed slot antenna was proposed by Luo [27], which operates as either a dual-band dual-linearly polarized antenna or a circularly polarized antenna, as shown in Figure 5.9. For circularly polarized operation, the cavity is designed with a square shape for exciting two orthogonal degenerate modes, that is, TE_{120} and TE_{210} modes. The cavity is fed at its corner and with a perturbation introduced by the cross slot due to its unequal lengths of arms, which ensures that two degenerate modes are successfully excited. As shown in Figure 5.10, at the operating frequency of 10 GHz, the *E*-field in the cross slot is orthogonally alternated with 90° phase delay for circularly polarized operation. Figure 5.11 presents the S-parameter, axial ratio as well as the gain of the circularly polarized antenna.

For a rectangular cavity, the TE_{210} and TE_{120} can be separated from each other, and a dual-band dual-linear polarization radiation is achieved accordingly, as illustrated in Figure 5.12. Furthermore, a quadrupole mode of TE_{220} at a higher frequency of 10.97 GHz can be observed but with a weak radiation, because at this mode the cavity has the opposite polarity at opposite ends of each slot arms, as can be seen in Figure 5.12c. Figure 5.13 gives the S-parameter and gain of the dual-band dual-polarized antenna. This type of SIC backed cross slot antenna shows the possibility of using multiple resonances in a single cavity to achieve multiband or wideband performance.

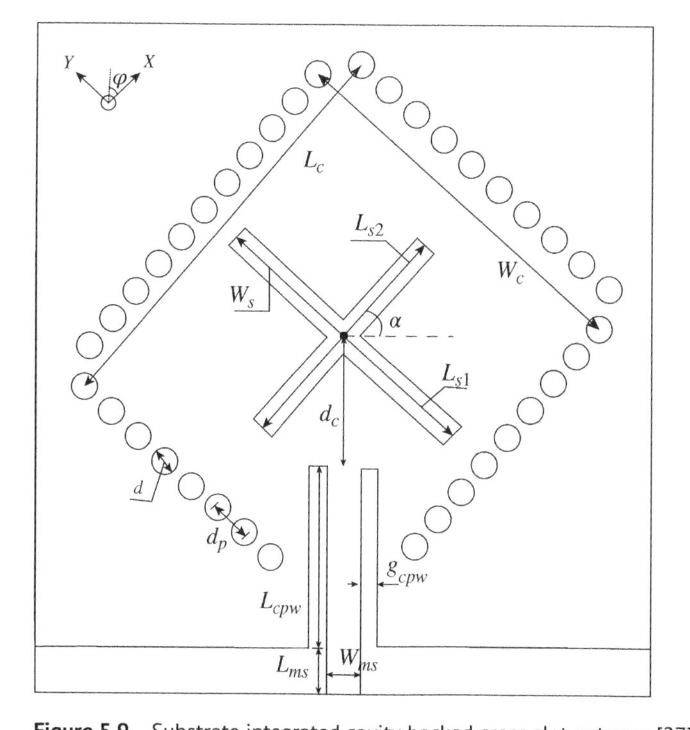

Figure 5.9 Substrate integrated cavity backed cross slot antenna [27].

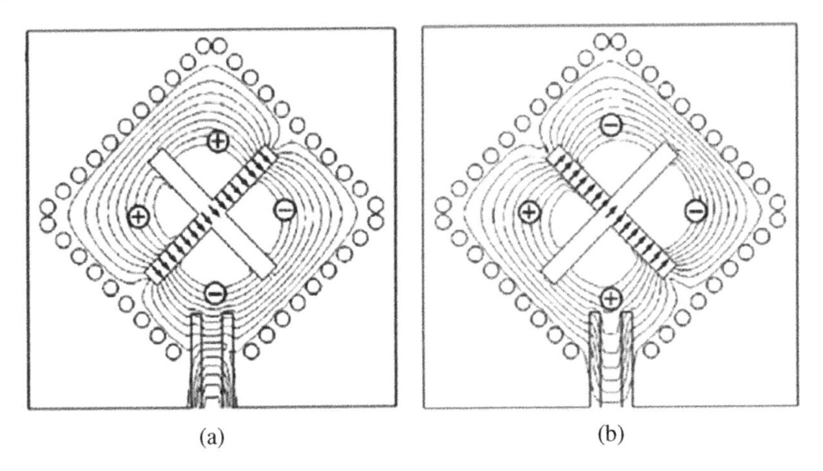

Figure 5.10 *E*-field distribution in the substrate integrated square cavity backed cross slot antenna with circular polarization at 10 GHz. (a) TE_{120} mode at 60° and (b) TE_{210} mode at 150°.

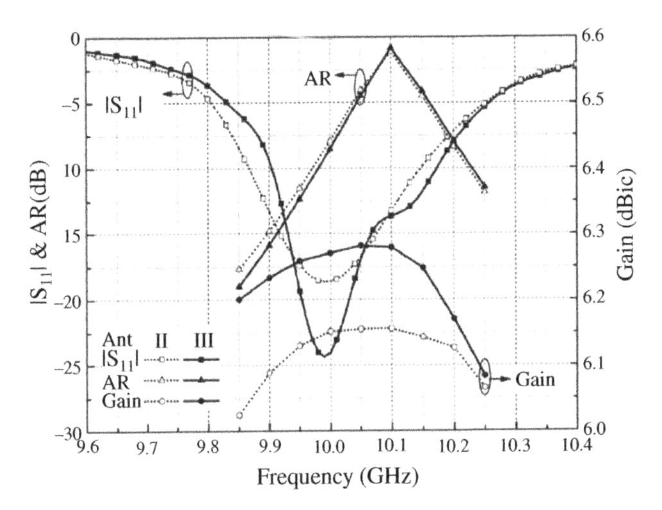

Figure 5.11 S-parameter of the circularly polarized substrate integrated cavity backed cross slot antenna.

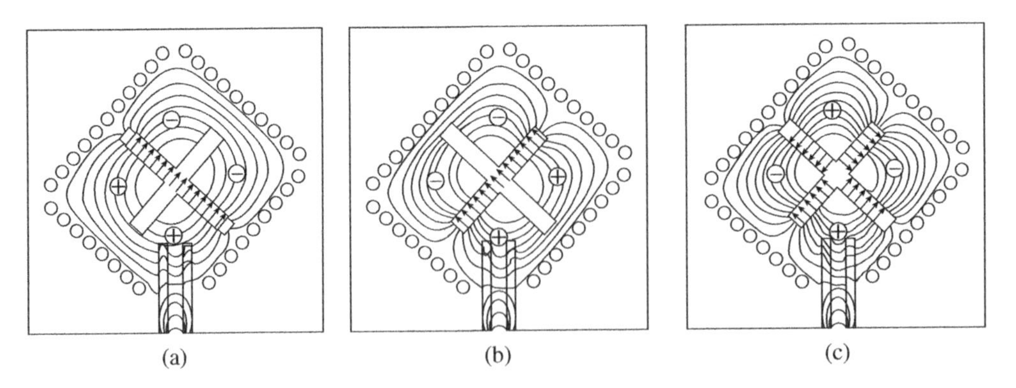

Figure 5.12 *E*-field distribution in the substrate integrated rectangular cavity backed cross slot antenna with dual-band dual-linear polarization. (a) TE_{210} mode at 9.5 GHz, (b) TE_{120} mode at 10.5 GHz, and (c) TE_{220} at 10.97 GHz.

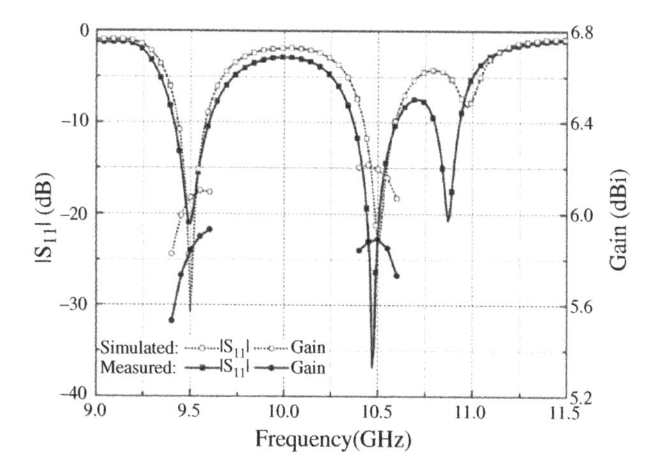

Figure 5.13 Reflection coefficient and gain of the substrate integrated rectangular cavity backed cross slot antenna with dual-band dual-linear polarization.

To enhance the bandwidth of the SIC backed slot antenna, a rectangular SIW cavity has been designed with two hybrid modes to be excited simultaneously [28], which is achieved by controlling the geometry of the cavity as well as the location of the slot. The proposed antenna is shown in Figure 5.14a, which has a similar rectangular cavity and slot as those used in the antenna shown in Figure 5.8a. However, the modes in this SIW cavity is a hybrid-mode case, while the one in the antenna shown in Figure 5.8a is only single dominate mode. To illustrate the hybrid-mode scheme, Figure 5.15 provides the two decomposition schemes of both the TE_{110} and TE_{120} modes in terms of different phase distributions. The two hybrid modes can generate two resonances as shown in Figure 5.14b, and it is seen clearly that the two closely located resonances widen the operating bandwidth accordingly.

It can be predicted that if more modes can be jointly generated in a single cavity, much wider bandwidth could be achieved. In 2017, Long et al. have presented the triple- or quad-resonance rectangular SIC-backed slot antennas as shown in Figures 5.16a,b [29]. The geometry of either the triple- or quad-resonance antennas are similar to each other, where a rectangular SIC is fed by a microstrip line from one end of the cavity and a transverse slot is cut near another end of the cavity. There are one- or two-grouped shorting vias inserted between the slot and the feeding port of the triple- or quad-resonance rectangular SIC-backed slot antennas, respectively. It is known that the rectangular cavity achieves multiple resonances, but the modes are usually separated with frequency gaps. The introduced shoring vias can squeeze several modes into a small frequency range, which can be operated jointly to result in a wide bandwidth.

To reveal the operating scheme, the resonant modes in the cavity are examined without and with the shorting vias, as shown in Figures 5.17 and 5.18, respectively. It can be seen from Figure 5.17 that the first three modes in the rectangular cavity are half-TE_{110} mode, odd TE_{210} mode, and even TE_{210} mode at the resonant frequencies of 6.55, 9.85, and 10.59 GHz, respectively. The half-, odd, and even modes are unusual modes in a conventional rectangular cavity, and they are attributed by the cutting slot, which functions as the radiator simultaneously. The slot provides an additional boundary into the cavity, which means the amplitudes and phases besides the two sides of the slot can be continuous or discontinuous. These modes are still separated with frequency gaps, and they are hardly excited jointly. With the presence of the shorting vias as shown in Figure 5.18, the first

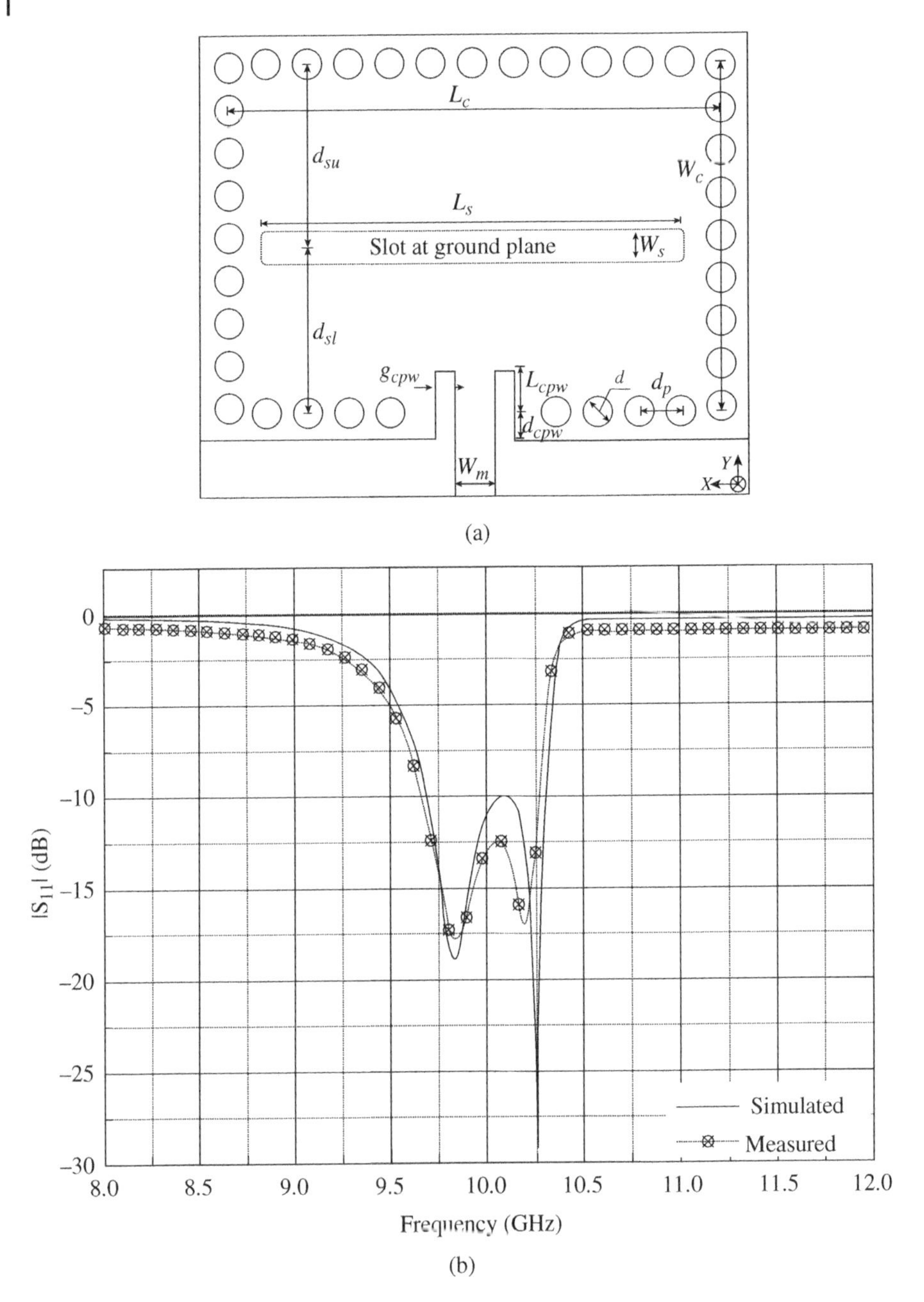

Figure 5.14 Hybrid-mode rectangular SIW cavity backed slot antenna [28]. (a) Geometry and (b) S-parameter.

three modes are kept, but their resonance frequencies are squeezed near to each other, that is, 9.09, 10.07, and 10.66 GHz, respectively. The final achieved operating bandwidth is 9.12–10.62 GHz or 15.2% in terms of $|S_{11}| < -10$ dB, gain >4 dBi, and efficiency >85%, as shown in Figure 5.19.

To further improve the bandwidth of the triple-resonance SIW cavity slot antenna, the fourth mode of the cavity, TE_{310}, is introduced together with the first three modes to achieve the quad-resonance antenna, as shown in Figure 5.20. Another group of the inserted vias shifts down the resonance frequency of TE_{310} to 11.12 GHz, the upper edge of the bandwidth. Thus, the design

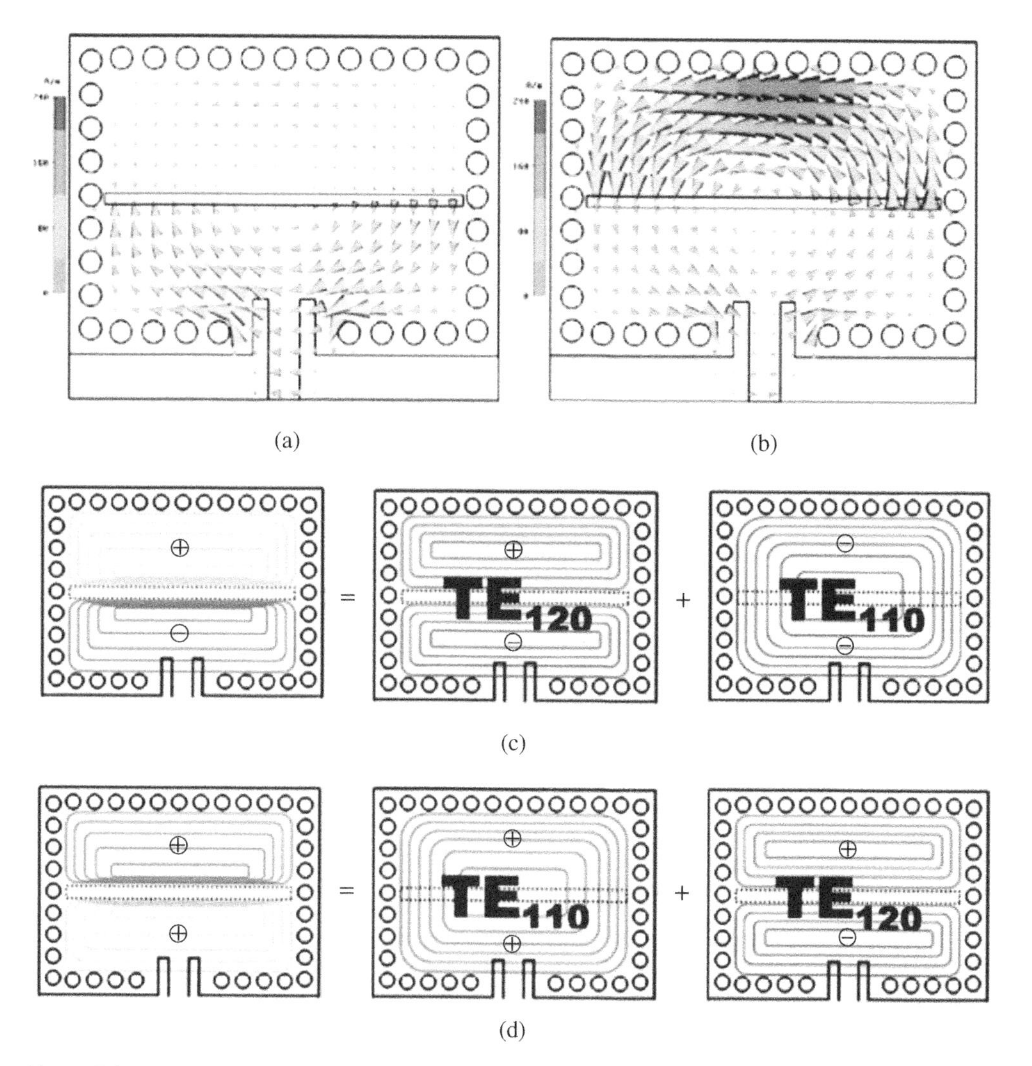

Figure 5.15 Field distributions in the SIW cavity of the hybrid-mode rectangular SIW cavity backed slot antenna at different frequencies. (a) *H*-field vector distribution at 9.84 GHz, (b) *H*-field vector distribution at 10.27 GHz, (c) dominant *E*-field isoline of a hybrid mode at 9.84 GHz, and (d) dominant *E*-field isoline of a hybrid mode at 10.27 GHz.

achieves a wider bandwidth of 9.36–11.26 GHz or 17.5%, in terms of $|S_{11}| < -10$ dB, gain >6 dBi, and efficiency >85%, as shown in Figure 5.21.

The radiator of either the triple- or the quad-resonance SIC backed antennas is a conventional slot, and the wideband radiation is achieved by tuning the resonant modes in the cavity with the shorting vias. The operating scheme is explained and illustrated from the perspective of the jointly operating multi modes in the cavity. On the other hand, the tuning of the multiple modes can be regarded as a special case of impedance matching because the shorting vias inserted in between the radiator and the feeding line function as a multi-order matching network. This reveals that the mechanisms of the multi-mode cavity-backed slot antenna and the inductive vias–based matching network are similar in the SIW transmission lines.

Furthermore, a penta-resonance SIW rectangular cavity-backed slot antenna is developed to jointly utilize the first five modes in a rectangular cavity by Wu et al. in 2019 [30]. The geometry is shown in Figure 5.22a, which has a cross slot etched onto the top of a rectangular SIW cavity.

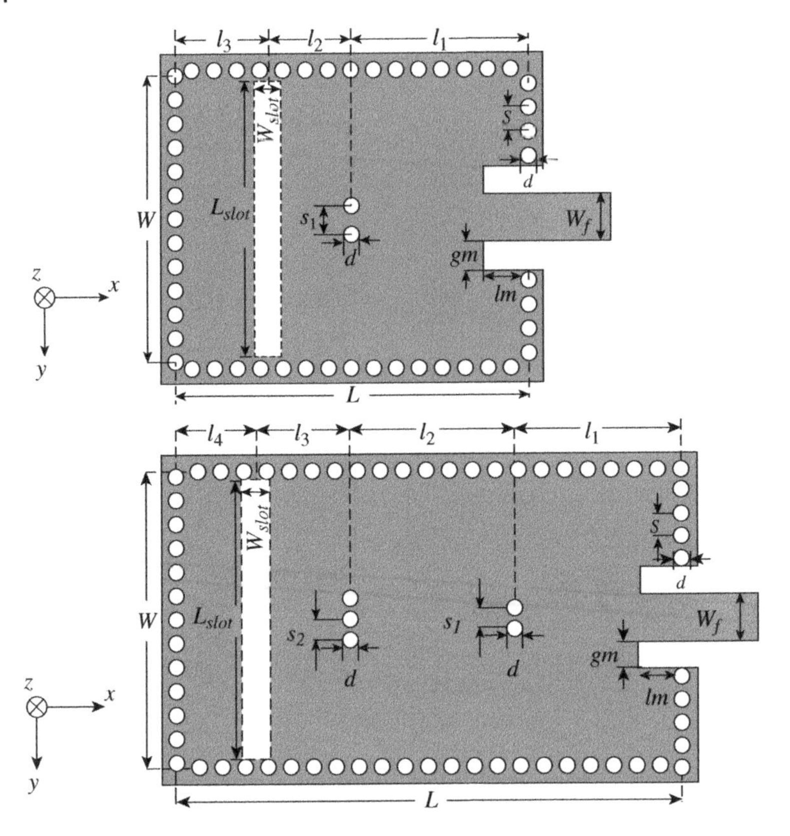

Figure 5.16 Configurations of (a) triple-resonance rectangular SIW cavity backed slot antenna and (b) quad-resonance rectangular SIW cavity backed slot antenna [29].

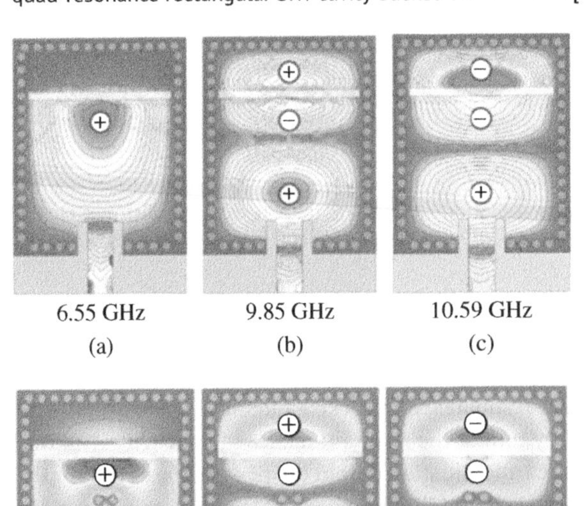

Figure 5.17 Electric field distributions of the triple-resonance SIW cavity backed slot antenna working in (a) half-TE_{110} mode, (b) odd TE_{210} mode, and (c) even TE_{210} mode.

9.09 GHz (a) 10.07 GHz (b) 10.66 GHz (c)

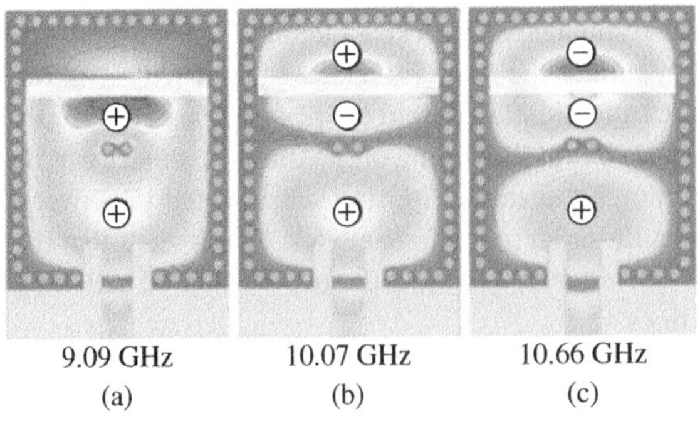

Figure 5.18 Electric field distributions of the triple-resonance SIW cavity backed slot antenna working in (a) half-TE_{110} mode, (b) odd TE_{210} mode, and (c) even TE_{210} mode.

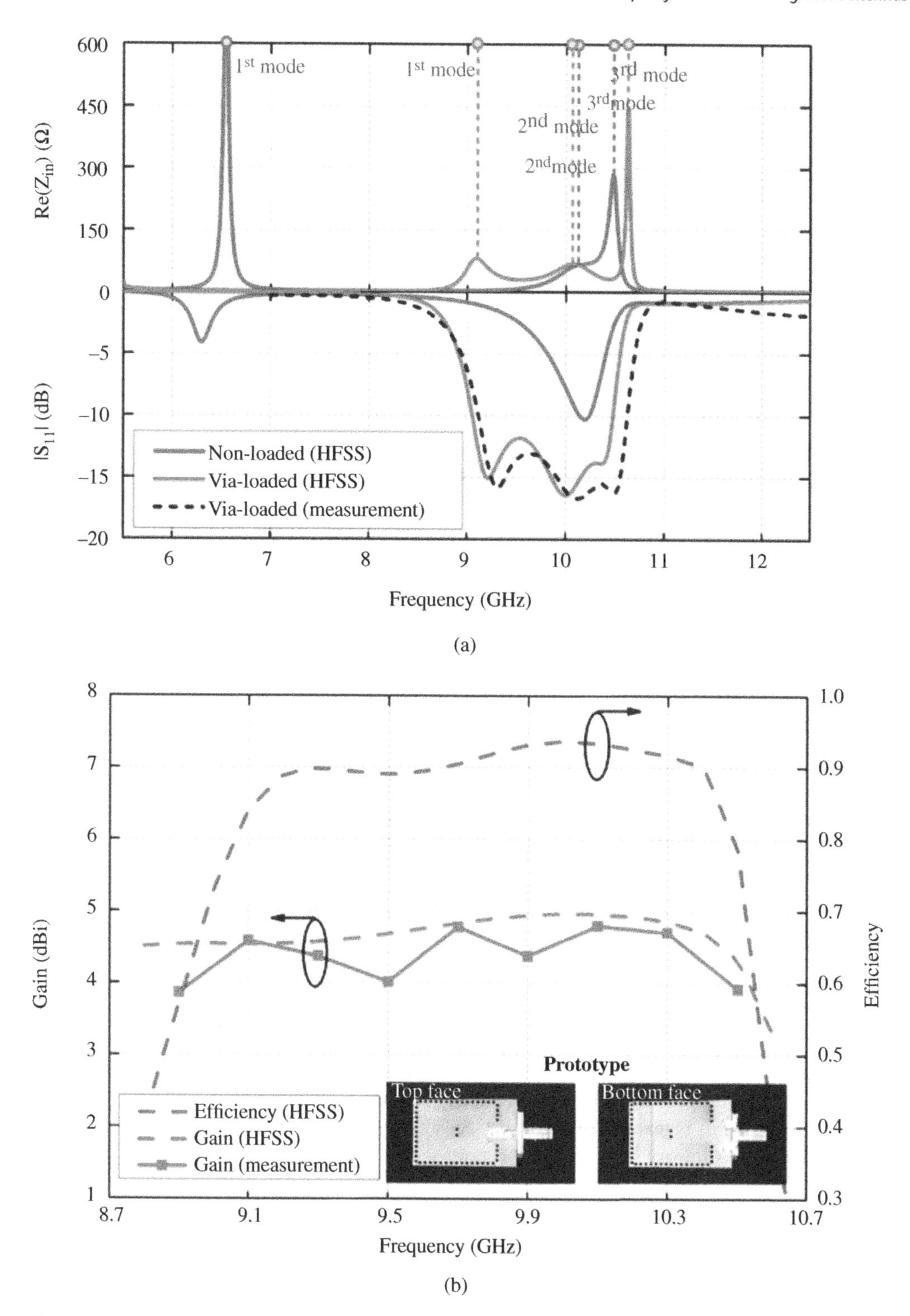

Figure 5.19 (a) Input impedance and $|S_{11}|$ and (b) gain and efficiency, for the triple-resonance SIW cavity backed slot antenna.

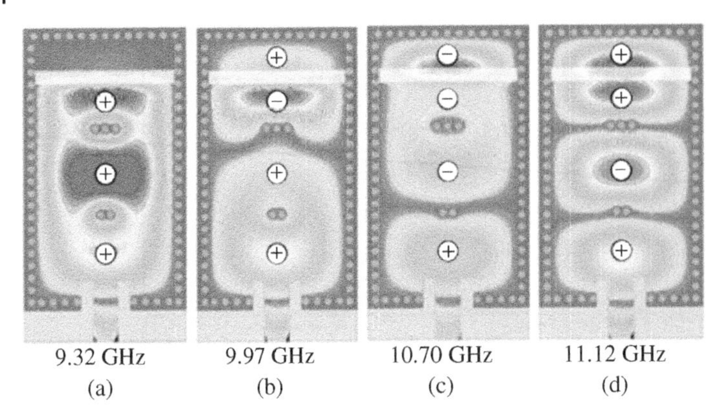

Figure 5.20 Electric field distributions of the proposed quad-resonance antenna (by HFSS) working in (a) half-TE_{110} mode, (b) odd TE_{210} mode, (c) even TE_{210} mode, and (d) TE_{310} mode.

In this case, the cavity is operated with five modes simultaneously, and the resonant frequencies are squeezed by inserting four shorting vias as two vias placed on the opposite side of the slot and another two placed in between the slot and the feeding port of the cavity. The five modes include half-TE_{110}, half-$TE_{120,}$ odd TE_{210}, TE_{210} mode, and even TE_{210} mode, appearing in sequence, as depicted in Figures 5.23a–e. The simultaneous operation of the five modes lead to an operating bandwidth of 8.89–10.95 GHz, or 20.8%, as shown in Figures 5.22b,c. It should be noted that although a much wider bandwidth is achieved here, the cross slot evidently reduces the co-to-cross polarization levels over the operating bandwidth. Thus, the application of such an antenna is rather limited.

In summary, the multi-mode operating scheme is an efficient method for the design of SIC backed slot antenna for either wideband or multiband operations.

5.2.4 Patch Loaded Substrate Integrated Cavity Antenna

In 2014, a *Q*-band SIW cavity loaded patch antenna was proposed by Yang et al. [31]. As shown in Figure 5.24a, the antenna is comprised of an SIC-backed rectangular patch and an SIW feeding line. The cavity is inserted with a metallic via in its center, which grounds the patch at its center, and a small gap is etched on the wider side of the patch near to the feeding line for impedance matching. The cavity resonates at its TE_{210} mode, well incorporated with the patch mode of (transvers magnetic) TM_{10}, as shown in Figure 5.24b. Hence, a multiple-resonance achieves a wide operating bandwidth of 35.3–41.3 GHz, or 15.6% in terms of $|S_{11}| < -10$ dB. The experiment results show that the maximum gain over the bandwidth is 6.5 dBi, and 3-dB beamwidths of *E*- and *H*-plane are 97.2° and 70.4°, respectively. The proposed cavity-backed patch antenna can be formed into a 4×4 array, as shown in Figure 5.26a. It can be seen that the array has a very compact form with two additional phase shifters formed by two sections with varied cross-sectional widths. The measured bandwidth of the array is from 41.2 to 44.8 GHz, or 8.7% in terms of $|S_{11}| < -10$ dB, and the gain is 17.3 dBi with an efficiency of 74.9%, as shown in Figure 5.25b. Note that the bandwidth of the antenna array is narrower than that of the antenna element because of the series feed scheme for the compact array size. The abovementioned SIC-loaded patch antenna and array demonstrate that with the help of SIW technology, the patch antenna can be successfully applied in the antenna design at mmW bands.

Since the width of the SIW is determined by its operating frequency, it is difficult to design the shunt fed antenna array with SIW lines because the spacing between two adjacent antenna elements cannot be kept less than a wavelength. Thus, alternating feeding lines are applied to overcome this problem. Figure 5.26 shows the design of SIC-backed patch antenna and array fed

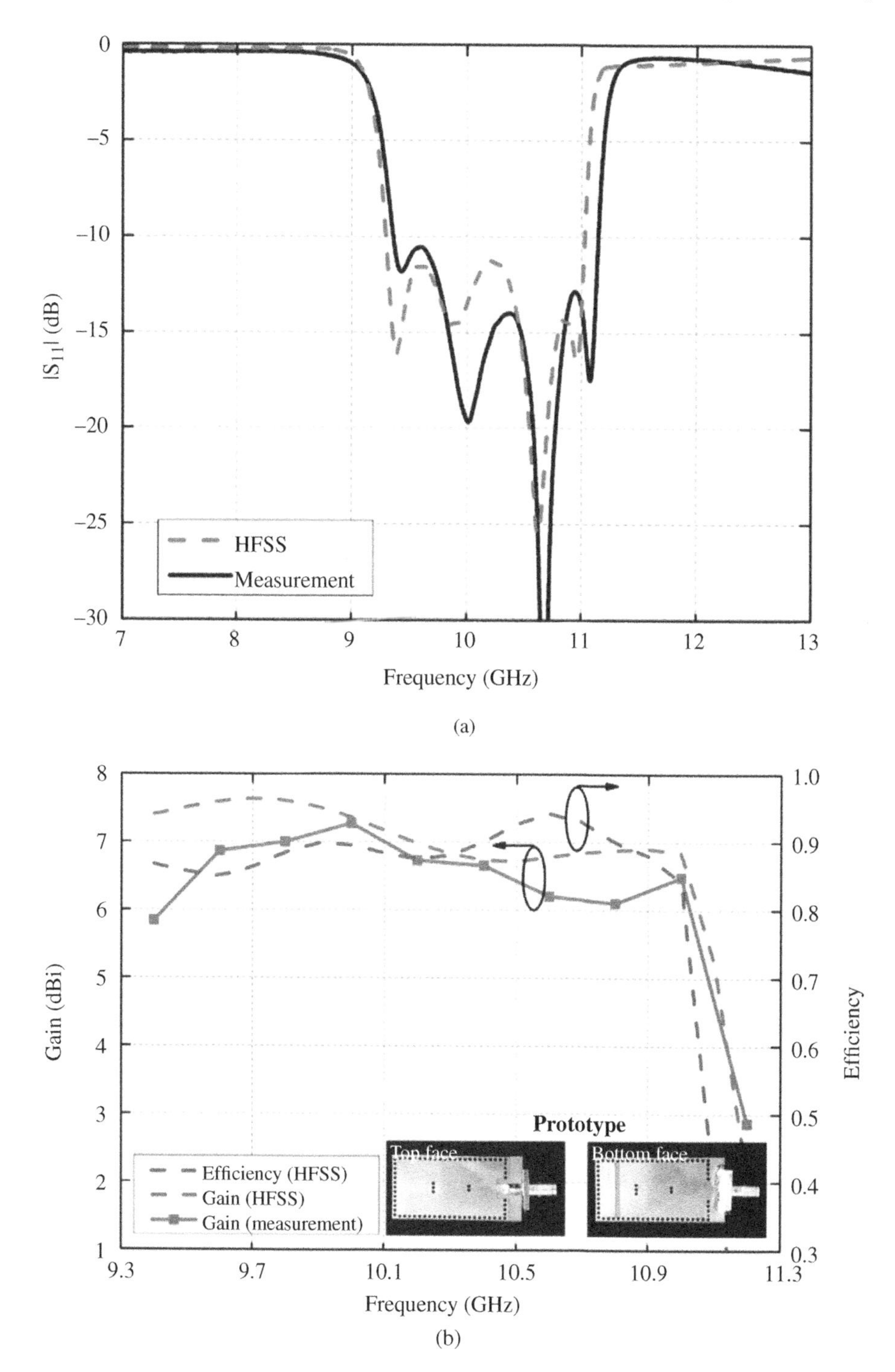

Figure 5.21 (a) Input impedance and $|S_{11}|$ and (b) gain and efficiency, of the quad-resonance SIW cavity backed slot antenna.

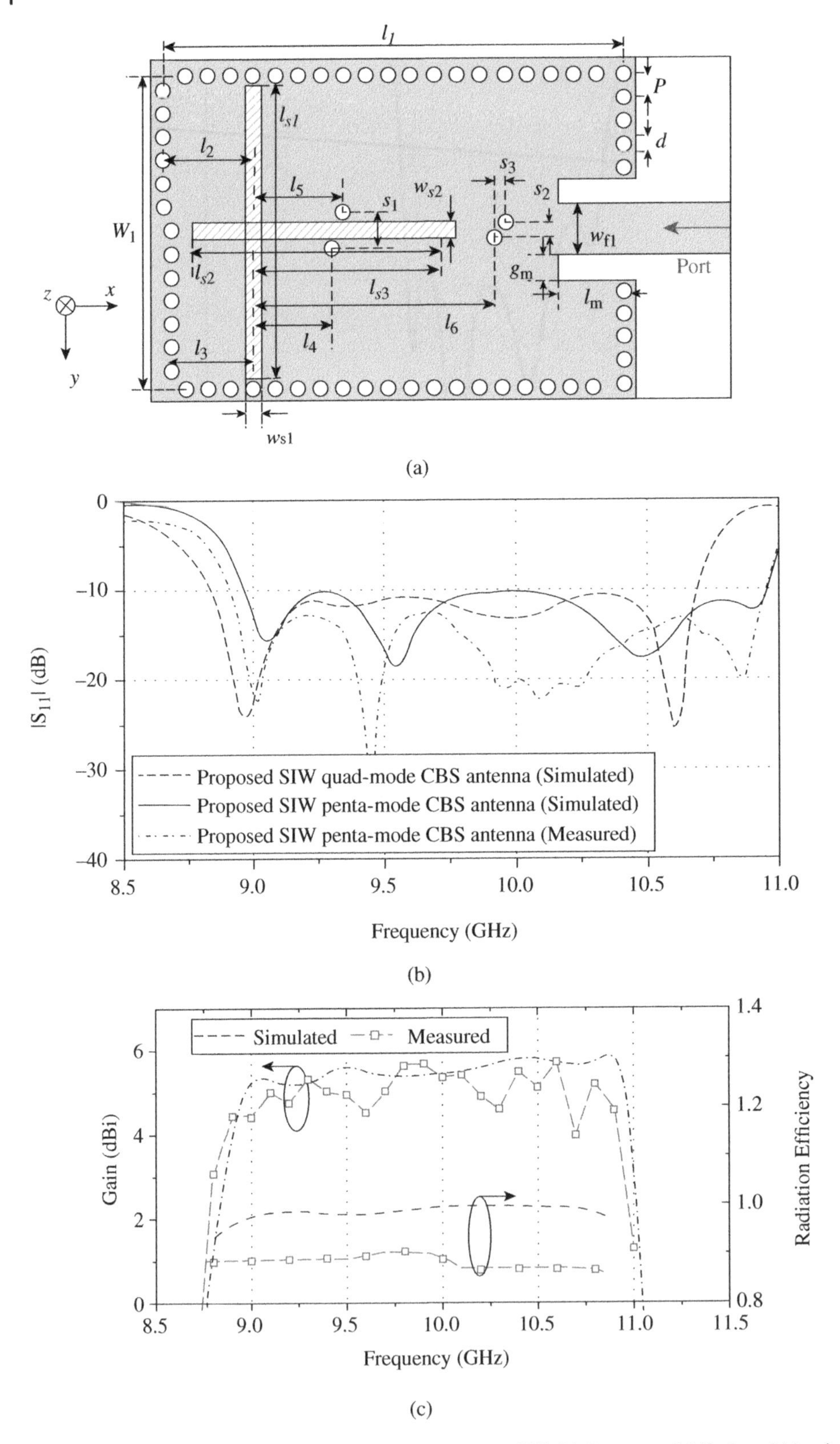

Figure 5.22 Penta-mode SIW cavity backed slot antenna [30]. (a) Geometry, (b) $|S_{11}|$, and (c) gain and efficiency.

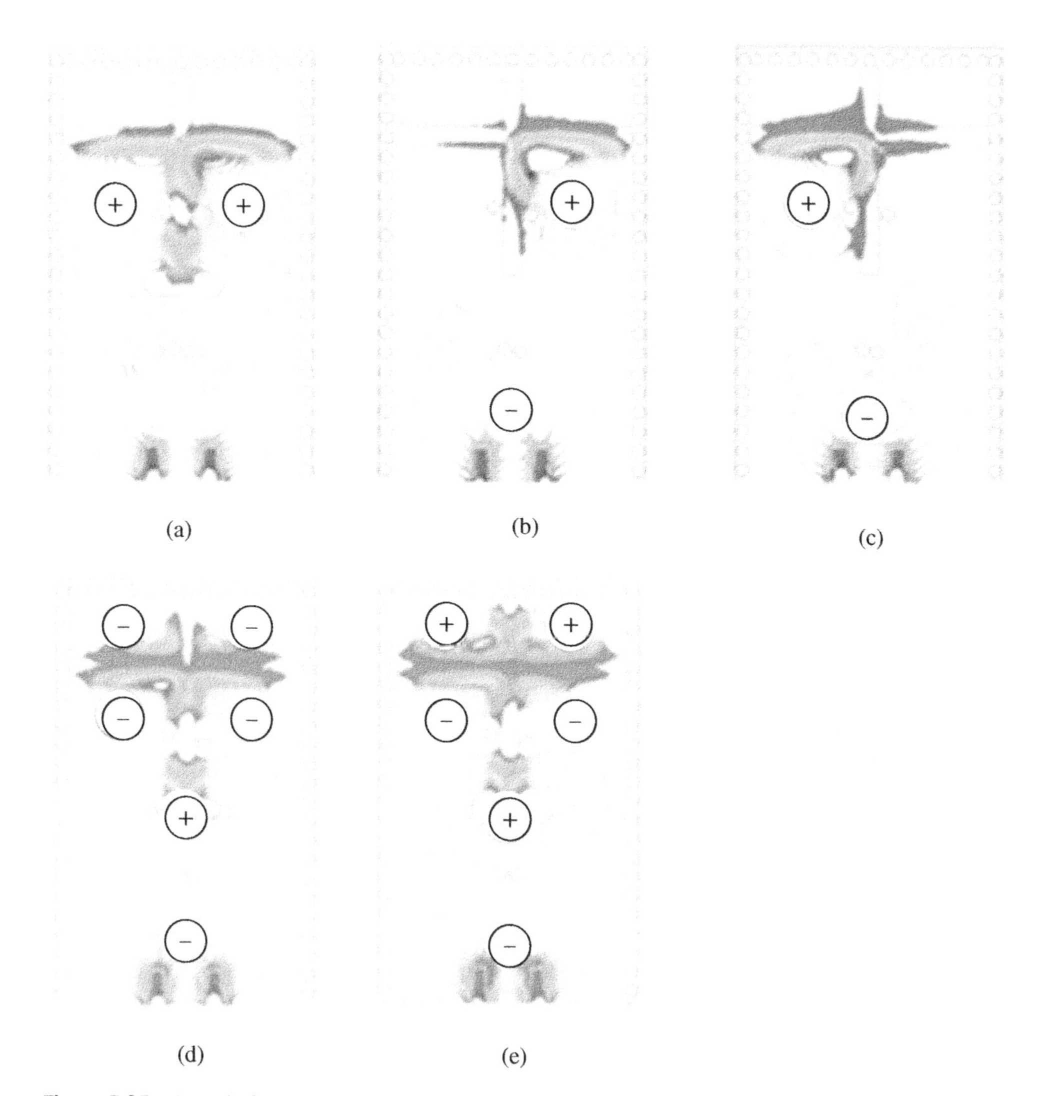

Figure 5.23 Electric field distributions of the penta-mode SIW cavity backed slot antenna at (a) 8.9, (b) 9.4, (c) 9.8, (d) 10.6, and (e) 10.9 GHz.

Figure 5.24 SIW cavity-backed rectangular patch antenna [31]. (a) Geometry and (b) *E*-field distribution and $|S_{11}|$.

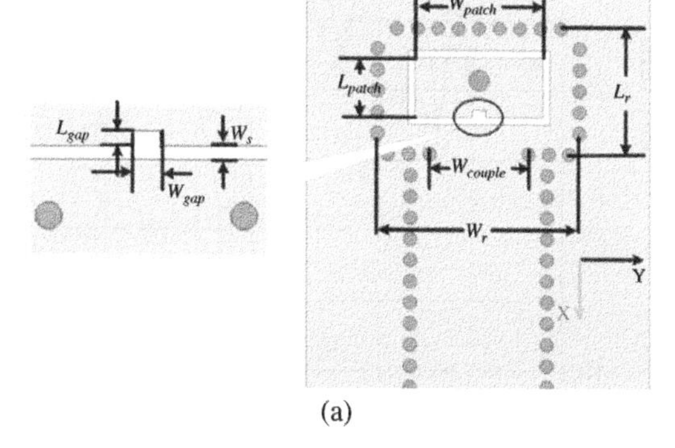

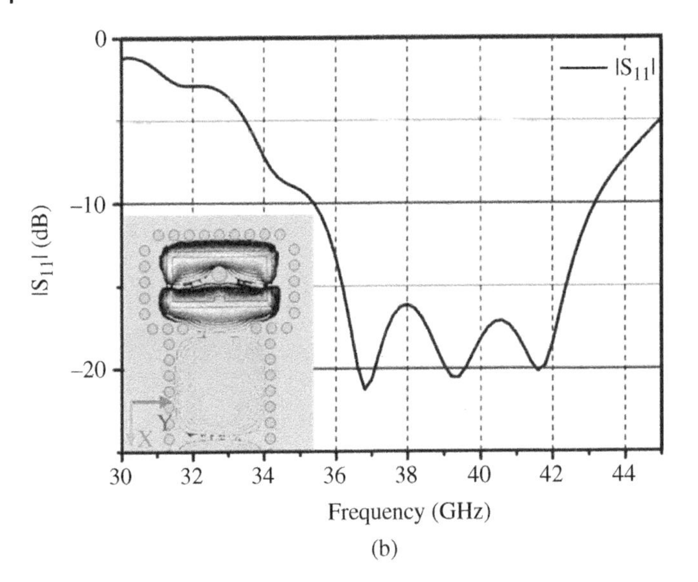

Figure 5.24 (Continued)

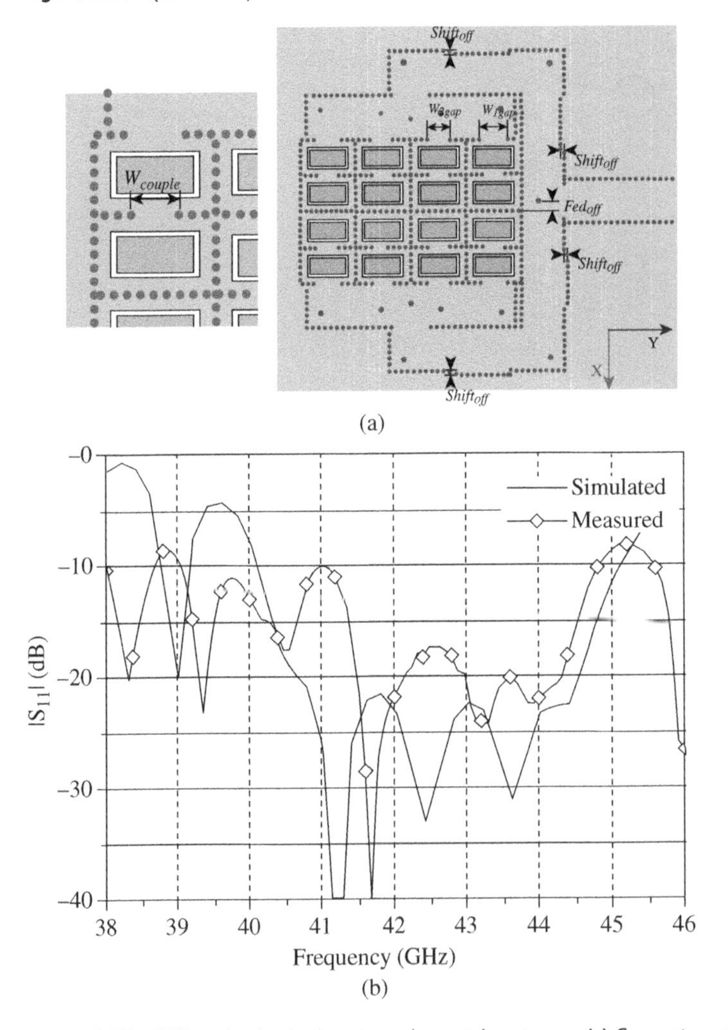

Figure 5.25 SIW cavity-backed rectangular patch antenna. (a) Geometry and (b) $|S_{11}|$.

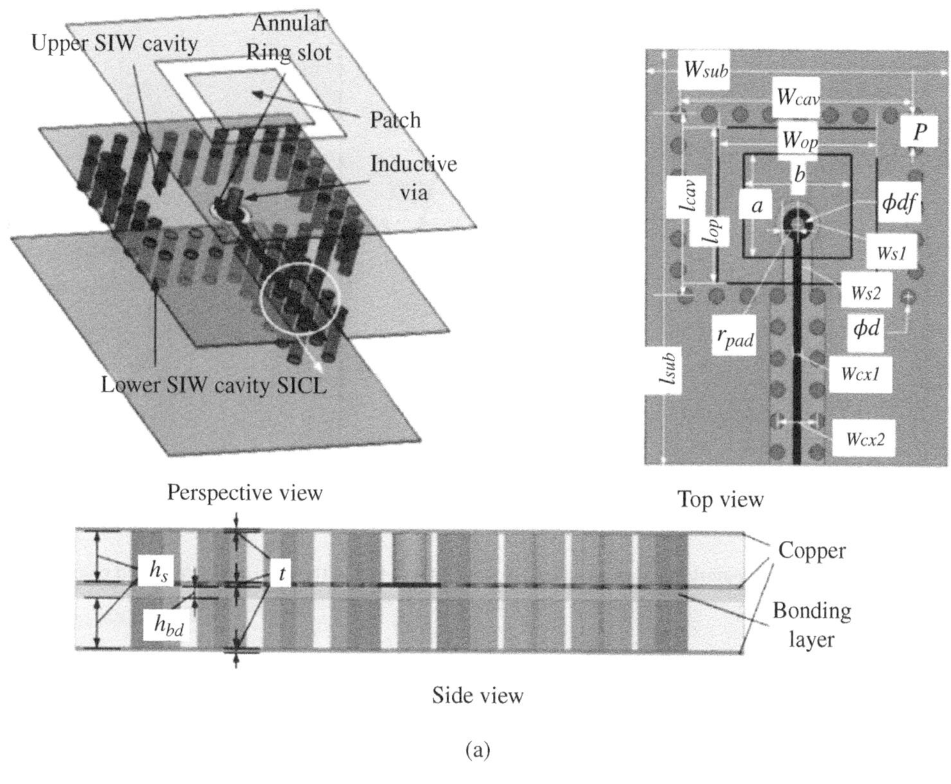

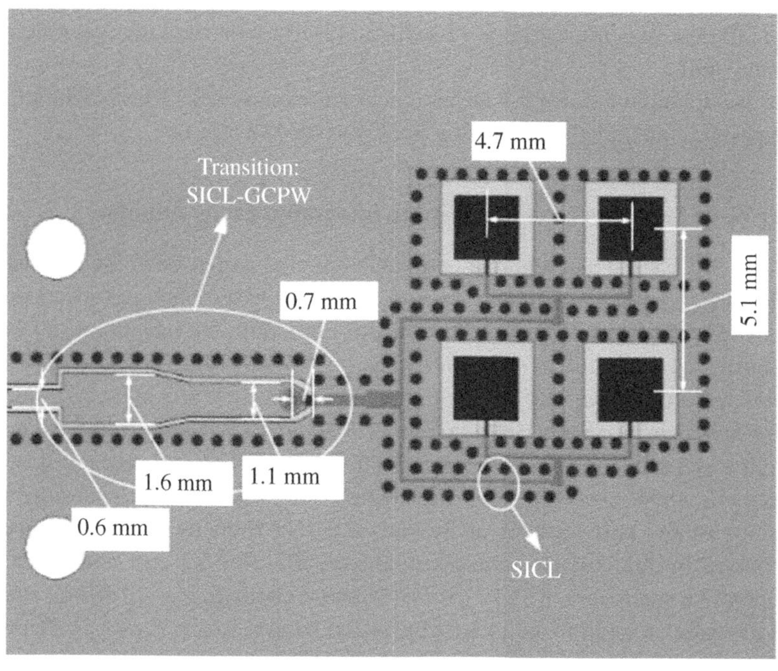

Figure 5.26 SIW cavity-backed rectangular patch antenna and array fed with substrate integrated coaxial lines [32]. (a) Geometry of antenna element and (b) geometry of a 4×4 array.

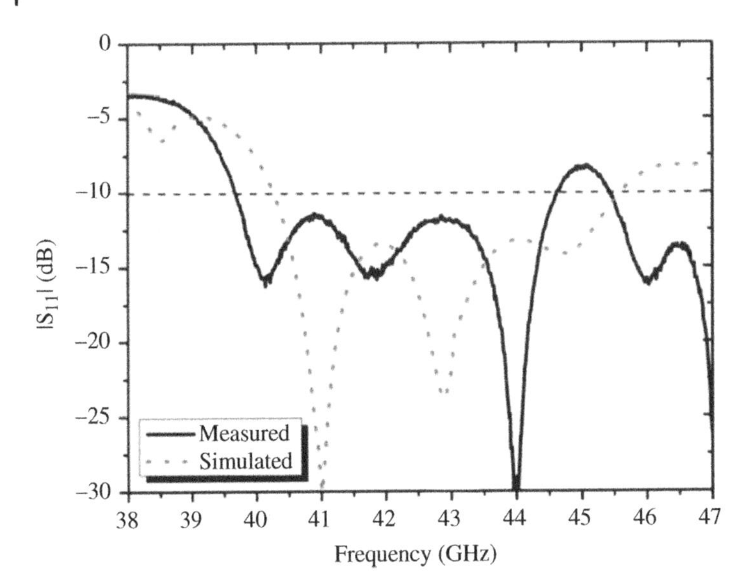

Figure 5.27 $|S_{11}|$ of the SIW cavity-backed rectangular patch array fed with substrate integrated coaxial lines.

with SICLs [32]. The SICL is a TEM transmission line, whose cross-section size is independent of operating frequency. Thus, the SICL is designed with a very narrow width in the shunt fed array design, easily fabricated between adjacent antenna elements, as shown in Figure 5.26b. As depicted in Figure 5.27, the measured results show that a bandwidth of 39.7–44.6 GHz, or 11.7% is achieved, and the maximum gain is 11.4 dBi at 42 GHz with the radiation efficiency of 88% within the bandwidth.

The SIC-backed patch antenna exhibits wideband performance as well as flexible design ability with the help of substrate integrated transmission lines, such as SIW, SICL.

5.2.5 Traveling-Wave Elements Loaded Substrate Integrated Cavity Antenna

The aforementioned SIC-backed slot antenna and patch antenna can achieve the relative bandwidths around 10–20%, which is reasonable for the resonance-based antennas also known as standing-wave antennas. To further break the bandwidth limitation, the traveling-wave antenna concept can be transferred to the design of SIC antennas.

The curl radiator was introduced into SIW fed antenna design by Wu et al. in 2017 [33]. The antenna geometry is shown in Figure 5.28, including a printed stacked curl-like element fed by an SIW through a slot implemented with four substrate layers. The curl element is a typical traveling-wave element, and it operates over a wide bandwidth. Figure 5.29 provides the surface current distributions in the curl element at 34 and 42.8 GHz showing that a traveling-wave operation is supported by the curl element. As predicted, the SIW-fed curl antenna element achieves a wide operation bandwidth of 35.7% (31–44.5 GHz) for both $|S_{11}| < -10$ dB and axial ratio (AR) < 3 dB, and the gain varies from 3.3 to 8.1 dBic in the bandwidth, as shown in Figure 5.30. Similarly, an S-shape dipole element was proposed to be fed by an SIW for wideband radiation by Zhang et al. in 2019 [34], as shown in Figure 5.31, which is implemented in two substrate layers.

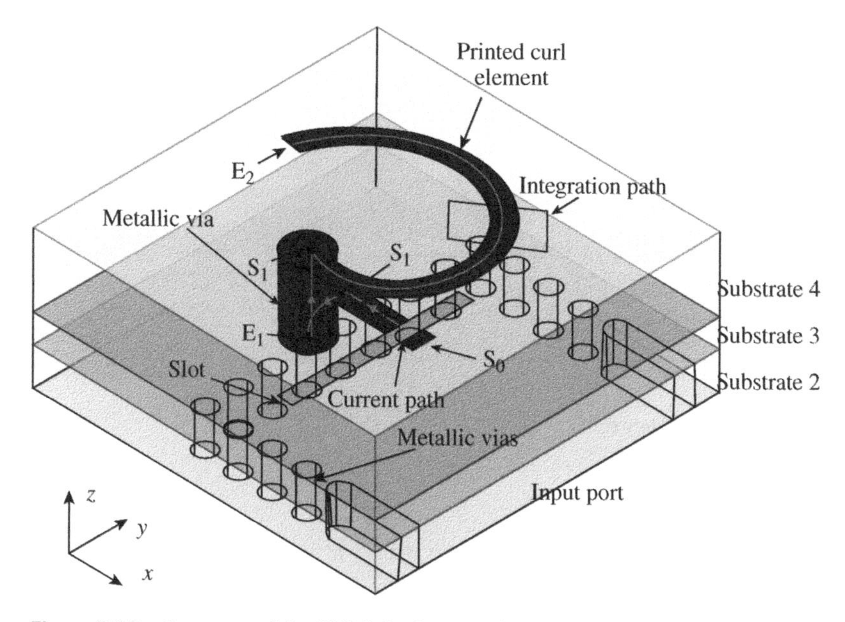

Figure 5.28 Geometry of the SIW-fed printed curl element [33].

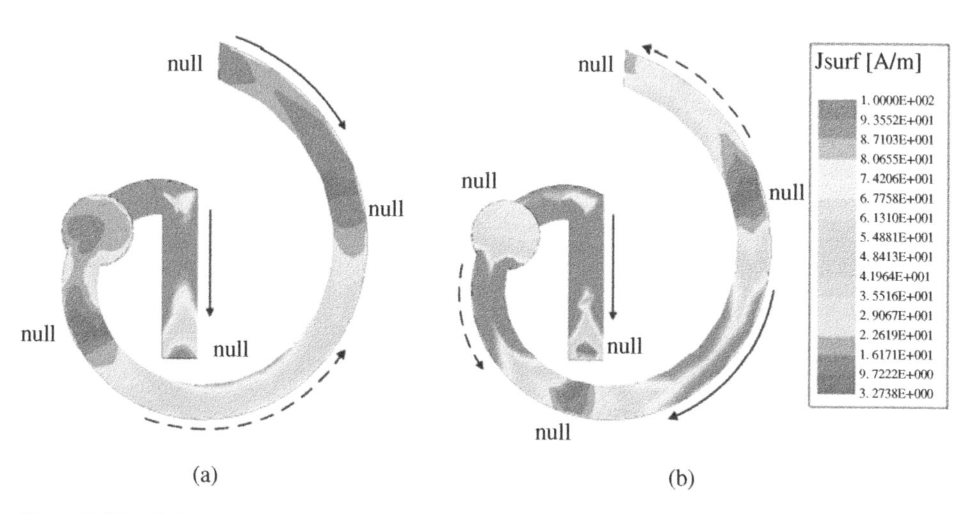

Figure 5.29 Surface current distributions in the curl element at (a) 34 and (b) 42.8 GHz.

The SIW fed S-shape dipole achieves a bandwidth of 36% in terms of both $|S_{11}| < -10$ dB and AR < 3 dB due to its traveling-wave operation scheme.

Although either the SIW fed stacked curl element antenna or the SIW fed S-shape dipole antenna are not comprised of any SIC, they still provide very valuable examples for SIC antenna design. In 2018, an SIC-fed two-arm spiral element antenna as shown in Figure 5.32 was proposed by Zhu et al. [35]. The antenna consists of three layers of substrate. Four identical two-arm spiral element antennas are printed on the top surface of substrate layer 1, and each of them is shorted to the metal ground through vias. In the substrate layer 2, an SIC is designed to operate at a higher mode of TE_{340},

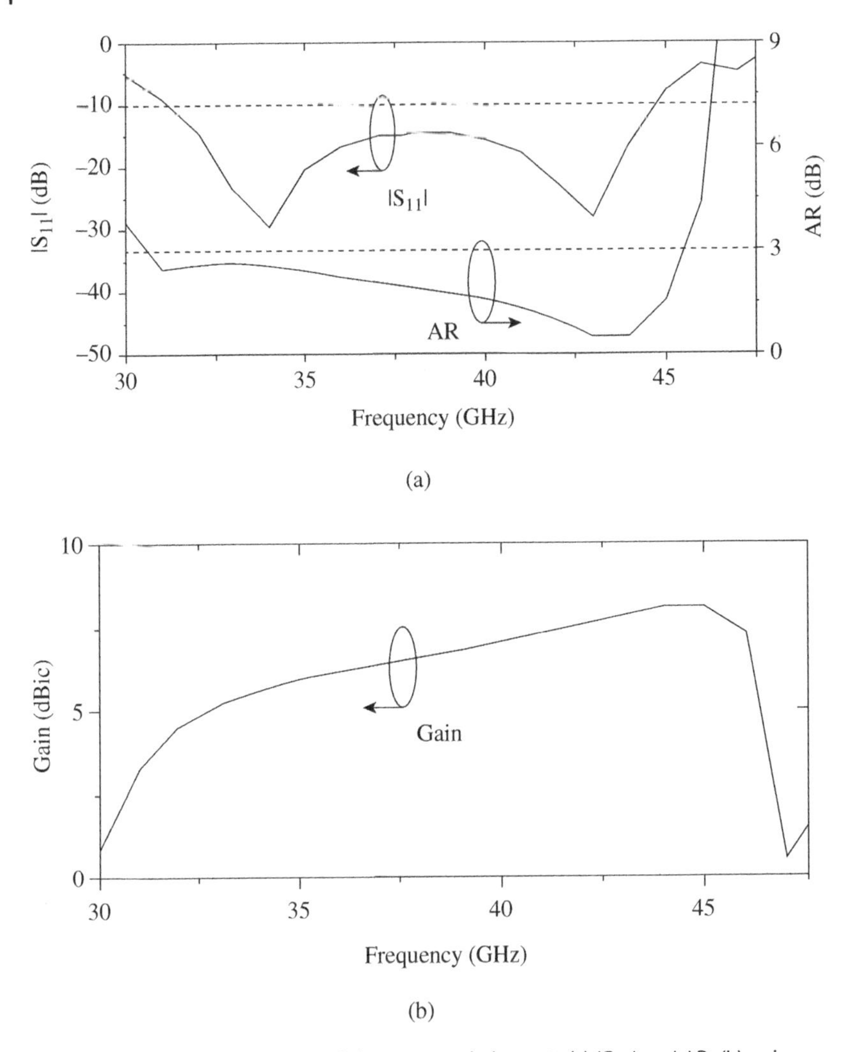

Figure 5.30 Simulated results of the proposed element. (a) $|S_{11}|$ and AR, (b) gain.

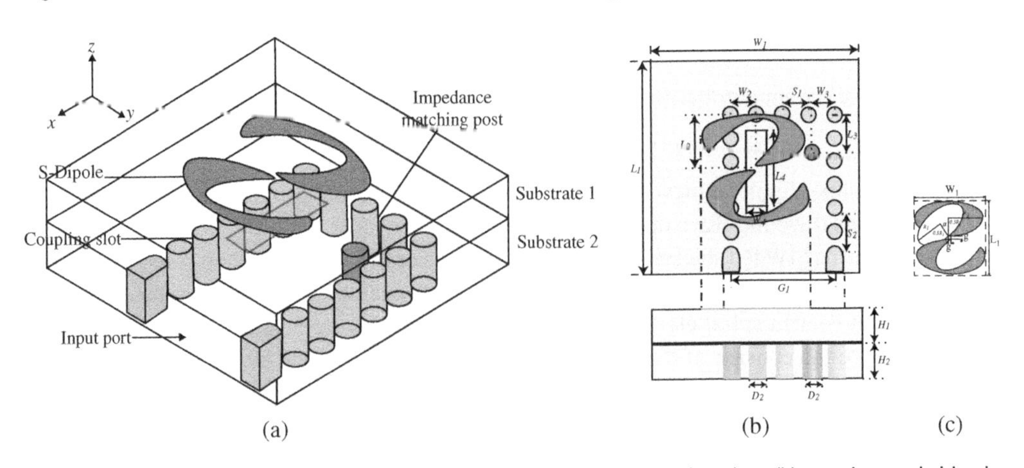

Figure 5.31 Geometry of the SIW fed S-shape dipole [34]. (a) Perspective view, (b) top view and side view, and (c) S-dipole.

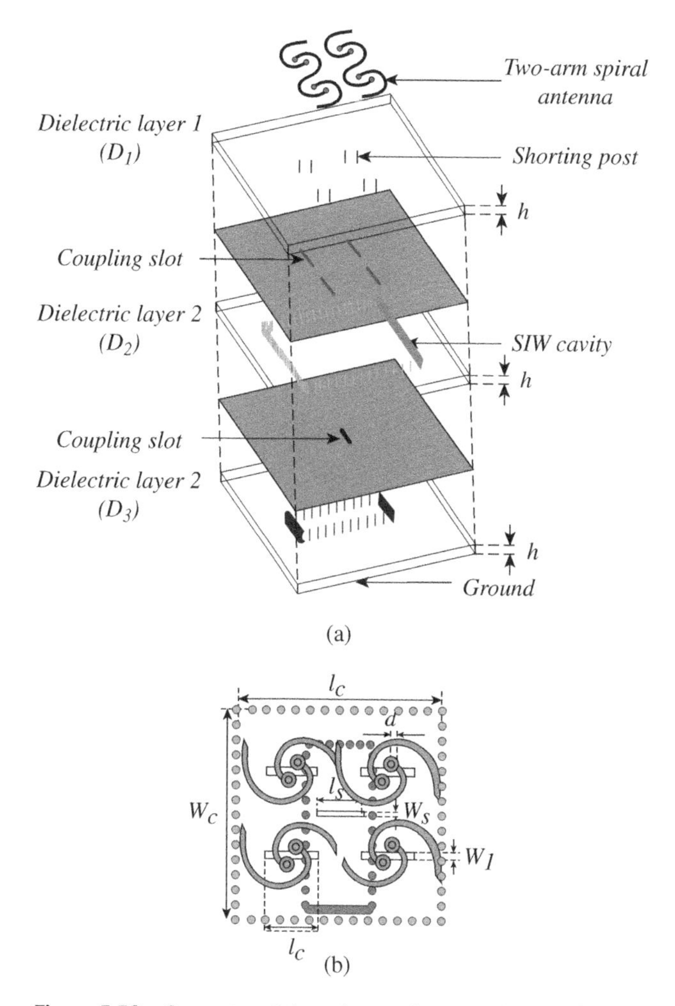

Figure 5.32 Geometry of the substrate integrated cavity fed two-arm spiral element antenna [35]. (a) Distributed 3-D view and (b) top view.

which ensures the cavity exciting the four two-arm spiral elements with equivalent amplitude and phase through the slots etched on top of the cavity. In the bottom substrate layer, a feeding SIW is used to excite the SIC through a transverse slot. The surface current distribution generated by spiral elements is demonstrated in Figure 5.33. The antenna achieves a bandwidth of 53–64 GHz, or 18% in terms of both $|S_{11}| < -10$ dB and AR < 3 dB, as shown in Figure 5.34. It should be noted that the SIC operates at its higher mode, which replaces a complicated feeding network with a more compact form. Meanwhile, the higher-mode cavity limits the bandwidth less than 20%.

In summary, the traveling-wave elements, such as curl element, S-shape dipole, and two-arm spiral element, are promising to or indeed enhance the bandwidth of SIC antennas as discussed above. It should be pointed out the properties or limitations of these designs, including: (i) only circular polarization can be supported due to the curl or spiral element configuration; (ii) multiple substrate layers are needed to form a stacked structure, which increases the cost and design complexity; and (iii) there is a trade-off between the element size and the relative operating bandwidth because on the one hand the wider bandwidth requires larger element size but on the other hand the larger element size limits the operating bandwidth of antenna arrays.

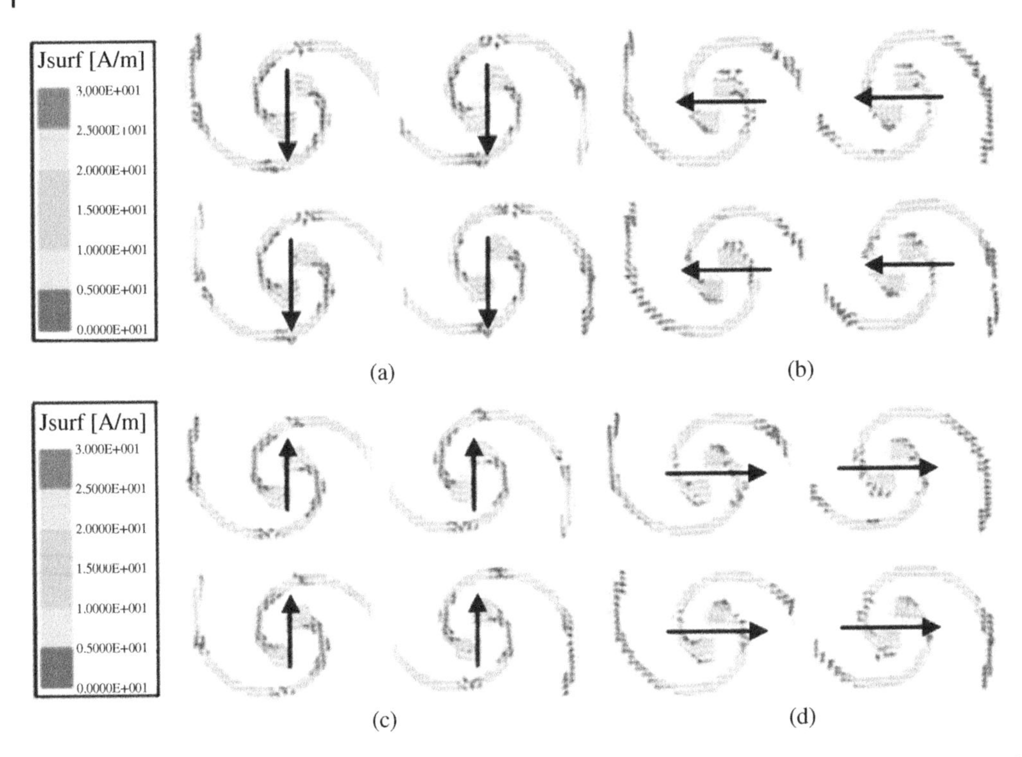

Figure 5.33 Surface current distributions generated by the spiral antenna at different phase of time. (a) 0°, (b) 90°, (c) 180°, and (d) 270°.

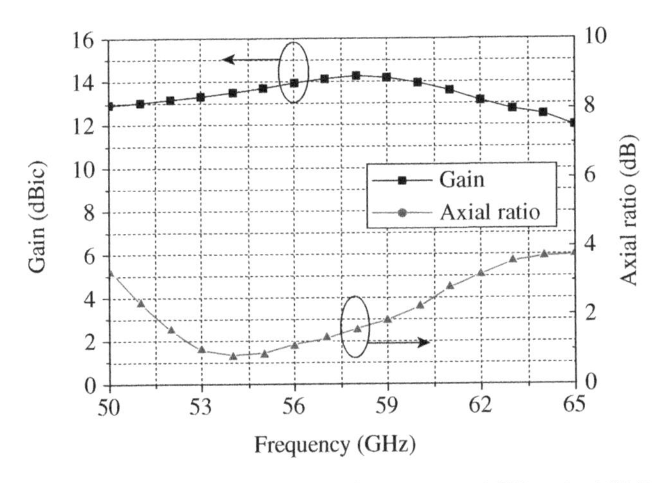

Figure 5.34 Gain and axial ratio of the proposed SIC-excited CP 2 × 2 subarray.

5.3 SIW Narrow Wall Fed SIC Antennas at *Ka*- and *V*-Bands

Besides the aforementioned broadband SIC antennas, another type of SIC antenna is introduced with the concept of the dielectric resonator antenna, to achieve wideband performance.

5.3.1 SIW Narrow Wall Fed SIC Antenna

The proposed antenna consists of a dielectric-loaded SIC and a feeding SIW with a short-circuit (S.C.) end as shown in Figure 5.35 [24]. The SIC is formed by metallic via-holes, a conductor ground, and an open-ended aperture on top of the substrate (RT/Duroid 6010, $\varepsilon_r = 10.2$, $\tan\sigma = 0.0023$). The cavity is applied in the dominant resonant mode only to ensure the wideband performance, and the energy radiates from the open-ended aperture to the free space at boresight. The SIC is coupled through an inductive window in the sidewall by the feeding SIW. An inductive via is positioned near the SIW narrow sidewall for impedance matching. The wideband feeding structure and the cavity ensure the antenna having a wideband radiation performance. The operating frequency of the proposed antenna is mainly determined by the resonant frequencies of the cavity. Therefore, the cavity is regarded as a resonator in the design as depicted in Figure 5.36.

The top of the dielectric-loaded cavity is open-ended while the rest of the walls are all perfectly electrically conducting. The substrate with high permittivity allows the open-ended side to be regarded as a magnetic wall as that of dielectric resonator antennas (DRAs) [36]. The main planes along the x- and y-axes as well as the open-ended plane are noted as planes 1, 2, and 3, respectively. The dimensions of the cavity are denoted as a, b, and c. In order to facilitate the following analysis, $a > b$ is assumed here without any loss of generality. c indicates the thickness of the substrate much smaller than a and b. Therefore, the first eigenmode of the resonator is $TE^z_{10\delta}$ and the following two eigenmodes are $TE^z_{20\delta}$, $TE^z_{01\delta}$ $(a > 2b)$ or $TE^z_{01\delta}$, $TE^z_{20\delta}$ $(a > b > a/2)$, respectively. The subscript δ denotes the number of variations in the standing wave pattern along the z-axis as used in [37]. The eigenmode resonant frequencies can be calculated by using the following equation:

$$f_e = \frac{1}{2\sqrt{\varepsilon\mu}}\sqrt{\left(\frac{m}{a}\right)^2 + \left(\frac{n}{b}\right)^2 + \left(\frac{l}{2c}\right)^2} \tag{5.7}$$

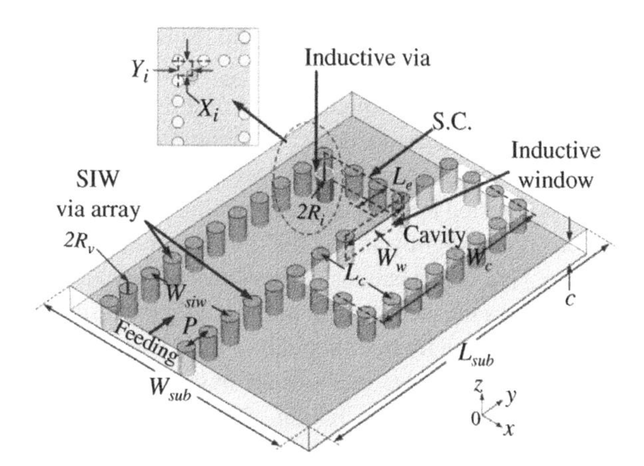

Figure 5.35 The SIW fed cavity antenna [24].

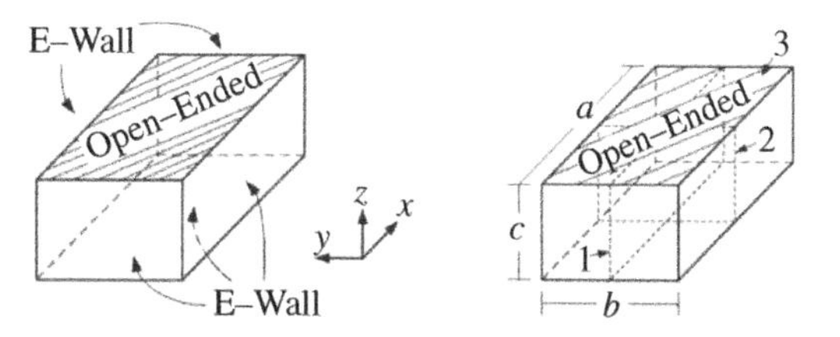

Figure 5.36 Dielectric-loaded cavity model of simplified resonator. (a) With boundary condition and (b) with dimensions and planes indications.

Freq.	Plane 1	Plane 2	Plane 3
$TE^z_{10\delta}$ 38.1 GHz			
$TE^z_{20\delta}$ 41 GHz			
$TE^z_{30\delta}$ 43 GHz			

Figure 5.37 The first three eigenmodes and the *E*-field in Planes 1, 2, and 3.

where f_e denotes the eigenmode frequency; ε and μ are the permittivity and permeability of the material respectively; and m, n, l refer to the number of variations in the standing wave pattern in the *x*-, *y*-, and *z*-axis, respectively. When $a = 4.6$ mm, $b = 1.9$ mm, and $c = 0.635$ mm (the thickness of RT/Duroid 6010), the resonant frequencies corresponding to these eigenmodes together with the electric field in planes 1, 2, and 3 are exhibited in Figure 5.37.

The cavity is excited by the coupling energy through the inductive window in the narrow side-wall of SIC. Using Eq. (5.2), the width (W_c) and length (L_c) of the dielectric-loaded cavity can be obtained from the simplified resonator size, a and b, respectively. The feeding SIW is designed to operate at single dominant mode under 43.5 GHz ($W_{siw} = 2.4$ mm, $R_v = 0.2$ mm, $p = 0.7$ mm). If the window is centered in the wider side wall (parallel to plane 1), only $TE^z_{10\delta}$ can be supported in this case. The mode $TE^z_{20\delta}$ would not be excited since the *E*-field of this mode is null in plane 2 as shown in Figure 5.37 while the *E*-field of the feeding SIW at the center of the window is the maximum. The resonance frequency of the $TE^z_{01\delta}$ mode is higher than that of the $TE^z_{10\delta}$ mode, nearly out of the dominant mode frequency band of the feeding SIW. Therefore, the proposed feeding scheme enables the dielectric-loaded cavity operating in single $TE^z_{10\delta}$ mode. In this case, b doesn't affect the resonant frequency because $a > b$. In practice, by tuning b, the characteristic impedance of the cavity is adjusted for impedance matching purpose. As the thickness c is much smaller than a, it is obvious that a slightly affects the resonant frequency of $TE^z_{01\delta}$ mode, which indicates the antenna operating frequency band is not sensitive to this parameter. In addition, the feeding inductive window destroys the electric wall of the cavity that will shift down the cavity's resonant frequencies. Therefore, the cavity size should be a little bit smaller than that of the calculation.

The feeding scheme includes an inductive window and an inductive via, which ensures the wideband impedance matching. The equivalent circuit of the feeding structure is shown in Figure 5.38. Both the inductive window and the inductive via can be modeled as a T-shape network, including a shunt inductance and two series capacitances. Z_0 and Z_0' represent the characteristic impedances of the feeding SIW and the cavity, respectively. In this case, the feeding

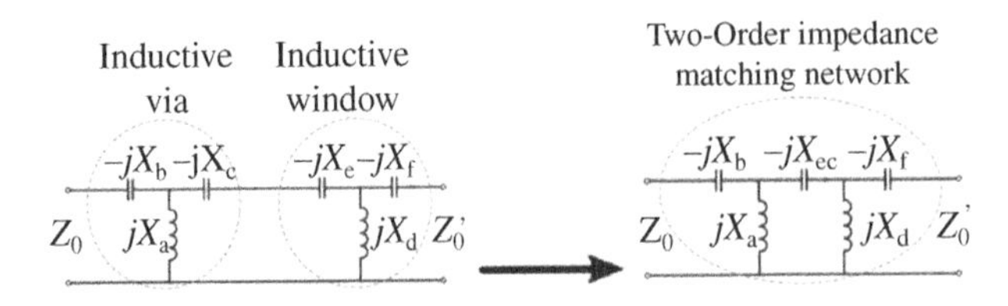

Figure 5.38 The equivalent circuit of the feeding structure.

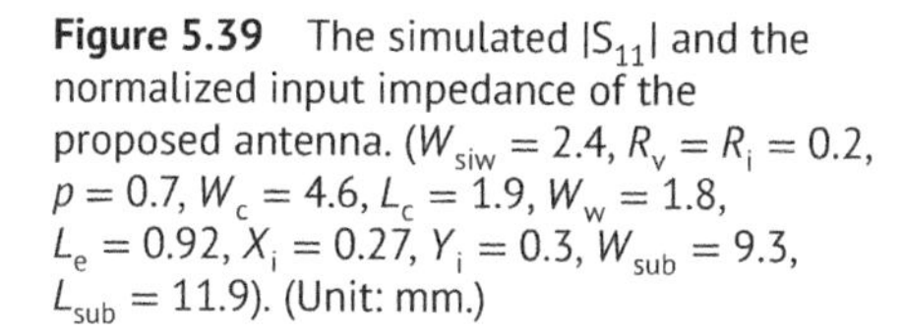

Figure 5.39 The simulated $|S_{11}|$ and the normalized input impedance of the proposed antenna. ($W_{siw} = 2.4$, $R_v = R_i = 0.2$, $p = 0.7$, $W_c = 4.6$, $L_c = 1.9$, $W_w = 1.8$, $L_e = 0.92$, $X_i = 0.27$, $Y_i = 0.3$, $W_{sub} = 9.3$, $L_{sub} = 11.9$). (Unit: mm.)

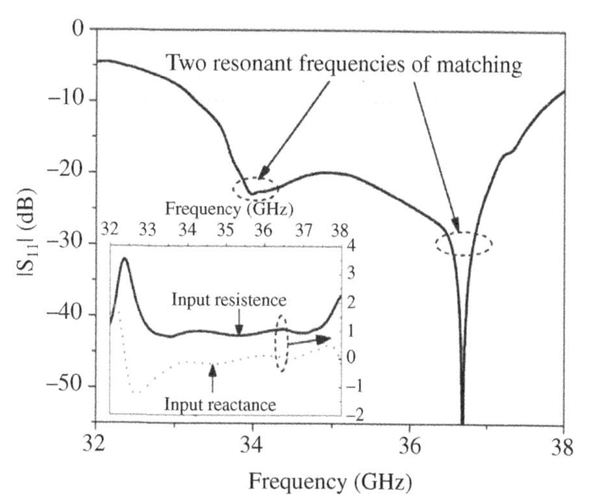

network can be regarded as a two-order impedance-matching network that can provide two resonant frequencies for the matching.

To validate the analysis, an antenna operating at 35.5 GHz was designed, simulated, and optimized using the full-wave electromagnetic field simulation software CST Microwave Studio. The initial size of the cavity and the feeding SIW is calculated using the given substrate parameters and the specified operating frequency. Then, all the parameters, especially the size and position of the inductive window and the inductive via are optimized for achieving a specific performance. From the simulated $|S_{11}|$ of the proposed antenna as shown in Figure 5.39, two resonances can be observed, as predicted. In addition, the feeding inductive window shifts down the resonant frequencies of the cavity so that the simulated impedance bandwidth ($|S_{11}| = -10$ dB) is 4.6 GHz or 13% (33.2–37.8 GHz) with the center frequency of 35.5 GHz; the bandwidth for −15 dB $|S_{11}|$ over the frequency range from 33.4 to 37.2 GHz, still greater than 10%.

As shown in Figure 5.40, the simulated *E*-field distribution at 35 GHz in the two main planes AA' and BB' are also provided for validating the above analysis. In the cavity, the *E*-field intensity

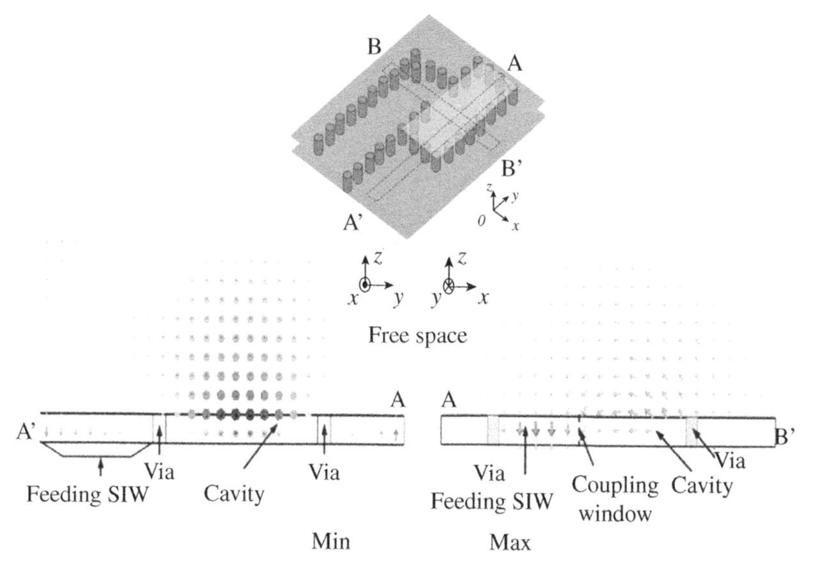

Figure 5.40 The simulated *E*-field distribution at 35 GHz on Planes AA′ and BB′. The arrow indicates the direction of the *E*-field. In *yz*-plane, the arrows are pointing into the paper.

distribution along the y-axis is approximated as a cosinusoid, while it shows no change the along x-axis, which is the same as that of Figure 5.37. The maximum E-field intensity along the z-axis is occurs at the interface between the substrate and the free space, which is consistent with regarding the interface as an equivalent magnetic wall. The results prove the cavity operating at a single $TE^z_{10\delta}$ mode.

The simulated radiation patterns at 35 GHz for the antenna are shown in Figure 5.41. The wide 3-dB beamwidths of 92.1° (E-plane) and 76.8° (H-plane) are similar to those of the rectangular DRA. The simulated ratio of co- to cross-polarization levels is about 20 dB. Figure 5.42 shows the simulated maximum gain larger than 6 dBi over the operating frequency band of 33.2–37.8 GHz, which suggests the high radiation efficiency of the antenna.

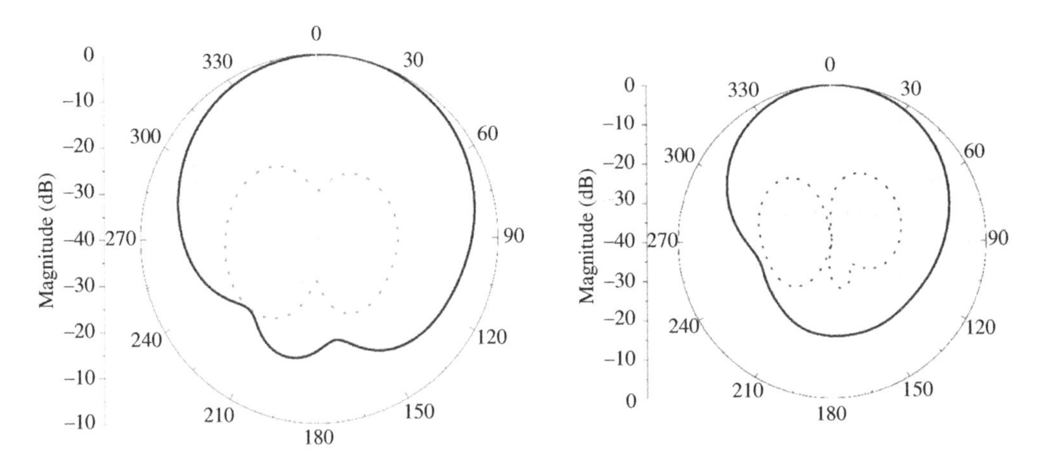

Figure 5.41 The simulated radiation patterns of the proposed antenna at 35 GHz. (a) E-plane (xz-plane) and (b) H-plane (yz-plane). (Solid line is the co-polarization. Dot line is the cross-polarization.)

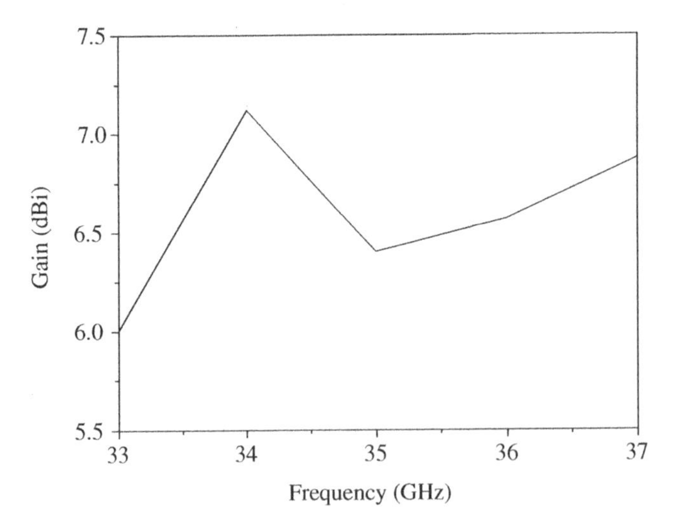

Figure 5.42 The simulated gain of the antenna.

5.3.2 SIW Narrow Wall Fed SIC Antenna Array at 35 GHz

Generally, there are series-fed arrays and shunt-fed arrays. Series-fed arrays suffer narrow-band performance and beam sweeping against the frequency. Therefore, a shunt-fed approach is incorporated with the antenna for a wideband performance and fixed beam direction, as shown in Figure 5.42.

The antenna array is composed of 2 × 2 antenna elements and a compact four-way tree-shape power divider. The power divider splits the power into four ways equally. All the elements are excited in-phase. The divider is similar to that proposed in [16]. The width of the SIW divider is equal to the feeding SIW for each element. The unique design consideration of the divider introduced herein is that the SIW and the antenna elements share part of via-holes to configure the side walls, which keeps the antenna array more compact. Since the antenna array is implemented with the via-holes, the spacing between the adjacent elements along the x-axis must increase or decrease discretely to fit the minimum pitch between the adjacent vias required by the standard PCB process. In this case, the final optimized spacing between adjacent antenna elements are 7.2 and 7.4 mm along the x- and y-axis, respectively, which are equal to 0.84λ and 0.863λ, where λ is the wavelength at 35 GHz in free space.

A 50-Ω microstrip line is introduced to connect with a small SubMiniature version A (SMA) connector during the measurement. A tapered SIW is used for impedance matching as illustrated in Figure 5.43. The overall size of the antenna prototype is 16.6 mm × 29.9 mm × 0.635 mm, and the optimized parameters are listed in the caption of Figure 5.43a. The antenna array was fabricated by a standard PCB process to validate the design.

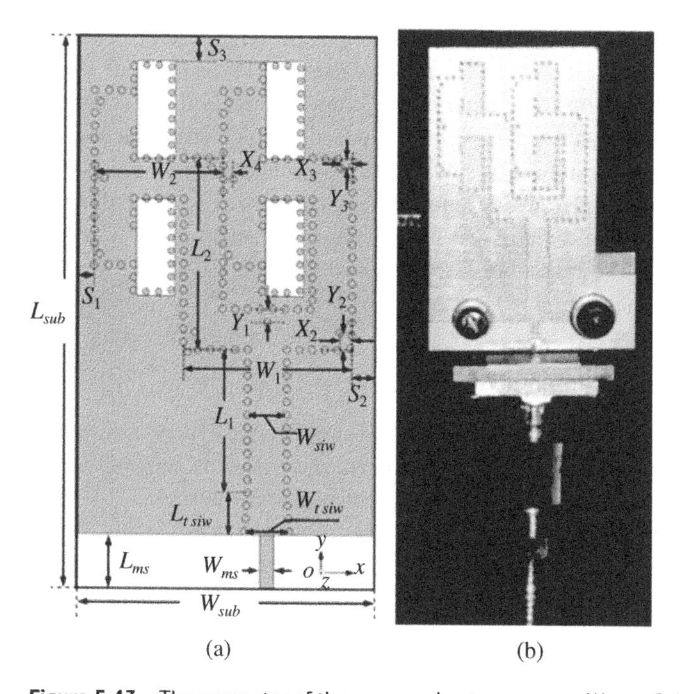

Figure 5.43 The geometry of the proposed antenna array. $W_{siw} = 2.4$, $W_{tsiw} = 2.5$, $L_{tsiw} = 2.1$, $W_1 = 9.4$, $L_1 = 7.7$, $W_2 = 7.2$, $L_2 = 10.3$, $R_v = R_i = 0.2$, $p = 0.7$, $W_c = 4.6$, $L_c = 1.7$, $W_w = 2.1$, $L_e = 1.1$, $X_i = 0.3$, $Y_i = 0.6$, $Y_1 = 0.7$, $X_2 = 0.65$, $Y_2 = 0.95$, $X_3 = 0.65$, $Y_3 = 0.5$, $X_4 = 0.55$, $S_1 = 1$, $S_2 = 1.2$, $S_3 = 1.6$, $W_{ms} = 0.8$, $L_{ms} = 2$, $W_{sub} = 16.6$, $L_{sub} = 29.9$. (Unit: mm.)

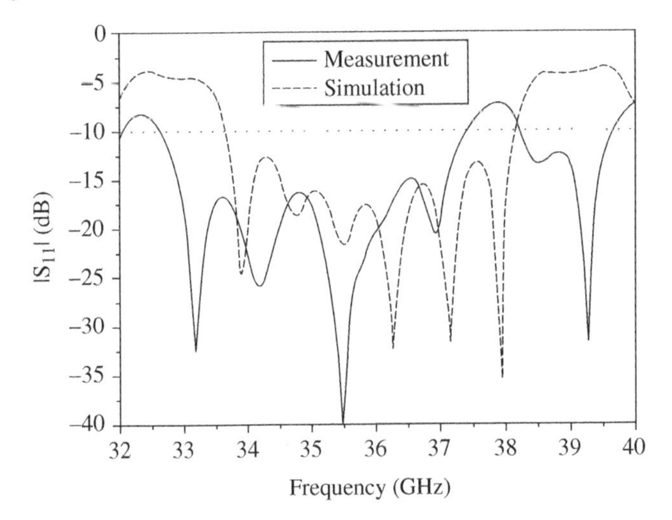

Figure 5.44 The reflection coefficients of the proposed four-cell antenna array.

Figure 5.44 compares the measured and simulated $|S_{11}|$ for the antenna prototype. It can be observed that the frequency range for −10 and −15 dB $|S_{11}|$ is from 32.7 to 37.4 GHz and 32.9 to 37 GHz or the bandwidth of 13.4% and 11.7%, respectively.

The measured radiation patterns at 35 GHz compared with the simulation are illustrated in Figure 5.45. The 3-dB beamwidths in the *E*- and *H*-plane are 29.1° and 32°, respectively. The side lobe levels are higher, as shown in Figure 5.45a, which may be caused by the large spacing between adjacent antenna elements. The measured ratio of co- to cross-polarization levels is about 15 dB in the main beam in the both *E*- and *H*-planes. The deteriorated cross-polarization performance in the *H*-plane may be caused by the additional radiation from the feeding microstrip line and/or the SMA connector. The microstrip line results in a significant cross radiation loss in such a frequency band because the *E*-field radiated from the microstrip is orthogonal to the radiation from the antenna. It can be improved when the antenna array is integrated with planar circuits fully implemented by SIW technology. The measured gain response is of 9.2–10.8 dBi over the frequency range from 33 to 37 GHz.

5.3.3 60-GHz SIW Narrow Wall Fed SIC Antenna Array

An SIW-fed SIC antenna array is designed for the unlicensed 60 GHz systems with frequency range from 57 to 64 GHz. This band is free of congestions, unlike lower frequency bands, more suitable for a high data-rate connection over a short range in mmW wireless local area networks.

The geometry of the 60 GHz antenna comprising 2×2 elements is shown in Figure 5.46, which is similar to the one shown in Figure 5.44. With the substrate Rogers/RO3006 ($\varepsilon_r = 6.0$, $\tan\delta = 0.0025$, thickness of 0.635 mm), the initial size of the cavity can be calculated using Eq. (5.7), that is, a = 1.72 mm and $b <$ 1.72 mm. In this case, the first two eigenmodes of the cavity are $TE^z_{10\delta}$and $TE^z_{01\delta}$, the resonant frequency of the second mode-$TE^z_{01\delta}$ is 71.2 GHz, which is out of the operating bandwidth of 57–64 GHz and ensures the cavity operating at dominant mode only. The radius of the via-holes, including the inductive via, is 0.15 mm, and the spacing between the adjacent vias is 0.6 mm. The width of the feeding SIW and the power divider is selected to ensure only the main mode $TE^z_{01\delta}$can be supported. The final optimized distance between the adjacent antenna elements are 4.44 mm and 3.49 mm along the *x*- and *y*-axes, respectively, which is equal to 0.888λ and 0.698λ, where λ is the wavelength at 60 GHz in free space. The overall antenna array

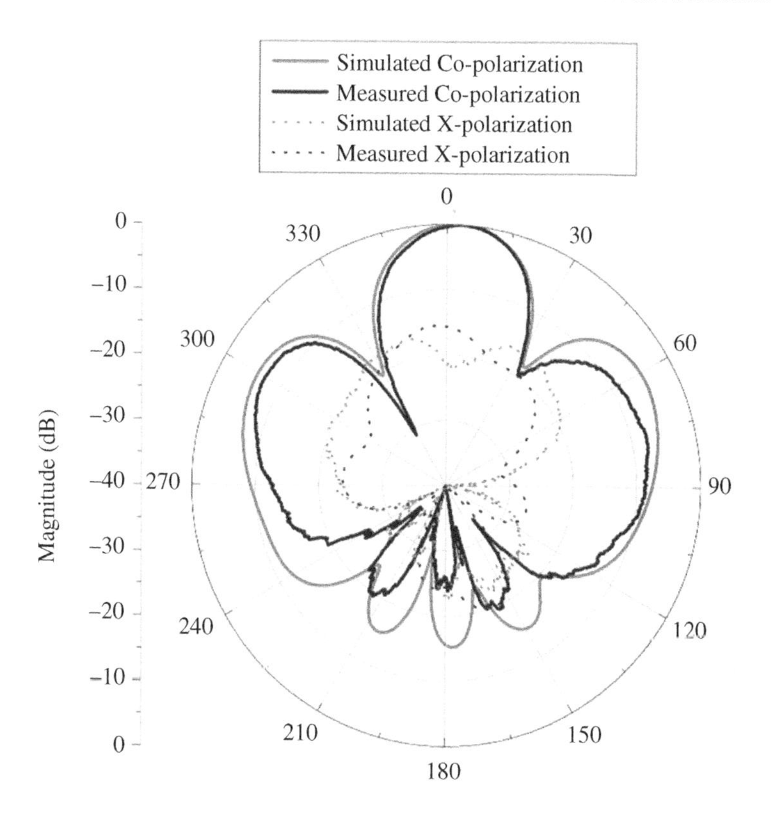

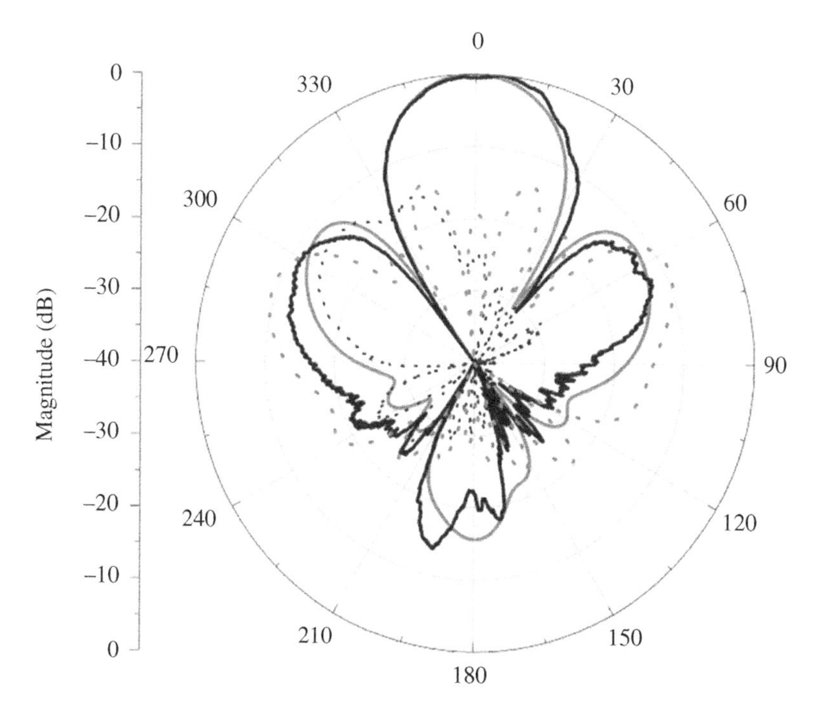

Figure 5.45 The measured normalized radiation patterns at 35 GHz. (a) *E*-plane (*xz*-plane) and (b) *H*-plane (*yz*-plane).

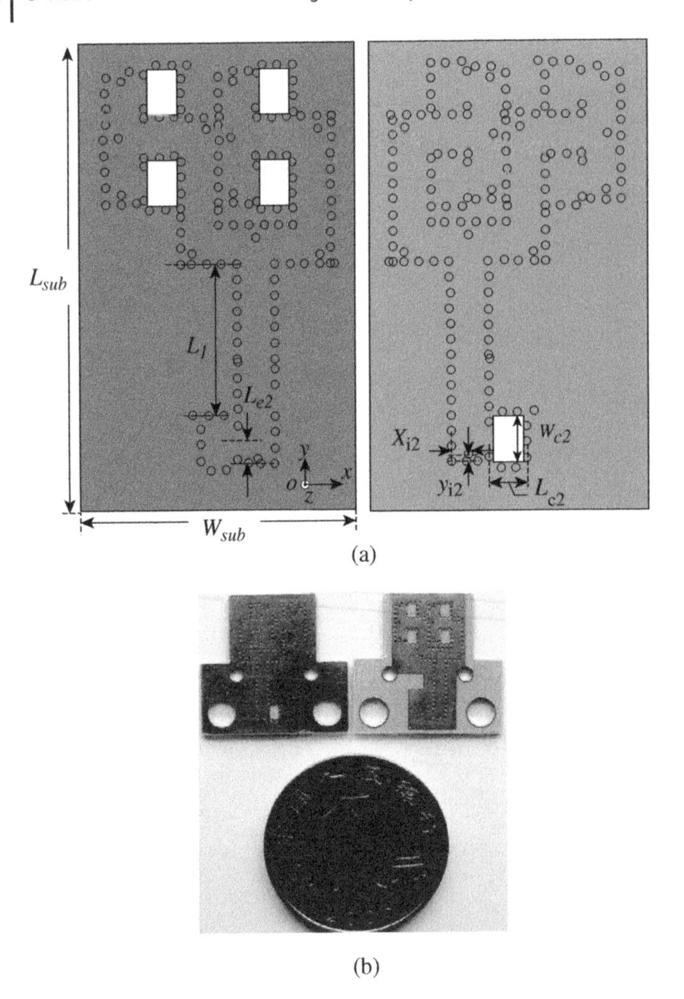

Figure 5.46 The geometry of the proposed 60-GHz antenna array. $W_{siw} = 1.5$, $W_1 = 9.4$, $L_1 = 7.7$, $W_2 = 7.2$, $L_2 = 10.3$, $R_v = R_i = 0.2$, $p = 0.7$, $W_c = 1.69$, $L_c = 1.14$, $W_w = 2.1$, $L_e = 1.1$, $X_i = 0.3$, $Y_i = 0.6$, $Y_1 = 0.7$, $X_2 = 0.65$, $Y_2 = 0.95$, $X_3 = 0.65$, $Y_3 = 0.5$, $X_4 = 0.55$, $S_1 = 1$, $S_2 = 1.2$, $S_3 = 1.6$, $W_{ms} = 0.8$, $L_{ms} = 2$, $W_{sub} = 10.88$, $L_{sub} = 17.76$. (Unit: mm).

size, including the transition, is 10.88 mm × 17.76 mm. The size is around 10.88 mm × 9.76 mm excluding the transition.

The measurement needs a waveguide-to-SIW transition. A dielectric-loaded cavity is also used in the transition to realize the wideband transmission where the inductive via and the inductive window are used for matching. The same phenomenon of two resonant frequencies can be observed in the simulated S-parameter of the transition as shown in Figure 5.47.

Figure 5.48a compares the simulated $|S_{11}|$ of the array with and without the waveguide-to-SIW transition. The design without the transition shows the frequency range from 56.62 to 64.25 GHz for −10 dB impedance bandwidth. The transition degrades the overall performance of the array, especially around 63.7 GHz with the $|S_{11}|$ a little higher than −10 dB.

Figure 5.48b exhibits the frequency range from 54.7 to 61.7 GHz for the measured $|S_{11}|$ less than −10 dB. The measured results show a similar trend to the simulation. The degraded performance

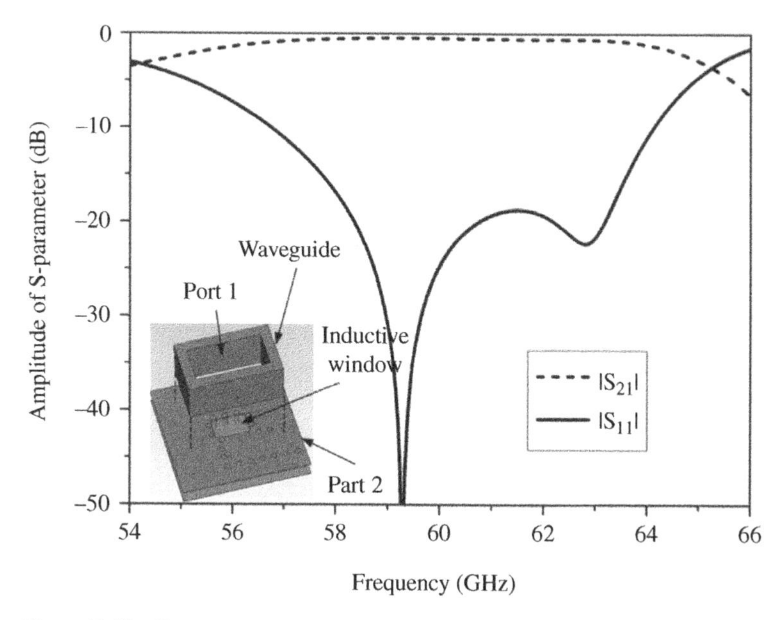

Figure 5.47 The simulated S-parameters of the waveguide-to-SIW transition.

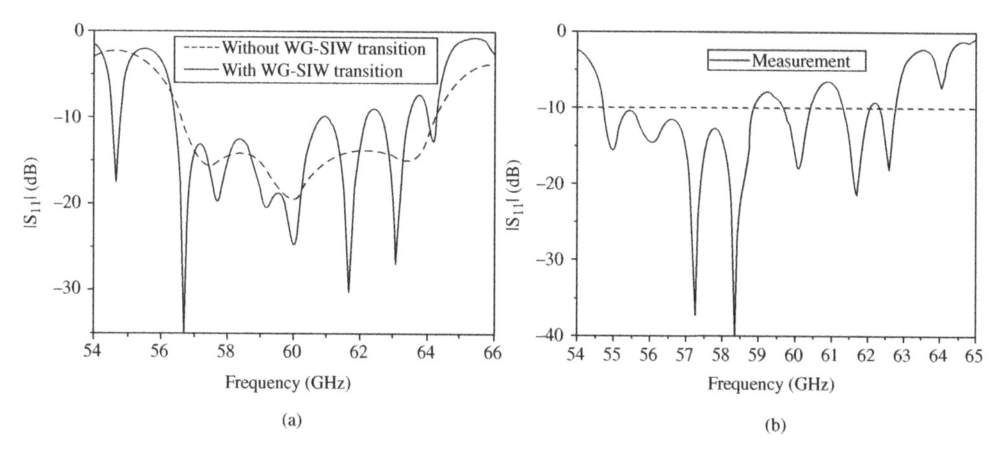

Figure 5.48 (a) Simulated and (b) measured reflection coefficients of the proposed 60 GHz antenna array.

with higher $|S_{11}|$ around 59 and 61 GHz may be caused by the fabrication error since the size of the via-holes nearly reaches the process limits. When the antenna is integrated into circuits, the transition can be omitted so that the antenna performance would be improved. In addition, the design should be tuned to accurately cover the desired bandwidth.

Figure 5.49 shows the normalized radiation patterns at 58 GHz. The 3 dB main beamwidths in the *E*- and *H*-plane are 27.8° and 32.2°, respectively. In Figure 5.49a, the side lobe levels in the *E*-plane are higher due to the larger spacing between the adjacent antenna elements and a higher measured co- to cross-polarization level of about 17 dB in main beams in both the *E*- and *H*-planes. Without the feeding microstrip lines in the array, the cross-polarization performance is kept acceptable in the *H*-plane. With the waveguide feeding structure, the measured back-lobe level is lower than the simulation, especially in the *H*-plane.

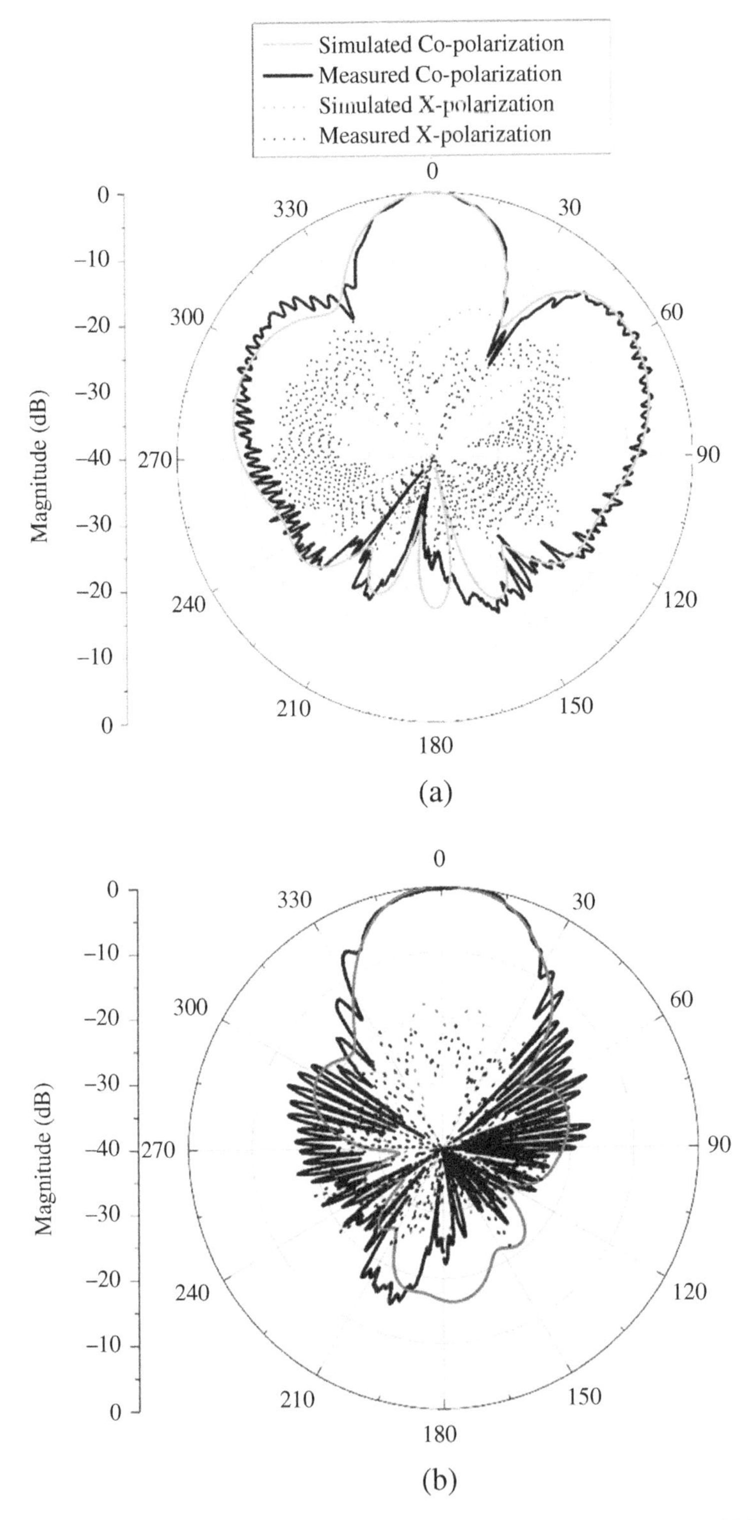

Figure 5.49 The measured normalized radiation patterns of the array at 58 GHz. (a) *E*-plane (*xz*-plane) pattern and (b) *H*-plane (*yz*-plane) pattern.

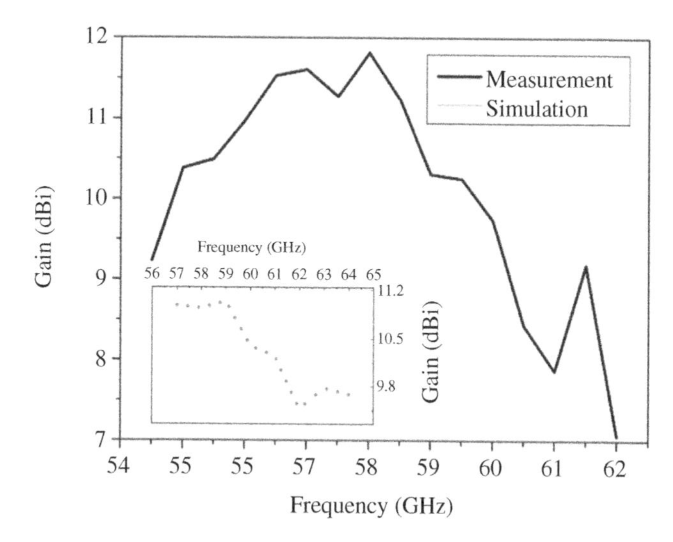

Figure 5.50 The maximum gain of the 2 × 2 antenna array.

Figure 5.50 compares the measured gain with the simulated realized gain. In the simulation, the gain is around 9.7–11 dBi from 57 to 64 GHz. In the desired band of 54.5 to 61.5 GHz, the measured gain is above 9 dBi, except at 61 GHz.

5.4 Summary

The chapter has reviewed the research progress in SIC antennas, including SIC-backed slot antennas, patch loaded SIC antennas, and traveling-wave element loaded SIC antennas. Typically, the SIW-fed narrow wall SIC antenna has been elaborated. The study has shown that the SIC antennas are capable of covering a wide operating frequency range from 10% to 35% using various broadband techniques, including the impedance matching method and multi-mode cavity, as well as the additional radiator loaded method. The inherent advantages, including low profile, low insertion loss, high efficiency, ease of integration, and low cost, make the SIC antennas very promising in mmW band wireless communications and other applications. Most of the SIC antennas can be fabricated with the conventional PCB process, and additional merits can be achieved when advanced processes are adopted for the fabrication of SIC antennas, such as LTCC and 3D-printing processes. Although only SIC antennas with fixed beams have been reviewed here, the SIC antenna is also a good candidate for scanning-beam or multiple-beam antenna design.

References

1 Lockie, D. and Peck, D. (2009). High-data-rate millimeter-wave radios. *IEEE Commun. Mag.* 10 (5): 75–83.

2 Andrews, J.G., Buzzi, S., Choi, W. et al. (2014). What will 5G be? *IEEE J. Sel. Areas Commun.* 32 (6): 1065–1082.

3 Ghosh, A., Thomas, T.A., Cudak, M.C. et al. (2014). Millimeter-wave enhanced local area systems: a high-data-rate approach for future wireless networks. *IEEE J. Sel. Areas Commun.* 32 (6): 1152–1163.

4 Busari, S., Mumtaz, S., Al-Rubaye, S., and Rodriguez, J. (2018). 5G millimeter-wave mobile broadband: performance and challenges. *IEEE Commun. Mag.* 56 (6): 137–143.

5 Hong, W., Jiang, Z., Yu, C. et al. (2017). Multibeam antenna technologies for 5G wireless communications. *IEEE Trans. Antennas Propag.* 65 (12): 6231–6249.

6 Tolbert, C., Straiton, A., and Britt, C. (1958). Phantom radar targets at millimeter radio wavelengths. *IRE Trans. Antennas Propag.* 6 (4): 380–384.

7 IEEE 802 LAN/MAN Standards Committee (2009). *IEEE Standard for Information technology – Local and metropolitan area networks – Specific requirements – Part 15.3: Amendment 2: Millimeter-wave-based Alternative Physical Layer Extension, IEEE Std 802.15.3c-2009* (Amendment to IEEE Std 802.15.3-2003), 1–200.

8 Hong, W. (2008). Research advances in SIW, HMSIW and FHMSIW. In: *China-Japan Joint Microwave Conference*, 357–358.

9 Bozzi, M., Georgiadis, A., and Wu, K. (2011). Review of substrate-integrated waveguide circuits and antennas. *IET Microwave Antennas Propag.* 5 (8): 909–920.

10 Zhu, F., Hong, W., Chen, J., and Wu, K. (2012). Ultra-wideband single and dual baluns based on substrate integrated coaxial line technology. *IEEE Trans. Microwave Theory Tech.* 60 (10): 3062–3070.

11 Xu, F. and Wu, K. (2005). Guided-wave and leakage characteristics of substrate integrated waveguide. *IEEE Trans. Microwave Theory Tech.* 53 (1): 66–73.

12 Deslandes, D. and Wu, K. (2003). Single-substrate integration technique of planar circuits and waveguide filters. *IEEE Trans. Microwave Theory Tech.* 51 (2): 593–596.

13 Yan, L., Hong, W., Hua, G. et al. (2004). Simulation and experiment on SIW slot array antennas. *IEEE Microwave Wirel. Compon. Lett.* 14 (9): 446–448.

14 Hao, Z.-C., Hong, W., Chen, J.X. et al. (2005). Compact super-wide bandpass substrate integrated waveguide (SIW) filters. *IEEE Trans. Microwave Theory Tech.* 53 (9): 2968–2977.

15 Tang, H.J., Hong, W., Chen, J. et al. (2007). Development of millimeter-wave planar diplexers based on complementary characters of dual-mode substrate integrated waveguide filters with circular and elliptic cavities. *IEEE Trans. Microwave Theory Tech.* 55 (4): 776–782.

16 Chen, P., Hong, W., Kuai, Z., and Xu, J. (2009). A substrate integrated waveguide circular polarized slot radiator and its linear array. *IEEE Antennas Wirel. Propag. Lett.* 8: 120–123.

17 Liu, J., Jackson, D.R., and Long, Y. (2012). Substrate integrated waveguide (SIW) leaky-wave antenna with transverse slots. *IEEE Trans. Antennas Propaga.* 60 (1): 20–29.

18 Cheng, Y.J., Hong, W., Wu, K., and Fan, Y. (2011). Millimeter-wave substrate integrated waveguide long slot leaky-wave antennas and two-dimensional multibeam applications. *IEEE Trans. Antennas Propaga.* 59 (1): 40–47.

19 Du, M., Xu, J., Dong, Y., and Ding, X. (2017). LTCC SIW-vertical-fed-dipole array fed by a microstrip network with tapered microstrip-to-SIW transitions for wideband millimeter-wave applications. *IEEE Antennas Wirel. Propag. Lett.* 16: 1953–1956.

20 Zou, X., Tong, C., Bao, J., and Pang, W. (2014). SIW-fed Yagi antenna and its application on monopulse antenna. *IEEE Antennas Wirel. Propag. Lett.* 13: 1035–1038.

21 Zhai, G., Cheng, Y., Yin, Q. et al. (2013). Super high gain substrate integrated clamped-mode printed log-periodic dipole array antenna. *IEEE Trans. Antennas Propaga.* 61 (6): 3009–3016.

22 Zhang, T., Zhang, Y., Cao, L. et al. (2015). Single-layer wideband circularly polarized patch antennas for Q-band applications. *IEEE Trans. Antennas Propag.* 63 (1): 409–414.

23 Hao, Z.C., Hong, W., Chen, A. et al. (2006). SIW fed dielectric resonator antennas (SIW-DRA). In: *2006 IEEE MTT-S International Microwave Symposium Digest*, 202–205.

24 Zhang, Y., Chen, Z.N., Qing, X., and Hong, W. (2011). Wideband millimeter-wave substrate integrated waveguide slotted narrow-wall fed cavity antennas. *IEEE Trans. Antennas Propaga.* 59 (5): 1488–1496.

25 Zhang, Y., Xue, Z., and Hong, W. (2017). Planar substrate-integrated endfire antenna with wide beamwidth for Q-band applications. *IEEE Antennas Wirel. Propag. Lett.* 16: 1990–1993.

26 Luo, G.Q., Hu, Z.F., Dong, L.X., and Sun, L.L. (2008). Planar slot antenna backed by substrate integrated waveguide cavity. *IEEE Antennas Wirel. Propag. Lett.* 7: 236–239.

27 Luo, G.Q., Hu, Z.F., Liang, Y. et al. (2009). Development of low profile cavity backed crossed slot antennas for planar integration. *IEEE Trans. Antennas Propag.* 57 (10): 2972–2979.

28 Luo, G.Q., Hu, Z.F., Li, W.J. et al. (2012). Bandwidth-enhanced low-profile cavity-backed slot antenna by using hybrid SIW cavity modes. *IEEE Trans. Antennas Propag.* 60 (4): 1698–1704.

29 Shi, Y., Liu, J., Long, Y., and Member, S. (2017). Wideband triple- and quad-resonance substrate integrated waveguide cavity-backed slot. *IEEE Trans. Antennas Propag.* 65 (11): 5768–5775.

30 Wu, Q., Yin, J., Yu, C. et al. (2019). Broadband planar SIW cavity-backed slot antennas aided by unbalanced shorting vias. *IEEE Antennas Wirel. Propag. Lett.* 18 (2): 363–367.

31 Yang, T.Y., Hong, W., and Zhang, Y. (2014). Wideband millimeter-wave substrate integrated patch antenna. *IEEE Antennas Wirel. Propag. Lett.* 13: 205–208.

32 Zhang, T., Zhang, Y., Hong, W., and Wu, K. (2015). Wideband millimeter-wave SIW cavity backed patch antenna fed by substrate integrated coaxial line. In: *IEEE International Wireless Symposium (IWS 2015)*, 1–4.

33 Wu, Q., Member, S., Hirokawa, J. et al. (2017). Millimeter-wave planar broadband circularly polarized antenna array using stacked curl elements. *IEEE Trans. Antennas Propaga.* 65 (12): 7052–7062.

34 Zhang, L., Wu, K., Wong, S.-W. et al. Wideband high-efficiency circularly polarized SIW-fed S-dipole array for millimeter-wave applications. *IEEE Trans. Antennas Propaga.* vol. 68, no. 3, pp. 2422–2427, 2020. .

35 Zhu, J., Liao, S., Li, S., and Xue, Q. (2018). 60 GHz wideband high-gain circularly polarized antenna array with substrate integrated cavity excitation. *IEEE Antennas Wirel. Propag. Lett.* 17 (5): 751–755.

36 Mongia, R.K., Ittipiboon, A., and Cuhaci, M. (1994). Measurement of radiation efficiency of dielectric resonator antennas. *IEEE Microwave Guided Wave Lett.* 4: 80–82.

37 Mongia, R.K. and Ittipiboon, A. (1997). Theoretical and experimental investigations on rectangular dielectric resonator antennas. *IEEE Trans. Antennas Propaga.* 45 (9): 1348–1356.

6

Cavity-Backed SIW Slot Antennas at 60 GHz

Ke Gong

College of Physics and Electronic Engineering, Xinyang Normal University, Xinyang 464000, Pepoe's Republic of China.

6.1 Introduction

A cavity-backed slot antenna (CBSA) eliminates the back radiation from the slot by placing the quarter-guided-wavelength cavity behind the slot antenna. As the short-ended cavity has high impedance near the resonance, most of the power radiates through the slot aperture. CBSAs have been used in various wireless communication systems due to their relatively high gain and uni-directional radiation patterns [1]. However, the traditional CBSAs have high profiles and are difficult to integrate with other planar circuits. SIW was introduced to implement the cavity in a planar substrate in 2008 [2]. Since then, such electromagnetically closed planar waveguide structure has been employed as a good candidate for design low-profile CBSAs in microwave and mmW bands [3, 4].

SIW based CBSAs have the following advantages compared with the conventional metallic waveguides:

- low profile with planar form;
- lightweight with small size;
- low fabrication cost by using printed-circuit technology;
- good for mass production; and
- easy to be integrated with other planar circuits on the same substrate.

Compared with conventional metallic waveguides, however, the SIW CBSAs also suffer from narrow operating bandwidth, substrate losses, and low power-handling capability.

This chapter discusses the cavity-backed SIW slot antennas by introducing: (i) the mechanism of the cavity-backed antenna; (ii) cavity-backed SIW slot antenna, including the feeding techniques, the backing SIW cavity, and the radiating slot; (iii) the wideband CBSAs, the dual-band CBSAs, the dual-polarized (DP) and circularly polarized (CP) CBSAs, and the miniaturized CBSAs; and (iv) SIW CBSA design examples operating at 60 GHz band.

Substrate-Integrated Millimeter-Wave Antennas for Next-Generation Communication and Radar Systems, First Edition.
Edited by Zhi Ning Chen and Xianming Qing.

6.2 Operating Principle of the Cavity-Backed Antenna

6.2.1 Configuration

A cavity-backed antenna (CBA) is generally composed of two parts: an excitation source and a metallic cavity. The two parts play an important role in designing the CBA. The excitation source may be designed in the forms of dipole, helical wire, and slot. Meanwhile, the metallic cavity with rectangular, circular, and elliptical configurations can be utilized to meet actual demands. Although the analysis and design methods vary for different kinds of CBAs, they all follow the same operating principle that a CBA can be considered as an open-ended resonator.

For analysis convenience, a CBA can be divided into three parts as shown in Figure 6.1, i.e., the excitation part, the cavity part, and the radiation part [5]. In Figure 6.1, the M_i is the matching component between the cavity and the feeding waveguide W_g, E_0 represents the excitation source with the electric current $\mathbf{J}_0$ and the magnetic current $\mathbf{M}_0$, V_r stands for the backing-cavity, S denotes the radiation aperture of the CBA, on which there are the equivalent electric current $\mathbf{J}_a$ and the equivalent magnetic current $\mathbf{M}_a$. From this perspective, the analysis of a CBA can be divided into three issues, namely the excitation, the cavity, and the radiation. Of them, the inner cavity is the key because the field distribution in the cavity V_r determines the field and current distribution of the aperture S. Thus, the far-field radiation characteristics can be worked out once the resonant cavity can be analyzed. The excitation is associated with the impedance matching between the excitation source and the feeding waveguide.

6.2.2 Analysis of the Backing-Cavity

In the general case, the CBAs feature a high-quality factor Q_A. Usually the backing-cavity of a CBA can be regarded as a resonant cavity, and the electromagnetic field distribution in the cavity can be obtained by using eigenmode analysis. Compared with a general closed resonant cavity, a CBA features some unique properties, such as the energy radiation, the electromagnetic perturbations, the resonant frequency deviation, and the mode spectrum narrowing.

The quality factor Q_A of a CBA can be obtained with the following formula,

$$\frac{1}{Q_A} = \frac{1}{Q_{wall}} + \frac{1}{Q_{die}} + \frac{1}{Q_{rad}}, \tag{6.1}$$

where Q_{wall}, Q_{med}, and Q_{rad} correspond to the quality factors associated with metallic loss of cavity wall, the dielectric loss in the cavity, and the radiation loss, respectively.

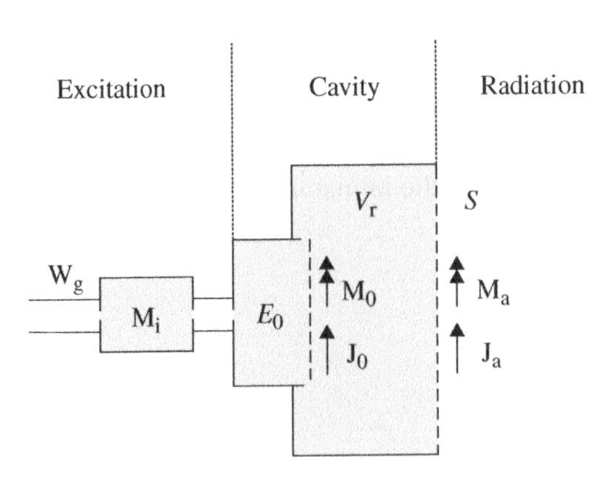

Figure 6.1 Division of a CBA for electromagnetic analysis.

In practical applications, the metallic loss and the dielectric loss of a CBA are very small, and most of the energy will radiate from its aperture. We have $\left(\frac{1}{Q_{wall}} + \frac{1}{Q_{die}}\right) \ll \frac{1}{Q_{rad}}$ and $Q_A \cong Q_{rad}$. Furthermore, the cavity resonant complex frequency in a CBA can be expressed as,

$$\dot{\omega}_0 = \omega_0' + j\omega_0'' = \omega_0'\left(1 + j\frac{1}{2Q_A}\right) \tag{6.2}$$

It can be seen form (6.2) that

$$Q_A = \frac{\omega_0'}{2\omega_0''} \tag{6.3}$$

For $\omega_0'' \neq 0$, and $\omega_0'' \ll \omega_0'$, the backing-cavity of a CBA operates at a quasi-periodic resonance. For a typical CBA, the radiation brings in a weak perturbation to the resonance of the backing-cavity, and the quality factor of $Q_A \geq 50 \sim 100$, that is, $\omega_0'' \leq 0.01 - 0.02\omega_0'$.

In the backing-cavity of a CBA, numerous resonant modes can be excited theoretically, and their resonant frequencies form an infinite discrete spectrum. The number of the resonant modes is recorded as N_r, and $N_r/\Delta\omega$ stands for the frequency spectrum density, which is given by [6],

$$\frac{N_r}{\Delta\omega} \approx \frac{V_r}{2\pi^2 c^3}\omega^2, \tag{6.4}$$

and

$$N_r \approx 4\pi\frac{V_r}{\lambda^3}\cdot\frac{\Delta f}{f_0}, \tag{6.5}$$

where V_r is the cavity's equivalent volume, $2\Delta f/f_0$ is the fractional bandwidth, and c is the velocity of light in vacuum.

As can be seen form the formulas (6.4) and (6.5), the frequency selectivity of a closed three-dimensional cavity becomes weak under the condition of V_r/λ^3; on the other hand, $N_r/\Delta\omega$ increases rapidly as the frequency increases. However, CBA is an open-ended resonator for energy radiation, which is different from a conventional resonant cavity. It means that the resonant mode in a CBA is not a real three-dimensional one instead of a two-dimensional resonant mode existing on the cross section of the cavity. Denoting the two-dimensional resonant surface by S_r, the frequency spectrum density of the backing-cavity in a CBA can be expressed as:

$$\frac{N_r}{\Delta\omega} \approx \frac{S_r}{\pi c^2}\omega. \tag{6.6}$$

The backing-cavity of a CBA is equivalent to an open-ended resonant cavity that supports a two-dimensional resonance on its cross section. The resonant mode is the same as the conventional waveguide resonance. Thus the waveguide mode theory can be employed to analyze the backing-cavity. For a CBA, the field distribution over the aperture can be obtained by utilizing the excitation theory of a waveguide, and then the far-field radiation can be characterized.

6.2.3 Design of the Backing-Cavity

In different CBAs, the backing-cavities may be designed in different shapes, such as rectangular, circular, and elliptical ones. Furthermore, the resonant modes and the electromagnetic field distributions in the backing-cavities will be different because their boundary conditions and excitation sources are different.

For a rectangular backing-cavity, the resonant frequency of the TE_{mnl} or TM_{mnl} mode is given by [7],

$$f_{mnl} = \frac{c}{2\pi\sqrt{\mu_r\varepsilon_r}}\sqrt{\left(\frac{m\pi}{a}\right)^2 + \left(\frac{n\pi}{b}\right)^2 + \left(\frac{l\pi}{d}\right)^2}, \tag{6.7}$$

where a, b, d are the length, width, and height of the resonant backing-cavity, respectively, and the indices m, n, and l refer to the numbers of variations in the standing wave pattern in the corresponding directions, respectively. μ_r and ε_r are the relative permeability and relative permittivity of the material filling the cavity, and c is the velocity of light in vacuum.

For a circular backing-cavity, the resonant frequency of the TE_{mnl} mode is [7],

$$f_{nml} = \frac{c}{2\pi\sqrt{\mu_r \varepsilon_r}} \sqrt{\left(\frac{p'_{nm}}{a}\right)^2 + \left(\frac{l\pi}{d}\right)^2}, \tag{6.8}$$

and the resonant frequency of the TM_{mnl} mode is [7],

$$f_{nml} = \frac{c}{2\pi\sqrt{\mu_r \varepsilon_r}} \sqrt{\left(\frac{p_{nm}}{a}\right)^2 + \left(\frac{l\pi}{d}\right)^2}, \tag{6.9}$$

where p'_{nm} and p_{nm} are the mth root of J'_n and J_n, respectively. The parameters a and d are the radius and height of the cylindrical backing-cavity, and l refers to the number of variations in the standing wave pattern in the axial direction.

6.3 Cavity-Backed SIW Slot Antenna

SIW technology provides a promising scheme for designing CBSAs with the merits of low-profile, low cost, easy integration with planar circuits and convenient fabrication such as PCB or LTCC process [8]. The SIW cavity can be realized using arrays of metallized vias through a single or multilayer substrate. The technology was incorporated in a CBA by Luo et al., in which the nonplanar metallic cavity is replaced by an SIW cavity structure [2]. The SIW CBSA presents unidirectional radiation characteristics with high gain while maintaining its low-profile planar configuration. After that, CBSAs based on SIW cavity have been proposed for microwave and mmW applications [3, 4].

6.3.1 Feeding Techniques

All the SIW CBSAs consist of the feeding part, the SIW backing-cavity, and the radiating slot. The CBSA can be excited by a microstrip line, a grounded coplanar waveguide (GCPW), an SIW, and a coaxial line. Feeding techniques as an important design part associate with the input impedance and radiation characteristics of the antenna.

A CBSA excited by a microstrip line and a GCPW are shown in Figures 6.2a–d, respectively. It can be seen from Figures 6.2a,c that the feed lines are designed on the same metallic layer with the radiating slot, and the whole structure remains planar. The drawback is the radiation from the feed line, which leads to an increased cross-polarization radiation. To overcome this, a microstrip line or a GCPW etched on the ground plane can be utilized as shown in Figures 6.2b,d. This method leads to the backward radiation from the feed line, leading to the poor front-to-back ratio.

The SIW feed arrangements are shown in Figures 6.2e,f. In Figure 6.2e, the feeding SIW is integrated in the same substrate with the CBSA, and the energy is coupled through an inductive window. Another slotted-SIW feeding method feeds the CBSA in a two- or multi-layer structure. In this configuration, the field is coupled from the feeding-SIW to the cavity through a transverse or longitudinal slot cut on the common ground plane, as shown in Figure 6.2f. The coupling slot is usually centered beneath the cavity, bringing to lower cross-polarization due to the symmetry of the configuration. The size and location of the slot decide the amount of energy coupling from the feeding SIW to the CBSA. The coupling slot can be either resonant or non-resonant. Although the resonant slot can provide another resonance to enhance the bandwidth, it is relatively sensitive to

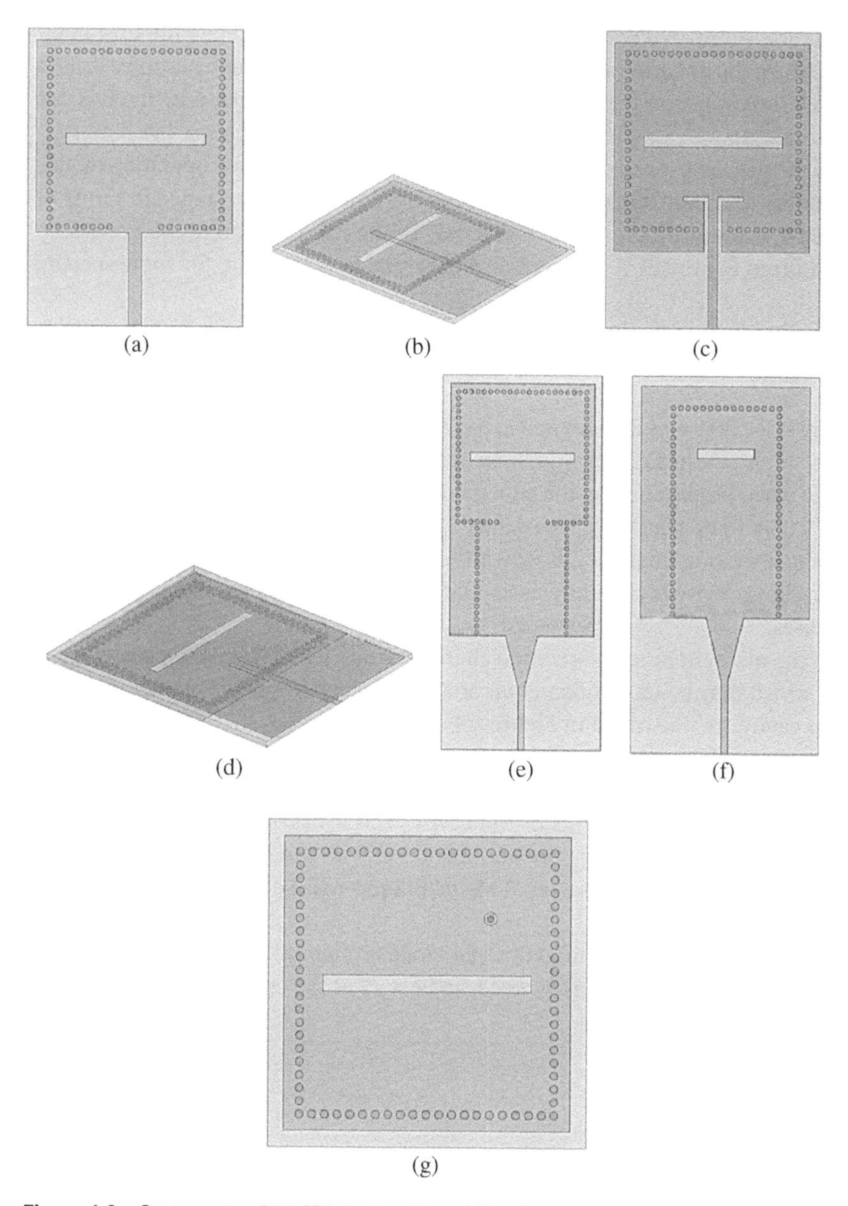

Figure 6.2 Rectangular SIW CBA fed by (a) and (b) microstrip line, (c) and (d) grounded CPW, (e) and (f) SIW, and (g) coaxial line.

assemble errors in the alignment of the different layers. In practical design, a non-resonant slot is normally employed, and the substrate parameters of the multilayers can be chosen separately for optimal CBSA's performance. Compared with the feeding structures shown in Figures 6.2a–d, the SIW feed scheme reduces the radiation loss, which is beneficial to the antenna efficiency, especially at mmW bands.

It should be pointed out that, in Figures 6.2c–f, a 50-Ω microstrip line is attached at the end of the feed structure to facilitate the testing.

The CBSA can also be excited by the coaxial line, and its center conductor and the outer conductor are connected to the top and bottom planes of the cavity, respectively. As shown in Figure 6.2g, the main advantage of this feed is that it can be placed at any desired location inside the SIW cavity to match its input impedance. Compared with other feed methods, it could eliminate the unwanted radiation and help to reduce the antenna size. The disadvantages are that the hole has to be drilled in the substrate and that the connector protrudes outside the bottom ground plane so that it is not completely a planar structure.

The off-center feed schemes are used to excite the SIW cavity with the expected operating modes or ensure the slot a double-resonant behavior and improve the antenna performance in terms of bandwidth and gain. Furthermore, different feed methods, like the coaxial line and the microstrip line, could be jointly utilized to provide a balanced or differential feed structure for some specific demands.

6.3.2 SIW Backing-Cavity

The SIW backing-cavity in a CBSA may be designed in rectangular or circular shape with a standard PCB or LTCC process, as shown in Figures 6.3a,b, respectively. Its metallized via arrays significantly suppress surface wave propagation in the substrate, and the low-profile cavity reduces the total thickness of the CBSA to be nearly $\lambda_0/30 \sim \lambda_0/40$. The cavity size and resonant mode have significant impact on the operating frequency, bandwidth, polarization, and so on. In comparison, the rectangular SIW backing-cavity is generally used, and the circular or elliptical backing-cavity is employed in some cases.

In order to maintain the inherent pure polarization characteristics, the SIW backing-cavity operates at a single mode, which may be the fundamental or adjacent higher mode. Some of resonant modes of a rectangular cavity are illustrated in Figure 6.4.

Hybrid modes or multi-modes can be excited simultaneously in a single SIW cavity, and their resonant frequencies can be adjusted by introducing some perturbations, for example, locating metallic via holes at certain positions of the cavity. The hybrid modes or multi-modes strategy is often used to design a CBSA with dual-band or wideband, dual-polarization or circular polarization, enhanced gain, and so on. This technique will be discussed in detail in Section 6.4.

For a rectangular SIW backing-cavity, the fundamental mode is often utilized in a CBAS design. As shown in Figure 6.4a, the length of the cavity needs to be $\lambda_g/2$ at the operational frequency to

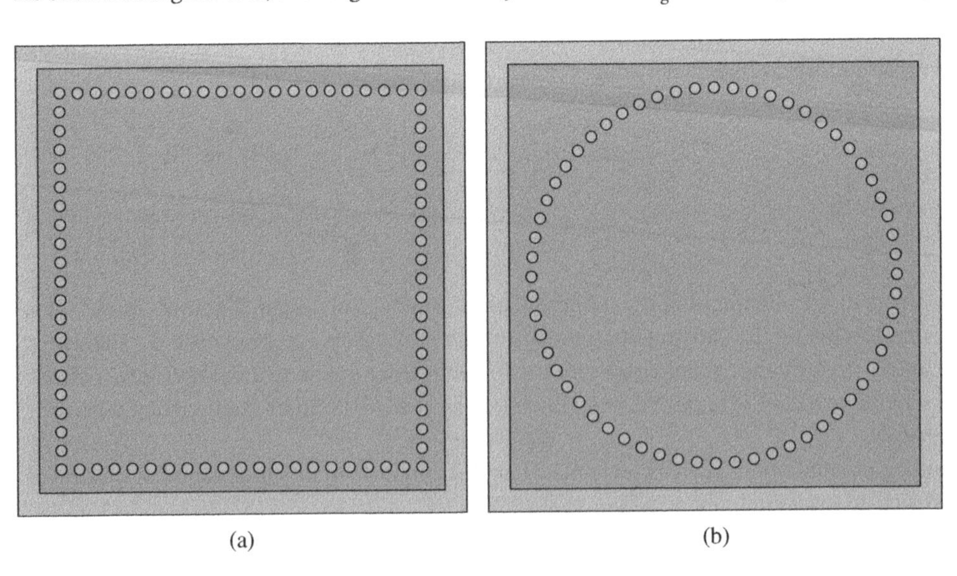

Figure 6.3 (a) Rectangular SIW backing-cavity and (b) circular SIW backing-cavity.

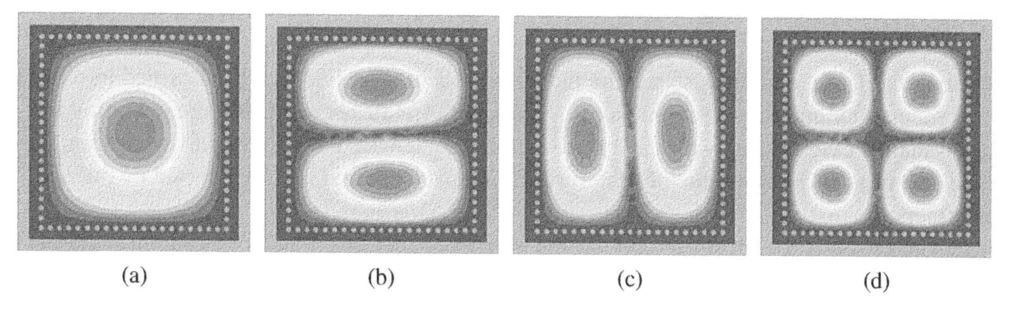

Figure 6.4 Rectangular SIW backing-cavity resonances at a single mode. (a) TE_{110}, (b) TE_{210}, (c) TE_{120}, (d) TE_{220}.

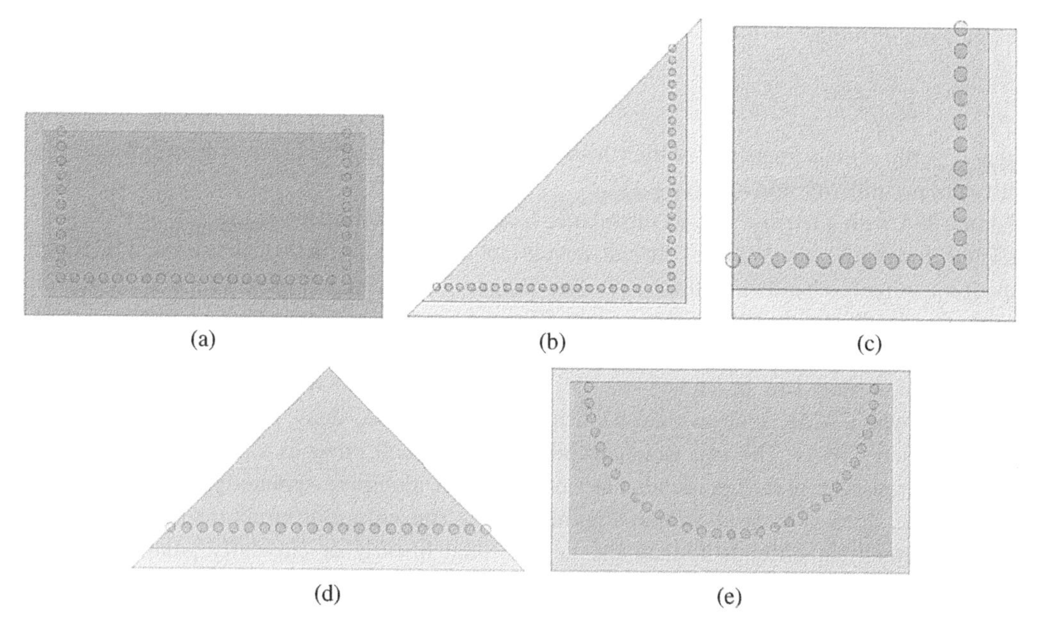

Figure 6.5 Rectangular SIW. (a) HM-cavity I, (b) HM-cavity II, (c) QM-cavity I, (d) QM-cavity II, and (e) circular SIW HM-cavity.

have an efficient radiation, where λ_g is the wavelength in free space divided by the square root of the dielectric constant of the cavity material. The SIW backing-cavity is divided into two halves along its symmetrical axis or diagonal line [9] to form the half-mode substrate integrated waveguide (HMSIW) cavity, shown in Figures 6.5a,b, which is with size reduction up to 50%. Furthermore, quarter-mode substrate integrated waveguide (QMSIW) cavities realized by cutting the SIW cavity along its two symmetrical axes or two diagonals achieve a 75% size reduction, as shown in Figures 6.5c,d. The applications of these miniaturized cavities will also be introduced later in Section 6.4.

In addition, the substrate parameters of the SIW backing-cavity have a certain impact on the performance of a CBSA. On the one hand, a high dielectric constant helps in the reduction of the antenna size, but it results in a narrower bandwidth and lower efficiency. On the other hand, a thick substrate can enhance the antenna operating bandwidth, but it may lead to the matching problem or increase the undesired feed radiation. The selection of the substrate needs to make a trade-off between the parameters and the performance.

6.3.3 Radiating Slot

According to different operating states, the radiating slots in the CBSA can be classified into two types: resonant slot and non-resonant slot [4]. For a CBSA with resonant slot, the operating frequency is mainly determined by the cavity size and the slot length. For a CBSA with non-resonant slot, the operating frequency is mainly determined by cavity size and slightly affected by the radiating slot length.

The resonant slot is the most used one in the SIW CBSA because it can offer a high radiation efficiency. The electric current flow on the ground metallic plane of a CBSA has two components. One is parallel to the slot line, which is creating a resonant condition, while the other is perpendicular to the slot line and responsible for far-field radiation. For a straight narrow slot as shown in Figure 6.6a, the resonance occurs when the slot has a half wavelength at the operating frequency.

For a narrow slot operating at the first resonate mode, its length l_s can be calculated by utilizing the following formula,

$$l_S = \frac{1}{2f_0\sqrt{\mu_r \varepsilon_e}}, \tag{6.10}$$

where f_0 is the central frequency of the CBSA, ε_r is the relative permeability of medium, ε_e is the equivalent permittivity, and $\varepsilon_e = (\varepsilon_r + 1)/2$.

For a CBSA with a narrow slot, its impedance bandwidth is around 3% or less [2]. By enlarging the width-to-length ratio (WLR) of the slot, a wide slot as shown in Figure 6.6b is used, which can provide an enhanced bandwidth up to 10%. An I-shaped slot, shown in Figure 6.6c, is also used for an inductive loading of the SIW backing-cavity. The slot length is approximately a half wavelength at the operating frequency.

Two resonant slots can be etched on the broad wall of the SIW backing-cavity to obtain some other type CBSAs, such as dual-band or wideband CBSA, dual- polarization, and circular polarization CBSAs. The two slots can be parallel to each other as shown in Figure 6.7a, and they can also be placed perpendicular to each other along the symmetry axis or diagonal direction, i.e., cross-slot, as shown in Figures 6.7b,c. The cross-slot structure is used as the radiating element in order to radiate the desired dual linearly or circularly polarized wave. Furthermore, the wide cross-slot can be employed to increase the bandwidth. The slots shown in Figure 6.7 may be designed with the same or different length to achieve dual-band or wideband characteristics.

Some other resonant slots with specific shapes have been implemented in the literature. Figure 6.8 shows the T-shape, H-shape, triangular-shape, spoon-shape, ramp-shape, and

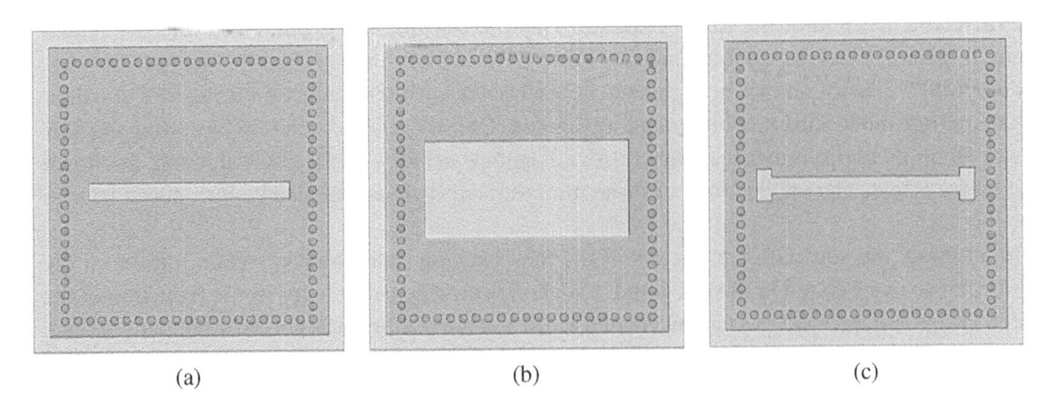

Figure 6.6 (a) Straight slot, (b) wide slot, and (c) I-shaped slot.

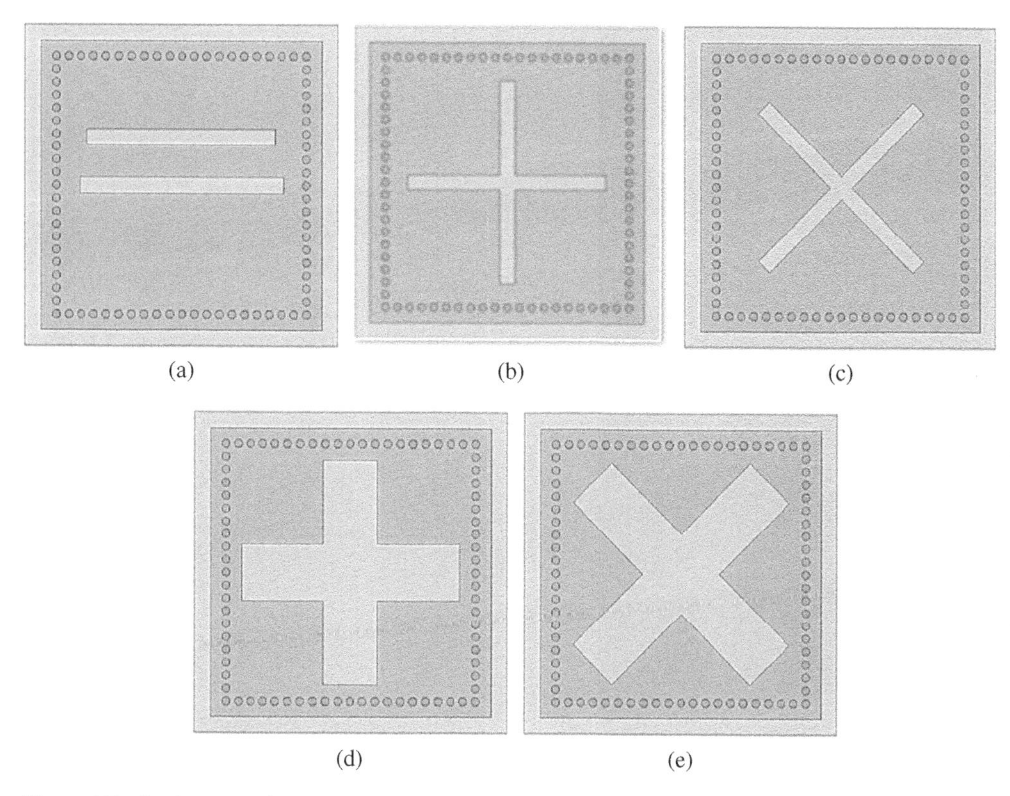

Figure 6.7 Dual-slot configurations. (a) Parallel slot, (b) Greek cross-slot, (c) Andrew's cross-slot, (d) wide Greek cross slot, and (e) wide Andrew's cross slot.

triangular-complementary-split-ring-shape, respectively. These slots can be employed to design wideband CBSA and miniaturized CBSA with multiple resonant modes or negative-order resonance mode.

Figure 6.9 shows some non-resonant slots, including the V-shaped slot, bow-tie-shaped slot, and dumbbell-shaped slot. These structures replace the straight slot with a modified slot, which creates additional hybrid current distribution in the SIW backing-cavity at higher frequency along with conventional mode. Under this condition, the CBSA can be considered as a multi-mode cavity with non-resonant slot, and the length of the modified slot can be much larger than the half-wavelength resonant length of a conventional slot antenna.

6.4 Types of SIW CBSAs

6.4.1 Wideband CBSAs

Due to the high-quality factor and single-resonance response, a conventional SIW CBSA suffers from a narrow bandwidth of about 1.7% [2], which limits their applications in broadband communication systems. To design a wideband SIW CBSA, there are five strategies that can be employed, they are: (i) substrate removal; (ii) dual- or multiple resonance of radiating slot; (iii) dual- or multiple resonant modes operating; (iv) wide slot; and (v) multilayer structure.

Substrate removal will decrease the Q of the slot and the backing cavity, and a greater fractionally bandwidth can be achieved. In [10], by partially removing the substrate under or next to the slot,

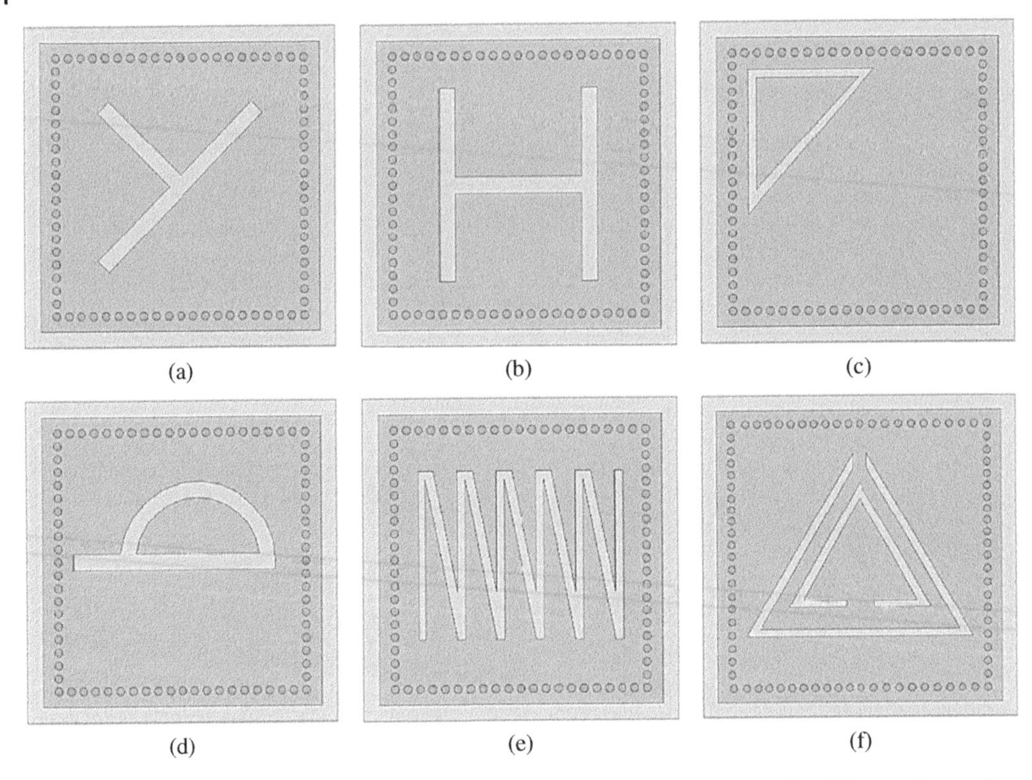

Figure 6.8 (a) T-shape slot, (b) H-shape slot, (c) triangular slot, (d) spoon-shape slot, (e) ramp-slot, and (f) triangular-complimentary-split-ring-slot (TCSRS).

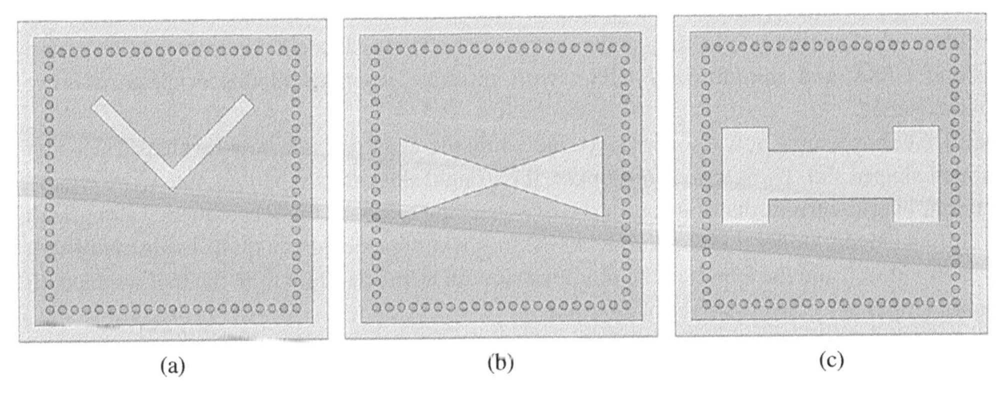

Figure 6.9 Non-resonant slot configuration. (a) V-shaped slot, (b) bow-tie-shaped slot, and (c) dumbbell-shaped slot.

the bandwidth was increased up to 2.16%. With a highly efficient impulse-radio ultra-wideband CBSA with air-filled SIW technology [11], an ultra-wide bandwidth of 29.4% has been achieved.

A radiating slot with a half wavelength resonates at multiple frequencies employing off-center feed or via-holes. In [12], a via-hole is placed above the slot to modify its electrical length and thus create an additional resonance at a higher frequency to improve the bandwidth by shortening the effective length of the slot. Its bandwidth reached up to 3.3%, and a 60% wider bandwidth has been achieved. A differentially fed CBSA with crossed slot has been discussed in [13], in which an impedance bandwidth of 19% has been achieved. Based on multimode resonance operation of SIW cavity and the multi-resonant character of radiating slot, a wideband dual-mode SIW cavity-backed triangular complementary split ring slot (TCSRS) antenna is designed and implemented at 28 and 45 GHz for 5G of wireless communications [14], it achieves a 16.67% impedance bandwidth at 28 GHz and 22.2% impedance bandwidth at 45 GHz.

Exciting the SIW backing-cavity with multiple modes is another method to widen the bandwidth of the CBSAs. This hybrid-mode method is suitable for both a resonant and non-resonant radiating slot. In [15], by dividing an SIW cavity into two halves with a non-resonant rectangular slot, two hybrid cavity modes (TE_{110} and TE_{120}) are generated, and the fractional bandwidth is improved to 6.3%. Compared with the rectangular slot, the bow-tie-shaped slot disturbs the current path of the higher-order mode and introduces a strong loading effect in the cavity, and as a result, the higher-order mode (TE_{120}) interacts with the dominant mode (TE_{110}) to generate two close hybrid modes to get a broadband response of 9.4% [16]. Similarly, high-order resonant modes (TE_{102}, TE_{301}, TE_{302}) inside the sub-cavity are excited with a modified dumbbell-shaped slot to achieve an impedance bandwidth of 26.7% [17]. In [18], shorting vias are introduced in the SIW cavity to realize triple- and quad-resonance CBSAs, with achieved bandwidths of 15.2% and 17.5%, respectively. Based on a similar principle, by loading the SIW cavity with unbalanced shorting vias, the quad- and penta-resonance CBSAs with cross-shaped slot in [19] achieve bandwidths of 20.0% and 20.8%, respectively. In [20], a coupled half-mode/quarter-mode SIW CBSA got a fractional impedance bandwidth of 11.7%. In [21], three high-order cavity modes (TE_{130}, TE_{310}, and TE_{330}) along with the slot mode led to a bandwidth over 26%. By implementing multiple slot elements [22, 23], additional resonance is generated, and the bandwidth could be increased to 15%.

A wide slot CBSA exhibits a wider impedance bandwidth along with enhanced gain and efficiency. The study of the slot WLR on its impedance bandwidth has been done in [24], it was found that the bandwidth increases from 3% to 11.6% with the slot WLR from 0.12 to 0.71. In [25], the bandwidth with high WLR has been further improved by introducing the rectangular patch at the center of the cavity, and the achievable bandwidth goes up to 15%. Reference [13] takes advantage of an off-center microstrip-fed wide slot to enhance the bandwidth to 19% with the aid of a thin covered dielectric substrate.

The multilayer structure with the coupled feeding element introduced in [26–28] is also conducive to widen the impedance bandwidth. In [26], the metamaterial-mushroom structure and the coupled feeding structure enable the bow-tie slot antenna to achieve an impedance bandwidth of 21.8%. In [28], four gradually widened SIW cavities are excited by the slot-coupled feeding structure to achieve an impedance bandwidth up to 30%. In [29], and by adding a multilayer dielectric cover, the bandwidth significantly increases to 40%.

6.4.2 Dual-Band CBSAs

Some design methods for wideband CBSA can be utilized to realize dual- or multi-band CBSA. Distinct from conventional antennas with a single radiating slot, some dual-band CBSAs are realized with dual-slot or specific slot shapes [30–37]. In [30], a dual-band dual-polarized CBSA has been proposed with cross-slot structure. In [31], the dual-band operation is achieved by simultaneously exciting the two hybrid modes (the mode of the slot and the mode of the patch inside the slot) of the triangular-ring slot, and the CP antenna achieves the overlapped bandwidths (of AR, impedance, and gain) wider than 8% and 5% at 9.5 GHz and 12 GHz, respectively. An SIW cavity-backed dumbbell-shaped slot antenna with dual-frequency bands presented bandwidths of lower and upper frequency bands at 9.5 and 13.85 GHz are about 2.0% and 1.46%, respectively [32]. In [33], a dual-band and polarization-flexible CBSA is proposed with bandwidths of 1.1% and 2% with respect to frequencies of 13.4 GHz and 17.9 GHz, respectively. The dual-band operation is achieved by simultaneously exciting the first high-order hybrid mode (superposition of TM_{310} and TM_{130}) and the second high-order mode (TM_{320}) in the SIW backing-cavity [34], and the 10-dB return loss bandwidths at resonant frequencies of 21 and 26 GHz are 3.7% and 2.6%, respectively. In [36], transverse dual-slot on SIW cavity is utilized as radiating elements, two passbands at 3.28 and 3.77 GHz were achieved with two radiation nulls.

6.4.3 Dual-Polarized and Circularly Polarized CBSAs

CBSAs with DP and CP radiation have been investigated. Single fed cross-slot is used in the SIW cavities to develop DP or CP radiation [30], and the achieved fractional axial ratio bandwidth is about 0.8%, which is much less than the fractional impedance bandwidth. In [38], the CP wave is generated at X-band by an annular slot with a shorting pin, and an impedance bandwidth of 18.74% and a CP bandwidth of 2.3% are obtained under the condition of VSWR 2:1 and axial ratio of less than 3 dB, respectively. A wideband dual-polarized CBSA was realized by differentially feeding a wide cross-shaped slot in [13]. With a dual-mode triangular-ring slot, a dual-band CP CBSA was achieved with the overlapped bandwidths about 8% and 5% at 9.5 and 12 GHz, respectively [14]. A semicircle slot CBSA based on the HMSIW cavity was introduced to realize CP radiation at 28 GHz in [39], showing a fractional AR bandwidth of 7.7% (26.60–28.55 GHz). An inset microstrip line is used to excite two orthogonal quarter-wave length patch modes with required phase difference for generating CP radiation in [40], and the fractional bandwidth for right-hand circular polarization (RHCP) antenna and left-hand circular polarization (LHCP) antenna is 1.74% (8.65–8.8 GHz) and 0.66% (8.63–8.688 GHz), respectively. In [41], two slots were placed perpendicular to each other for circular polarizations with 3-dB AR bandwidth of 14 MHz (2.421–2.435 GHz). By employing two annular exponential slots carved on the surface of a circular SIW cavity with single feed, dual-band CP radiation is achieved for mmW applications at 37.5 and 47.8 GHz in [42]. The vertically linear polarization (VLP) and horizontally linear polarization (HLP) are realized by adopting the TE_{430} and TE_{340} modes with different signal schemes in [43], and a polarization-diverse CBSA array provided a 3-dB AR bandwidth for the RHCP from 10.75 to 10.83 GHz. Besides, a differential dual-polarized SIW CBSA is designed in [44] with cross-slot by exciting two orthogonal degenerate modes (diagonal TE_{120} and TE_{210}). In [45], bow-tie slots are used to excite closely placed hybrid modes to implement dual-band dual-polarization CBSAs in X-band.

6.4.4 Miniaturized CBSAs

For an SIW CBSA, the length of the conventional cavity is about a half-guided wavelength, which is determined by the permittivity of the dielectric inside the cavity and lateral cavity length. Therefore, a dielectric with high permittivity is desired for the miniaturization of the conventional cavity. However, the use of the high-permittivity material results in narrow bandwidth and low efficiency. The following two schemes are employed to realize a miniaturized CBSA: (i) replacing the SIW cavity with the folded-SIW cavity, namely, HMSIW or QMSIW cavity; and (ii) exciting negative-order resonance mode in an SIW cavity.

A folded-SIW cavity structure for the miniaturized CBSA in [41] achieves a 72.8% size reduction. In [40, 46], compact CBSAs are realized using an HMSIW technique. The specific V-, T-, and cross-shaped radiating slot divide the SIW mode into HMSIW or QMSIW mode to get a miniaturized structure, and they have been used to design compact self-triplexing and self-quadruplexing SIW CBSAs [47–49]. On the other hand, the negative-order resonances are used to reduce the dimensions of the cavity [50–52]. As the negative-order resonances occur at lower frequency than positive resonance, cavity size reduction can be achieved. In [50], a miniaturized composite left/ right-handed (CRLH) SIW CBA is developed by etching an interdigital slot on the broad wall of an SIW cavity. By employing the first negative-order resonance mode in a compact SIW cavity resonator, an ultra-miniature SIW CBSA is realized with 87% miniaturization for 2.1 GHz applications [51].

6.5 CBSA Design Examples at 60 GHz

As an example, a CBSA is designed at 60 GHz and the corresponding 2×4 array is implemented using single layer PCB process for mmW applications [24]. In the example, the backing-cavity is not only a reflector for the radiating slot but also a radiating element for the whole antenna. The impedance bandwidth, radiation patterns and gain of the antenna are analyzed against the WLR of the radiating slot.

6.5.1 SIW CBSA with Different Slot WLR

Figure 6.10 depicts the geometry of the SIW CBSAs with different slot WLR, similar to the one shown in Figure 6.10a. A transverse slot with various WLR (W_s/L_S) = 0.12, 0.4, 0.71, respectively, is etched onto the broad wall of an SIW cavity with the size of $W_s \times d_1$ and fed by the SIW with a width of W_{SIW}. An inductive window is employed to couple the energy. d_0 is the distance from the center of the slot to the short-circuited end. D_1 and P are the diameter of the via-hole and the spacing between adjacent via-holes, respectively. The slotted-cavity and the feeding SIW are formed with the conditions $D_1/P \geq 0.5$ and $D_1/\lambda_0 \leq 0.1$ [52].

The slot is with the overall length of approximate a half operating wavelength, and the slotted backing-cavity resonates at its dominant mode frequency [53]. Their initial dimensions can be obtained from Eqs. (6.7) and (6.10). The antenna is designed onto a piece of a 0.635 mm thick RO3006 substrate with a dielectric constant of $\varepsilon_r = 6.15$ and loss tangent of $\tan\delta = 0.0025$.

The CBSA exhibits a dual-resonant characteristic, and its operating bandwidth can be enhanced by increasing the slot WLR. Figure 6.11 shows the simulated return losses of the antennas with WLRs of 0.2, 0.4, 0.6, and 0.71, respectively. When the WLR varies from 0.2 to 0.71, the impedance matching with two resonances is maintained by tuning the parameters including L_S, d_1, W_{siw1}, d_0,

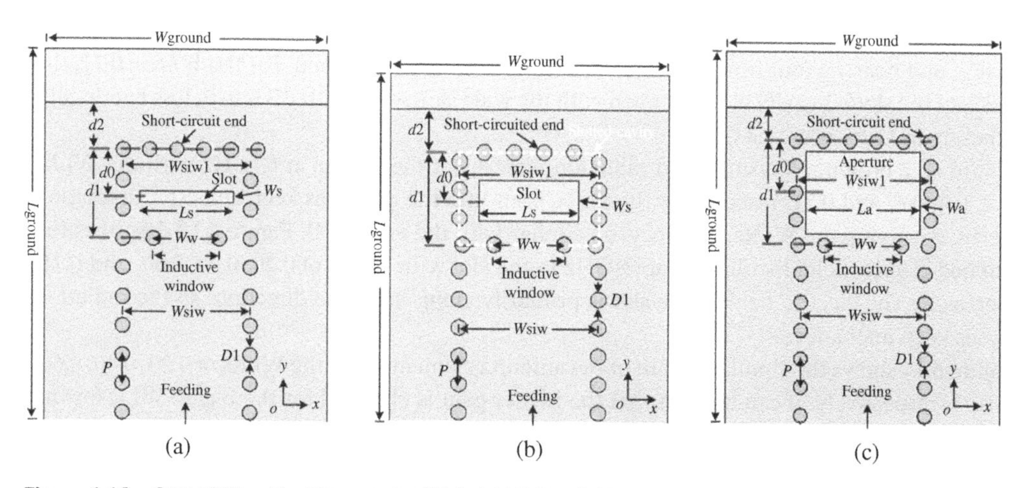

Figure 6.10 SIW CBSA with different slot WLR. (a) WLR = 0.12, (b) WLR = 0.4, and (c) WLR = 0.71.

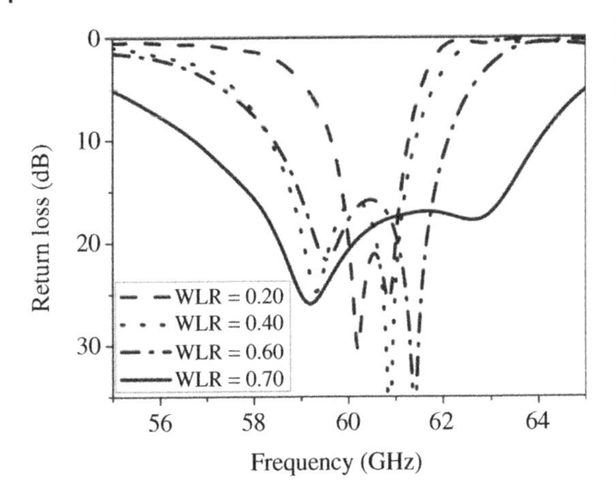

Figure 6.11 Simulated return loss of the SIW cavity-backed slot antennas with different slot WLRs.

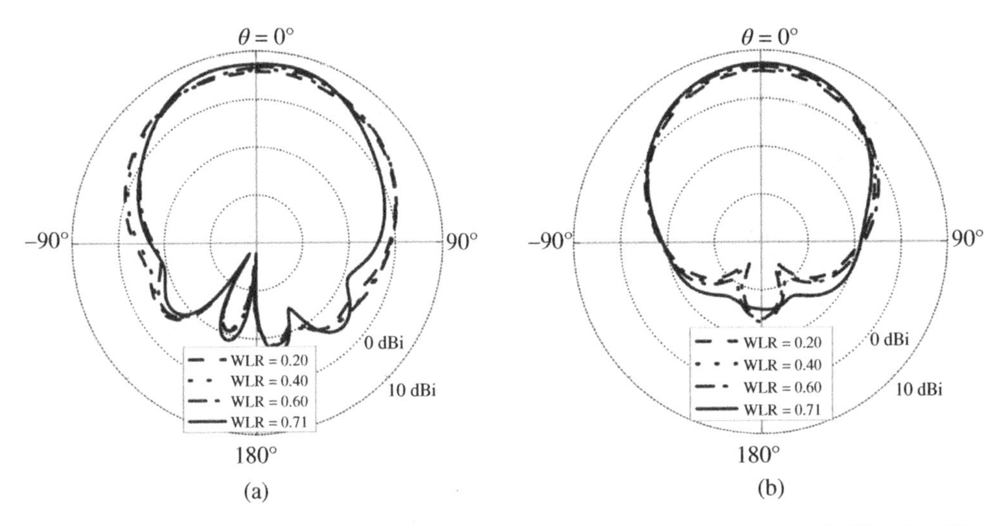

Figure 6.12 Simulated radiation patterns of the SIW cavity-backed slot antennas at 60 GHz with different WLRs in (a) *E*-plane (*yz*-plane) and (b) *H*-plane (*xz*-plane).

and W_w, and nearly a four times enhancement in bandwidth is obtained. For the $WLR = 0.71$, the top side of the slotted-cavity is nearly open with the wide slot, and the 10-dB return loss bandwidth reaches up to 11.6% or 57–64 GHz.

Figure 6.12 presents the simulated radiation patterns of the CBSAs at 60 GHz with the WLRs of 0.2, 0.4, 0.6, and 0.71, respectively. It can be seen that the antennas keep consistent radiation patterns as varying the WLRs, and they are similar with the one in [2]. Figure 6.13 describes the simulated electric field distribution at 60 GHz in the slot with WLRs of 0.20, 0.40, 0.60, and 0.71, respectively. The electric field in the slot is primarily along the same direction, so the radiation patterns keep unchanged.

Figure 6.14 shows the simulated gain of the antenna elements with the WLRs of 0.20, 0.40, 0.60, and 0.71, respectively. It can be seen that the higher gain is obtained for the bigger WLR for the same ground plane size of 10 mm × 6 mm. The simulation results show that the maximum radiation efficiency of the antenna elements increase from 76% to 94% when the slot WLR varies from 0.20 to 0.71.

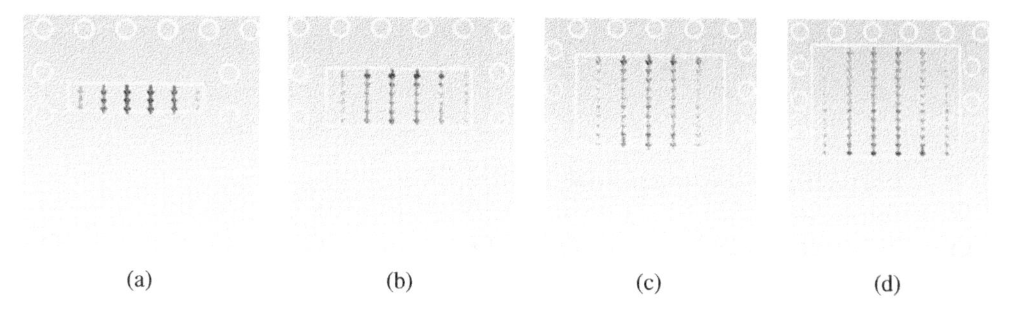

Figure 6.13 Simulated electric field distribution at 60 GHz in the slots of the SIW cavity-backed slot antennas with the WLR of (a) 0.20, (b) 0.40, (c) 0.60, and (d) 0.71.

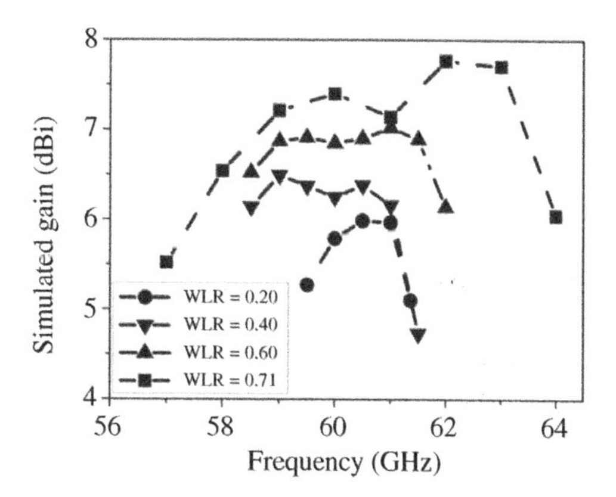

Figure 6.14 Simulated gain of the SIW cavity-backed slot antennas with different slot WLRs.

6.5.2 Array Examples with Different WLRs of Slot

Three SIW CBSA arrays with the different WLRs are exemplified on the RO3006 substrate at the 60 GHz band. The geometry of these 2 × 4 antenna arrays is shown in Figure 6.15, consisting of eight elements and a compact eight-way tree-shape power divider, in which the feeding SIW is modified with the inductive via-hole. The power divider splits the power into eight ways equally and excites all the elements in phase. The width of the SIW divider is equal to the feeding SIW for each element, and the divider shares part of via-holes with the elements to keep the antenna arrays more compact. Taking into account the mutual coupling between the elements, the parameters of the arrays should be optimized. The antenna arrays are excited by a WR-15 waveguide. For measurement convenience, a stepped waveguide-to-SIW transition similar to that in [54] is employed.

Figure 6.16 shows the photographs of the prototypes with the WLRs of 0.12, 0.5, and 0.71, respectively. For the array with the WLR of 0.12, the distances between the adjacent elements are $L_x = 3.3$ mm ($0.66\lambda_0$) and $L_y = 3.15$ mm ($0.63\lambda_0$) along the x- and y-axis, respectively, where λ_0 is the wavelength at 60 GHz in free space. For the arrays of WLR = 0.5 and 0.71, the distances are $L_x = 4$ mm ($0.8\lambda_0$), $L_y = 3.15$ mm ($0.63\lambda_0$), and $L_x = 3.45$ mm ($0.69\lambda_0$), $L_y = 3.37$ mm ($0.674\lambda_0$), respectively. The other geometrical parameters of the fabricated arrays are listed in Table 6.1.

Figure 6.17 shows the simulated and measured return loss of the antenna arrays with the waveguide-to-SIW transition. The measured return losses follow the trend of the simulated ones well with a slight frequency shifting, which is caused by the fabrication tolerance and the permittivity fluctuation. For the arrays of WLR = 0.12 and 0.50, the measured impedance bandwidths for 10-dB return loss are approximately 3.3% and 6.5%, respectively, which are consistent with the simulations. For the arrays of WLR = 0.71, the simulated impedance bandwidth for 10-dB return

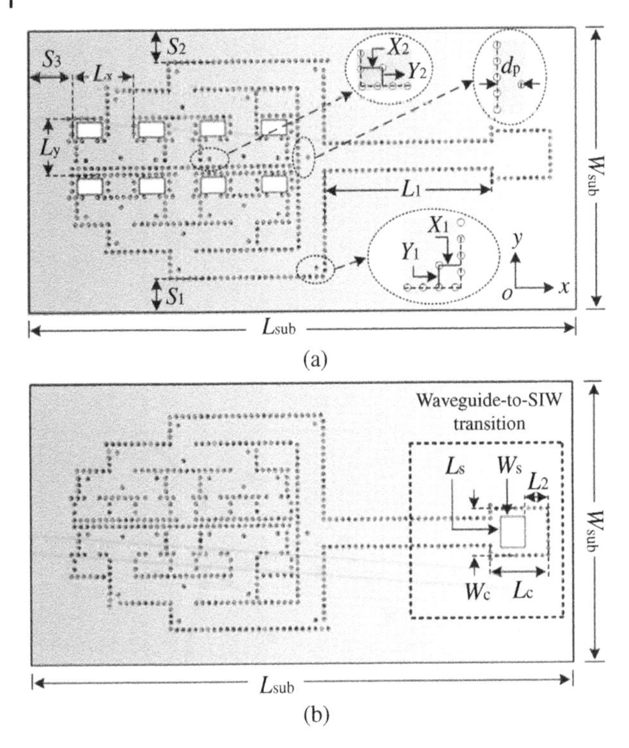

Figure 6.15 Geometry of the 2 × 4 antenna array prototypes. (a) Top view and (b) bottom view.

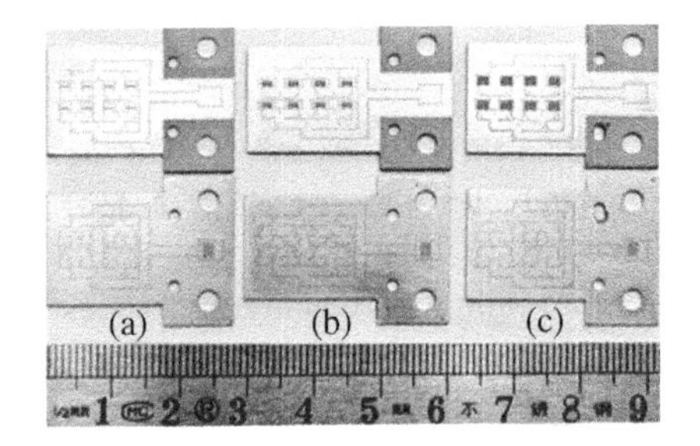

Figure 6.16 Photograph of the fabricated 2 × 4 arrays. (a) WLR = 0.12, (b) WLR = 0.5 and (c) WLR = 0.71.

Table 6.1 Geometrical parameters of the arrays.

Dimensions (mm)		
$X_1 = 0.45$	$Y_2 = 0.4$	$S_2 = 1.8$
$Y_1 = 0.45$	$d_p = 0.53$	$S_3 = 1.8$
$X_2 = 0.5$	$S_1 = 1.8$	$L_1 = 8$

loss reaches 11.6%, whereas the measured return loss exhibits a slight deterioration. It degrades to 9 and 8.2 dB around 57.3 and 59.2 GHz, respectively. The array of WLR = 0.71 is more sensitive to the fabrication tolerance than those of WLR = 0.12 and WLR = 0.50, because the slot/aperture with a bigger WLR is closer to the inductive window and the via-holes. In practical applications, the antenna can be directly integrated into circuits, and it can be expected that the return loss of the antenna without transition would be improved.

Figure 6.18 shows the simulated and measured normalized radiation patterns of the arrays with the slot WLRs of 0.12, 0.50, and 0.71, respectively, and it can be seen that they exhibit similar

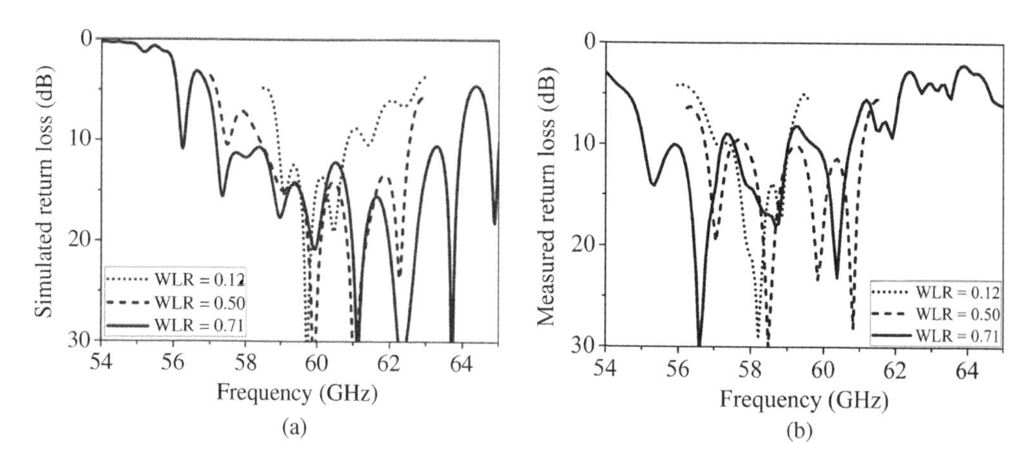

Figure 6.17 (a) Simulated return loss and (b) the measured return loss of the arrays with the waveguide-to-SIW transition.

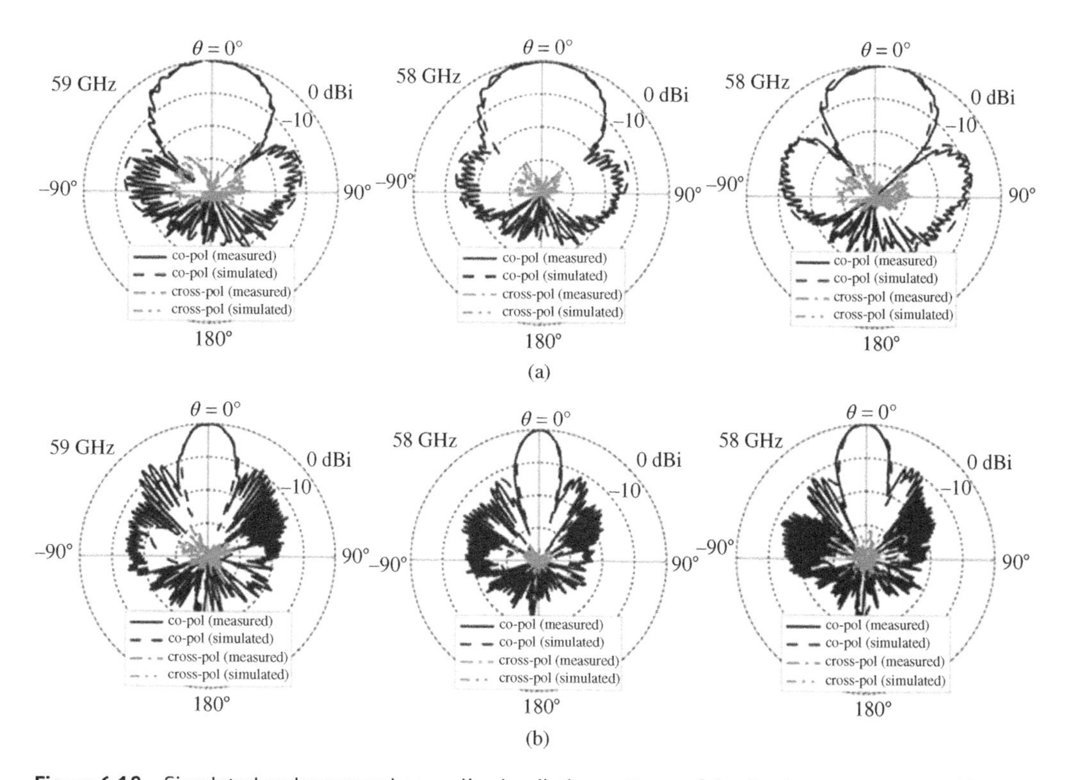

Figure 6.18 Simulated and measured normalized radiation patterns of the 2 × 4 antenna arrays with different WLRs in band. (a) *E*-plane and (b) *H*-plane. (from left to right: WLR = 0.12, 0.50 and 0.71, respectively).

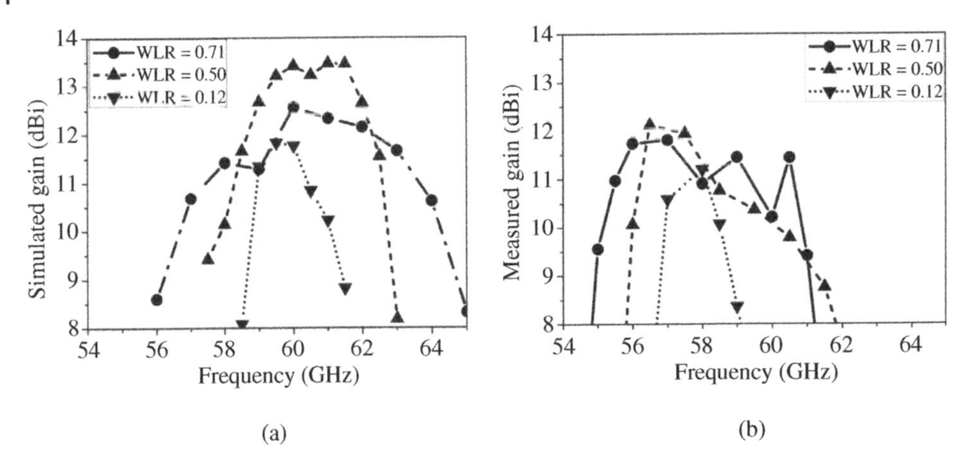

Figure 6.19 (a) Simulated gain and (b) measured gain of the arrays with different WLRs.

radiation patterns. The 3-dB beamwidths in the E- and H-plane are about 48° and 18°, respectively. Furthermore, low cross-polarization levels are observed in the both the E- and H-planes.

The measured and simulated gains of these arrays are compared and shown in Figure 6.19. The 1–2-dB drop in the measured gains is caused by the dielectric loss of the substrate. For the three arrays, the fluctuation of the measured gains is less than 2 dB over their operating bandwidth. The radiation efficiency reached up to 73% for the 2×4 antenna array of WLR = 0.71 at V-band.

6.6 Summary

This chapter has discussed the SIW technique in a planar configuration to realize a low-profile CBSA. First, the operating principle of a CBA has been briefly introduced. Second, SIW CBSA has been described in terms of the feeding techniques, the SIW backing-cavity, and the radiating slot, respectively. The feeding methods for the SIW CBSA, including microstrip line, GCPW, SIW, and coaxial line, have been addressed along with the discussion of their advantages and disadvantages. Single-, dual-, and multi-mode SIW cavities, HMSIW cavities, and QMSIW cavities have been introduced for all kinds of SIW CBSAs. The cavity size and resonant mode have significantly affected the performance of the antenna. The radiating slots in SIW CBSAs have been classified as resonant and non-resonant slots. The resonant slot has relatively yielded higher radiation efficiency. Single narrow or wide straight slot, parallel dual-slot, perpendicular cross-slot have been widely used as the resonant slots. Besides, the I-, T-, H-, triangle-, spoon-shaped slots, and TCSRS, have been utilized for resonant slots. The non-resonant slots have been usually designed in V-, bow-, and dumbbell-shapes. Then, wideband, dual-band, dual- and circular-polarization, and miniaturized SIW CBSAs, have been discussed as design examples associated with their design methods.

References

1 Galejs, J. (1969, ch. 7). *Antennas in Inhomogeneous Media*, 1e, 84–103. New York: Pergamon.

2 Luo, G.Q., Hu, Z.F., Dong, L.X., and Sun, L.L. (2008). Planar slot antenna backed by substrate integrated waveguide cavity. *IEEE Antennas Wirel. Propag. Lett.* 7: 235–239.

3 Bozzi, M., Georgiadis, A., and Wu, K. (2011). Planar slot an circuits and antennas. *IET Microw. Antennas Propag.* 5 (8): 909–920.

4 Luo, G.Q., Wang, T.Y., and Zhang, X.H. (2013). Review of low profile substrate integrated waveguide cavity backed antennas. *Int. J. Antennas Propag.*: 746920.
5 Kumar, A. and Hristov, H.D. (1989). *Microwave Cavity Antennas*, 11–13. Norwood, MA: Artech House.
6 Vainshtein, L.A. (1966). *Open Resonators and Waveguide*. Moscow: Sovjetskoe Radio Publishing.
7 Pozar, D.M. (1998). *Microwave Engineering*, 2e, 313–322. New York: Wiley.
8 Deslandes, D. and Wu, K. (2001). Integrated microstrip and rectangular waveguide in planar form. *IEEE Microw. Compon. Lett.* 11 (2): 68–70.
9 Lai, Q., Fumeaux, C.H., Hong, W., and Vahldieck, R. (2009). Characterization of the propagation properties of the half-mode substrate integrated waveguide. *IEEE Trans. Microw. Theory Tech.* 57 (8): 1996–2004.
10 Yun, S., Kim, D.Y., and Nam, S. (2012). Bandwidth and efficiency enhancement of cavity-backed slot antenna using a substrate removal. *IEEE Antennas Wirel. Propag. Lett.* 11: 1458–1461.
11 Brande, Q.V., Lemey, S., Vanfleteren, J., and Rogier, H. (2018). Highly efficient impulse-radio ultra-wideband cavity-backed slot antenna in stacked air-filled substrate integrated waveguide technology. *IEEE Trans. Antennas Propag.* 66 (5): 2199–2209.
12 Yun, S., Kim, D.Y., and Nam, S. (2012). Bandwidth enhancement of cavity-backed slot antenna using a via-hole above the slot. *IEEE Antennas Wirel. Propag. Lett.* 11: 1092–1095.
13 Paryani, R.C., Wahid, P.F., and Behdad, N. (2010). A wideband, dual-polarized substrate-integrated cavity backed slot antenna. *IEEE Antennas Wirel. Propag. Lett.* 9: 645–648.
14 Choubey, P., Hong, W., Hao, Z.C. et al. (2016). A wideband dual-mode SIW cavity-backed triangular-complimentary split-ring-slot (TCSRS) antenna. *IEEE Trans. Antennas Propag.* 64 (6): 2541–2545.
15 Luo, G.Q., Hu, Z.F., Li, W.J. et al. (2012). Bandwidth-enhanced low-profile cavity-backed slot antenna by using hybrid SIW cavity modes. *IEEE Trans. Antennas Propag.* 60 (4): 1698–1704.
16 Mukherjee, S., Biswas, A., and Srivastava, K.V. (2014). Broadband substrate integrated waveguide cavity-backed bow-tie slot antenna. *IEEE Antennas Wirel. Propag. Lett.* 13: 1152–1155.
17 Cheng, T., Jiang, W., Gong, S., and Yu, Y. (2019). Broadband SIW cavity-backed modified dumbbell-shaped slot antenna. *IEEE Antennas Wirel. Propag. Lett.* 18 (54): 936–940.
18 Shi, Y., Liu, J., and Long, Y. (2017). Wideband triple- and quad-resonance substrate integrated waveguide cavity-backed slot antennas with shorting vias. *IEEE Trans. Antennas Propag.* 65 (11): 5768–5775.
19 Wu, Q., Yin, J., Yu, C. et al. (2019). Broadband planar SIW cavity-backed slot antennas aided by unbalanced shorting vias. *IEEE Antennas Wirel. Propag. Lett.* 18 (2): 363–376.
20 Deckmyn, T., Reniers, A., and Seolders, A.C. (2017). A novel 60 GHz wideband coupled half-mode/quarter-mode substrate integrated waveguide antenna. *IEEE Trans. Antennas Propag.* 65 (12): 6915–6926.
21 Han, W., Yang, F., Ouyang, J., and Yang, P. (2015). Low-cost wideband and high-gain slotted cavity antenna using high-order modes for millimeter-wave application. *IEEE Trans. Antennas Propag.* 63 (11): 4624–4631.
22 Varnoosfaderani, M.V., Lu, J., and Zhu, B. (2014). Matching slot role in bandwidth enhancement of SIW cavity-backed slot antenna. In: *Proc. 3rd Asia-Pacific Conf. Antennas Propag.*, 244–247.
23 Chen, Z. and Shen, Z. (2014). A compact cavity-backed endfire slot antenna. *IEEE Antennas Wirel. Propag. Lett.* 13: 281–284.
24 Gong, K., Chen, Z.N., Qing, X. et al. (2012). Substrate integrated waveguide cavity-backed wide slot antenna for 60-GHz bands. *IEEE Trans. Antennas Propag.* 60 (12): 6023–6026.

25 Yang, T.Y., Hong, W., and Zhang, Y. (2014). Wideband millimeter-wave substrate integrated waveguide cavity-backed rectangular patch antenna. *IEEE Antennas Wirel. Propag. Lett.* 13: 205–208.

26 Kang, H. and Park, S.O. (2016). Mushroom meta-material based substrate integrated waveguide cavity backed slot antenna with broadband and reduced back radiation. *IET Microw. Antennas Propag.* 10 (14): 1598–1603.

27 Chen, Z., Liu, H., Yu, J., and Chen, X. (2018). High gain, broadband and dual-polarized substrate integrated waveguide cavity-backed slot antenna array for 60 GHz band. *IEEE Access* 6: 31012–31022.

28 Cai, Y., Zhang, Y., Ding, C., and Qian, Z. (2017). A wideband multilayer substrate integrated waveguide cavity-backed slot antenna array. *IEEE Trans. Antennas Propag.* 65 (7): 3465–3473.

29 Yang, H., Lu, J., Lin, C. et al. (2016). Design of wideband cavity backed slot antenna with multilayer dielectric cover. *IEEE Antennas Wirel. Propag. Lett.* 15: 861–864.

30 Luo, G.Q., Hu, Z.F., Liang, Y. et al. (2009). Development of low profile cavity backed crossed slot antennas for planar integration. *IEEE Trans. Antennas Propag* 57 (10): 2972–2979.

31 Zhang, T., Hong, W., Zhang, Y., and Wu, K. (2014). Design and analysis of SIW cavity backed dual-band antennas with a dual-mode triangular-ring slot. *IEEE Trans. Antennas Propag.* 62 (10): 5007–5016.

32 Mukherjee, S., Biswas, A., and Srivastava, K.V. (2015). Substrate integrated waveguide cavity-backed dumbbell-shaped slot antenna for dual frequency applications. *IEEE Antennas Wirel. Propag. Lett.* 14: 1314–1317.

33 Lee, H., Sung, Y., Wu, C.T.M., and Itoh, T. (2016). Dual-band and polarization flexible cavity antenna based on substrate integrated waveguide. *IEEE Antennas Wirel. Propag. Lett.* 15: 488–491.

34 Li, W., Xu, K.D., Tang, X. et al. (2017). Substrate integrated waveguide cavity-backed slot array antenna using high-order radiation modes for dual-band applications in *K*-band. *IEEE Trans. Antennas Propag.* 65 (9): 4556–4565.

35 Mukherjee, S., Biswas, A., and Srivastava, K.V. (2015). Substrate integrated waveguide cavity backed slot antenna with parasitic slots for dual-frequency and broadband application. *Proc. 9th Eur. Conf. Antennas Propag.*: 7–11.

36 Niu, B.J., Tan, J.-H., and He, C.-L. (2018). SIW cavity-backed dual-band antenna with good stopband characteristics. *Electron. Lett.* 54 (22): 1259–1260.

37 Mukherjee, S. and Biswas, A. (2016). Design of dual band and dual-polarized SIW cavity backed bow-tie slot antennas. *IET Microw. Antennas Propag.* 10 (9): 1002–1009.

38 Kim, D., Lee, J.W., Cho, C.S., and Lee, T.K. (2009). X-band circular ring-slot antenna embedded in single-layer SIW for circular polarization. *Electron. Lett.* 45 (13): 668–669.

39 Wu, Q., Wang, H., Yu, C., and Hong, W. (2016). Low-profile circularly polarized cavity-backed antennas using SIW techniques. *IEEE Trans. Antennas Propag.* 64 (7): 2832–2839.

40 Razavi, S.A. and Neshati, M.H. (2013). Development of a low-profile circularly polarized cavity-backed antenna using HMSIW technique. *IEEE Trans. Antennas Propag.* 61 (3): 1041–1047.

41 Yun, S., Kim, D.Y., and Nam, S. (2015). Folded cavity-backed crossed-slot antenna. *IEEE Antennas Wirel. Propag. Lett.* 14: 36–39.

42 Wu, Q., Yin, J., Yu, C. et al. (2017). Low-profile millimeter-wave SIW cavity-backed dual-band circularly polarized antenna. *IEEE Trans. Antennas Propag.* 65 (12): 7310–7315.

43 Li, W., Tang, X., and Yang, Y. (2019). Design and implementation of SIW cavity-backed dual-polarization antenna array with dual high-order modes. *IEEE Trans. Antennas Propag.* 67 (7): 4889–4894.

44 Srivastava, G. and Mohan, A. (2019). A differential dual-polarized SIW cavity-backed slot antenna. *IEEE Trans. Antennas Propag.* 67 (5): 3450–3454.

45 Mukherjee, S. and Biswas, A. (2016). Design of dual band and dual-polarised dual band SIW cavity backed bow-tie slot antennas. *IET Microw. Antennas Propag.* 10 (9): 1002–1009.

46 Caytan, O., Lemey, S., Agneessens, S. et al. (2016). Half-mode substrate-integrated-waveguide cavity-backed slot antenna on cork substrate. *IEEE Antennas Wirel. Propag. Lett.* 15: 162–165.

47 Kumar, A. and Raghavan, S. (2018). A self-triplexing SIW cavity-backed slot antenna. *IEEE Antennas Wirel. Propag. Lett.* 17 (5): 772–775.

48 Kumar, A. and Raghavan, S. (2018). Design of SIW cavity-backed self-triplexing antenna. *Electron. Lett.* 54 (10): 611–612.

49 Priya, S., Dwari, S., Kumar, K., and Mandal, M.K. (2019). Compact self-quadruplexing SIW cavity-backed slot antenna. *IEEE Trans. Antennas Propag.* 67 (10): 6656–6660.

50 Dong, Y. and Itoh, T. (2010). Miniaturized substrate integrated waveguide slot antennas based on negative order resonance. *IEEE Trans. Antennas Propag.* 58 (12): 3856–3864.

51 Saghati, A.P. and Entesari, K. (2017). An ultra-miniature SIW cavity-backed slot antenna. *IEEE Antennas Wirel. Propag. Lett.* 16: 313–316.

52 Xu, F. and Wu, K. (2005). Guided-wave and leakage characteristics of substrate integrated waveguide. *IEEE Trans. Microw. Theory Techn.* 53 (1): 66–73.

53 Gong, K., Chen, Z.N., Qing, X. et al. (2013). Empirical formula of cavity dominant mode frequency for 60-GHz cavity-backed wide slot antenna. *IEEE Trans. Antennas Propag.* 61 (2): 969–972.

54 Kai, T., Hirokawa, J., and Ando, M. (2004). A stepped post-wall waveguide with aperture interface to standard waveguide. *Proc. Antennas Propag. Soc. Int. Symp.* 2: 1527–1530.

7

Circularly Polarized SIW Slot LTCC Antennas at 60 GHz

Yue Li

Department of Electronic Engineering, Tsinghua University, Beijing 100084, People's Republic of China

7.1 Introduction

With the high demand for wireless personal area networks (WPAN) [1] and the 5G wireless communications [2], antennas at mmW bands are widely studied. For example, for the 60-GHz WPAN with the regulated spectrum of 57–64 GHz, wideband antennas are required for high-speed and short-range data transmission. As another requirement, circularly polarized (CP) antennas are preferred due to the merits of polarization matching without considering the antenna orientation [3, 4]. However, it is quite challenging for mmW CP antenna arrays to achieve both wide impedance and axial-ratio (AR) bandwidths simultaneously, especially considering the fabrication error and array configuration.

In this chapter, we introduce the emerging techniques utilized in the 60-GHz CP antenna array designs. In Section 7.2, the state-of-the-art design methods are reviewed and summarized. Two cases are addressed: one is the element selection for mmW CP antenna array, discussing the advantages and design benefits for each element candidate; the other one is the key technique for AR bandwidth enhancement, which is usually much more difficult in mmW array design. As a feasible example, a 60-GHz CP antenna array with AR bandwidth enhancement is provided in Section 7.3. The proposed wideband CP antenna array is fabricated using the process of LTCC, with the benefits of multiple layer structure, flexible metallization, and low fabrication error for applications.

7.2 Key Techniques of mmW CP Antenna Array

With the design of the wideband CP antenna array in mmW band, there are two issues that need to be clarified first: the selection of antenna element suitable for CP antenna array and the configuration of the selected elements for the desired AR bandwidth. With the understanding of these two issues, the CP element selection and AR bandwidth enhancement techniques are reviewed and compared.

7.2.1 Antenna Element Selection

There are three considerations for the element selection. First, the element is suitable for CP performance for wider AR bandwidth. Second, the element can be fabricated under the limit of

Substrate-Integrated Millimeter-Wave Antennas for Next-Generation Communication and Radar Systems, First Edition.
Edited by Zhi Ning Chen and Xianming Qing.

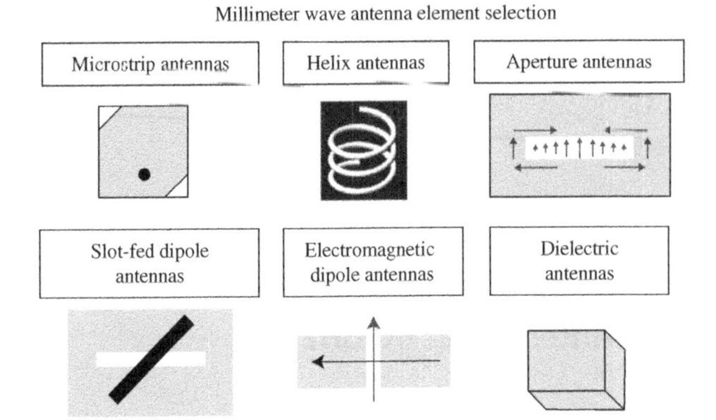

Figure 7.1 A summary of CP antenna element selection for mmW applications.

fabrication tolerance. That is to say, the element should be simple and without using complex and elaborated structures. Third, we also need the compatibility with existing mmW fabrication processes: LTCC, complementary metal oxide semiconductor (COMS), PCB, micro-electromechanical systems (MEMS), just to name a few. As a significant part of the mmW CP antenna array, the antenna element candidates include microstrips [5–14], helixes or spirals [15–18], apertures or slots [19–24], and other types, such as slot-coupled dipoles [25], electromagnetic dipoles [26, 27], and dielectric resonators [28], as exhibited in Figure 7.1. In the following, the advantages for each candidate are discussed with the performance comparison.

Microstrip antennas are the most preferred due to the requirements for low profile and multiple-layer structure [3, 4]. It is quite easy to achieve circular polarization using single or double feeds, such as corner-truncated patch for single feed and 90° hybrid network for double feed structures. The general principle is to excite two orthogonal modes with identical magnitude and 90° phase difference. However, for antenna array design with large numbers of elements, the single feed structure is most utilized. There are usually four ways to excite an asymmetrical microstrip antenna with single feeds, as illustrated in Figure 7.2. As shown in Figure 7.2a, the microstrip is excited by a slot etched on the waveguide, which is an irreplaceable transmission line for mmW applications, with the merits of low dissipation loss and low cross talk [7]. However, the waveguide feeding transmission line is bulky and difficult to realize the sequential rotation feeding network (SRFN) for wideband AR performance. Instead, as shown in 7.2b, it is quite easy to achieve sequential rotation feeding structure with microstrip lines [12]. By connecting a probe inside the patch cavity as depicted in Figure 7.2c, the freedom of feeding position is provided for improved impedance matching [13]. There is no need to connect the feeding strip with the microstrip antennas, the coupled method also works for the purpose of impedance matching [14], as described in Figure 7.2d. However, note that the microstrip lines suffer from more loss than waveguide when the operating frequency goes higher or the array becomes larger. Therefore, for higher frequency or a quite large array, the waveguide is the best choice for CP antenna array feeding structure, and the coupled case is more useful.

For the requirement of wideband AR performance, helix antennas may be the preferred choice for their intrinsic wideband CP quality. As a structure based on a traveling wave concept, the three-dimensional (3-D) helix or two-dimensional (2-D) spiral are wideband antenna types for both AR and impedance bandwidth [3, 4]. However, how to integrate such a 3-D structure with existing mmW fabrication process is a big challenge. As illustrated in Figure 7.3, the 3-D helix structure has already been positioned within multiple layers, based on the process of LTCC or PCB. It is indeed a progress in achieving wideband CP array using the multiple layered helix or spiral structures.

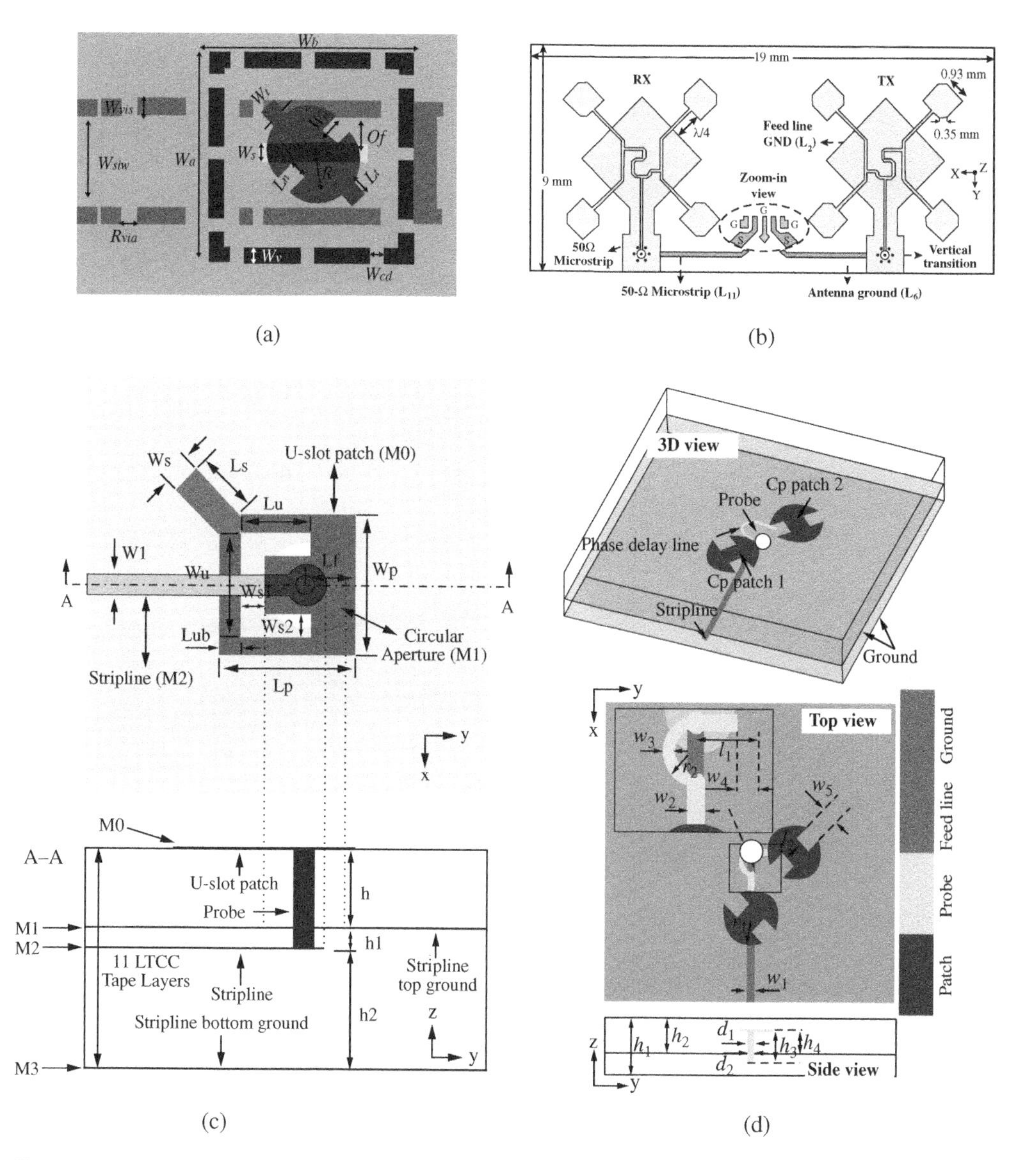

Figure 7.2 Selected microstrip antennas as the elements of mmW CP antenna array with different feeding methods. (a) Slot-coupled microstrip in [7], (b) strip-fed microstrip in [12], (c) probe-fed microstrip in [13], and (d) strip-coupled (not directly connected) microstrip in [14].

In Figure 7.3a, the 3-D helix element is designed within multiple layer PCB, with the loop part in each layer and vertical part as vias [15]. In Figure 7.3b, the 2-D spirals are designed in a rectangular shape with less quality deterioration [16]. In Figure 7.3c,d, the stacked and dual-fed spirals are also positioned in the flatland arrangement with good CP performance [17, 18]. For the helix or spiral elements, the advantage is their intrinsic CP quality, but the dimensions are usually large due to the traveling wave principle.

Another useful element for mmW CP antenna array is aperture or slot [3, 4], as shown in Figure 7.4. The aperture element is with the merits of wide bandwidth and high gain [20]. As shown in Figure 7.4a, the higher gain is determined by the size of the aperture, which is usually

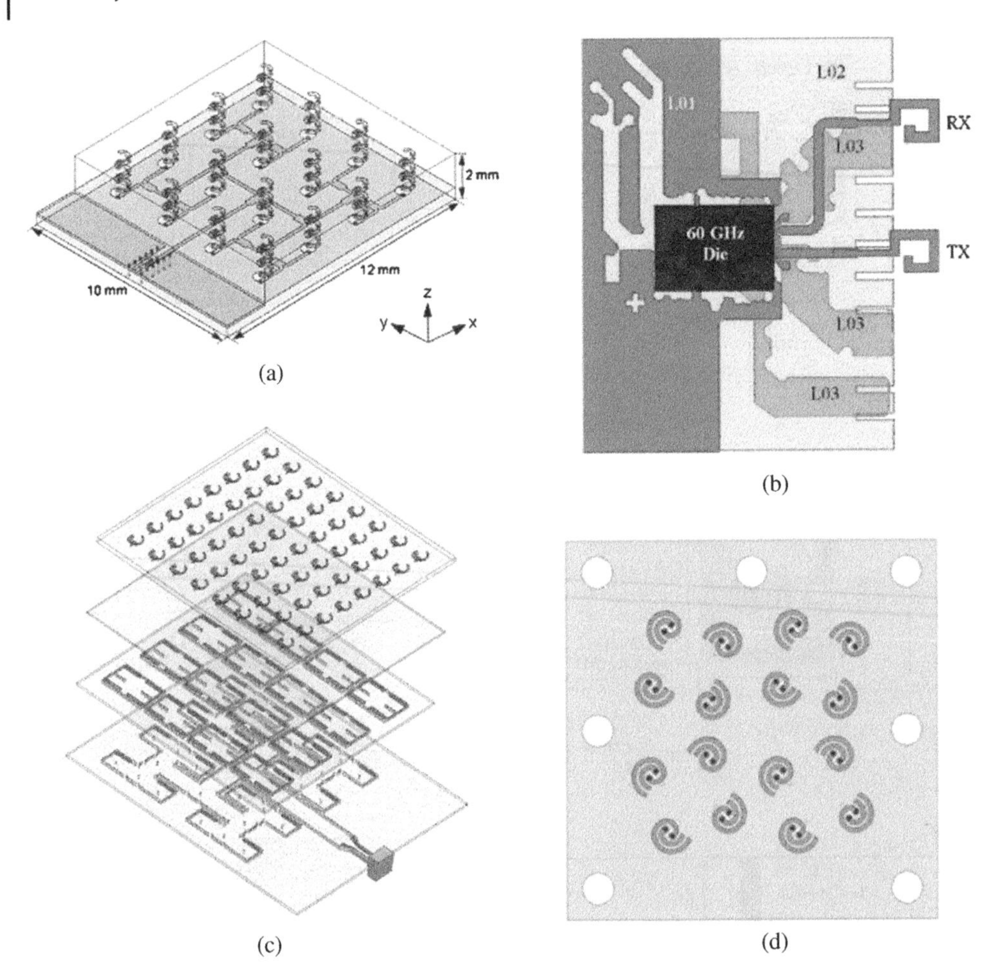

Figure 7.3 Examples of helix or spiral antennas as the elements of mmW CP antenna array. (a) Single-fed 3-D helix in [15], (b) planer rectangular spiral in [16], (c) planar stacked spiral in [17], and (d) dual-fed spiral in [18].

with a planar structure and easily achieved using an SIW technique [20]. However, as depicted in Figure 7.4b, for the low-profile case with a back cavity, the bandwidth decreases when the cavity becomes more closed to the aperture [22]. In the literatures [23, 24], the cavity aperture can be excited by strips or dipoles, as shown in Figure 7.4c,d. However, for the cavity case, the dimensions of the overall element look quite large. As a result, the planar apertures are usually selected for the purpose of low profile and convenient integration with waveguide transmission lines. The cases for Figures 7.4c,d can also be treated as cavity-loaded structures for impedance and AR bandwidth enhancement, which are also suitable for array configurations.

Besides the microstrip, helix, and aperture elements, there are still several types of array elements feasible for mmW CP antenna array configurations. For example, the slot-coupled dipoles are utilized for CP application [25]. As illustrated in Figure 7.5a, the electric field from the slot is divided into two orthogonal components: one is parallel to the dipole and the other one is perpendicular

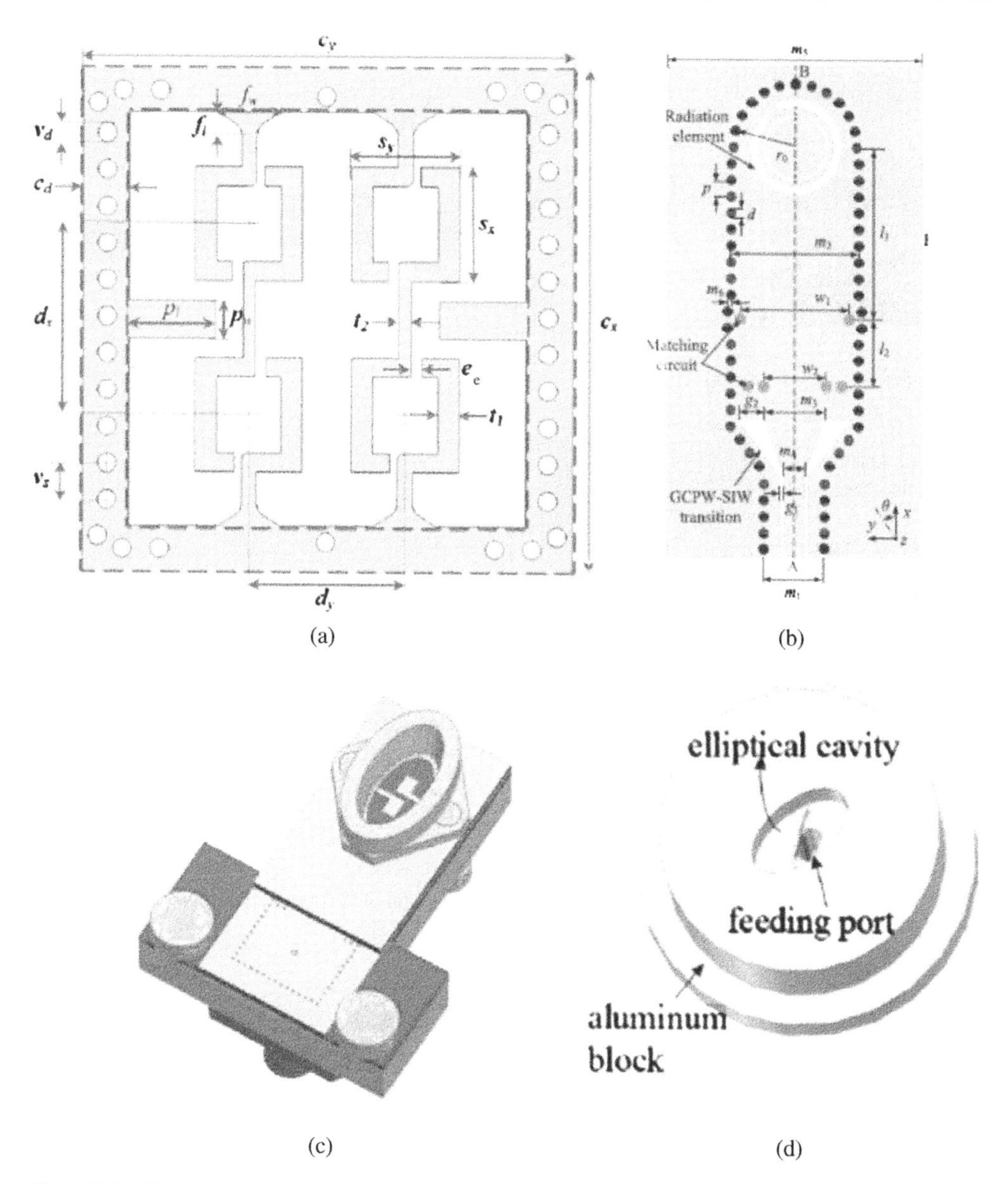

Figure 7.4 Examples of aperture or slot antennas as the elements of mmW CP antenna array. (a) Differential-fed aperture in [20], (b) cavity-backed slots in [22], (c) strip-fed cavity aperture in [23], and (d) dipole-fed cavity aperture in [24].

to the dipole. The parallel one is with coupling to the dipole and the perpendicular one is without such effect. By tuning the angle of dipole, the two orthogonal components are with equal magnitude and 90° phase difference [25]. A similar concept is proposed in [26] to design the electric dipole and magnetic dipole, as shown in Figure 7.5b. By properly tuning the two dipoles, the orthogonal electric fields can achieve equal magnitude and 90° phase difference for CP radiation. In [28], the dielectric resonator is used for mmW applications. As shown in Figure 7.5c, the CP radiation is generated by the asymmetrical dielectric resonator structure.

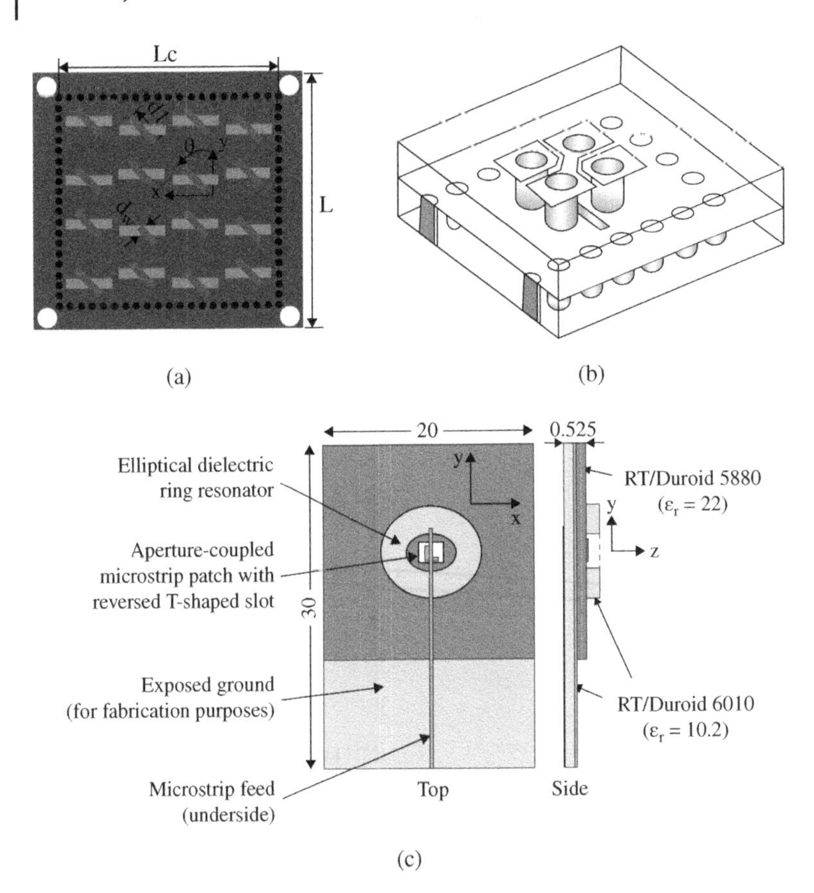

Figure 7.5 Examples of other types of elements of mmW CP array. (a) Slot-coupled dipole in [25], (b) electromagnetic dipole in [26], and (c) dielectric resonator in [28].

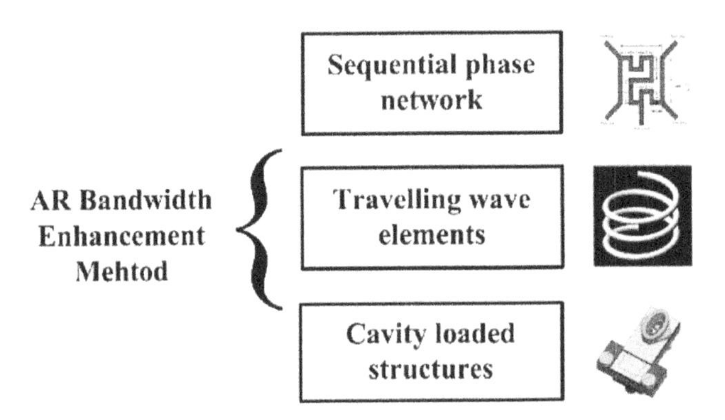

Figure 7.6 AR bandwidth enhancement methods.

7.2.2 AR Bandwidth Enhancement Methods

Three methods for AR bandwidth enhancement are summarized here. As shown in Figure 7.6, the wideband AR can be achieved using (i) SRFN, (ii) traveling wave elements, or (iii) cavity loading structures.

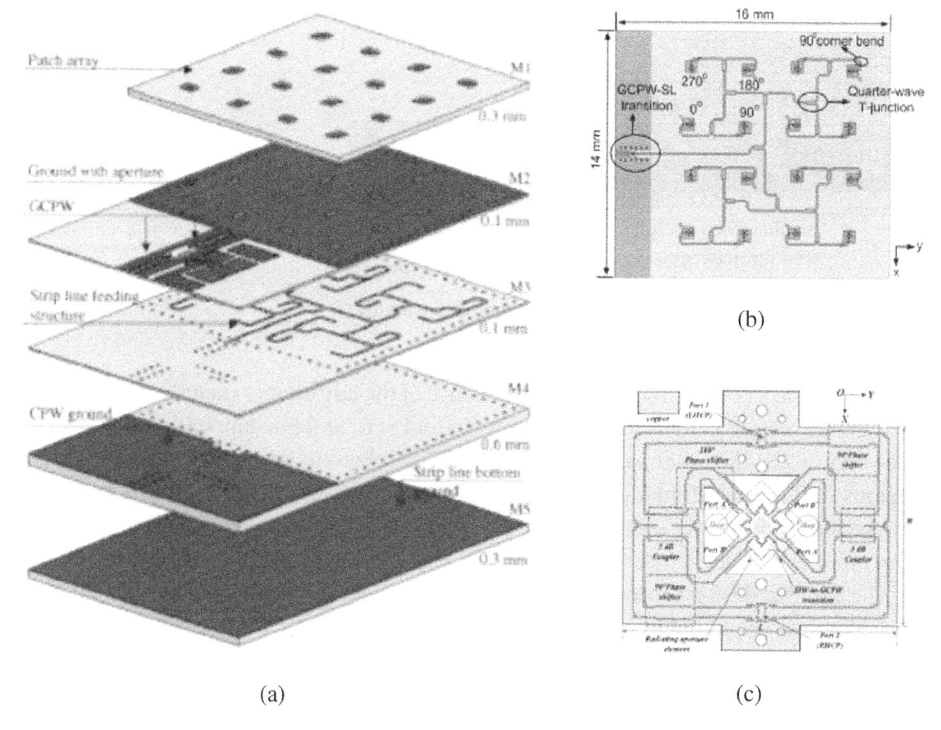

Figure 7.7 Examples of different SRFNs. (a) Slot-coupled SRFN in [11], (b) probe-fed SRFN in [13], and (c) SIW-fed SRFN in [14]. Both (a) and (b) are based on LTCC, while (c) is based on PCB process.

As illustrated in Figure 7.3, the helix element operates as the traveling wave antenna with wideband AR bandwidth [15–18], for example, up to 37.8% 3-dB AR bandwidth in [18]. For the cavity loading method, as used in [23, 24] and shown in Figure 7.4c,d, the achieved 3-dB AR bandwidth is up to 22.7%. It is worth mentioning that the overall bandwidth should be determined not only by 3-dB AR bandwidth but also the impedance bandwidth, e.g., −10 dB reflection coefficient and the 3-dBic gain bandwidth.

Next, we will focus on the SRFN design for wideband CP antenna array, as shown in Figure 7.7. The SRFN is a well-known method to enhance the AR performance, by feeding the four elements with a phase configuration of 0°, 90°, 180°, and 270° and identical amplitude [11–14, 29–31]. The 3-dB AR bandwidth is optimized up to 26% [14], covering the required bandwidth of 57–64 GHz for 60-GHz WPAN.

7.3 Wideband CP LTCC SIW Antenna Array at 60 GHz

As an example, we introduce the design process of a CP antenna array at 60 GHz in detail. The mmW band of 57–64 GHz is available for WPAN in short-range communication applications. CP antenna arrays are used for both high gain against the high pathloss and high-quality wireless communication links. As reviewed above, the slot-coupled dipole antenna is a candidate but with a relatively narrow bandwidth. In this section, a wideband design by replacing the dipole with a rectangular patch is demonstrated. A 4 × 4 array is formed, and via-fences are used to improve the isolation between elements for array applications [32]. To lower the feeding loss, we use an SIW as

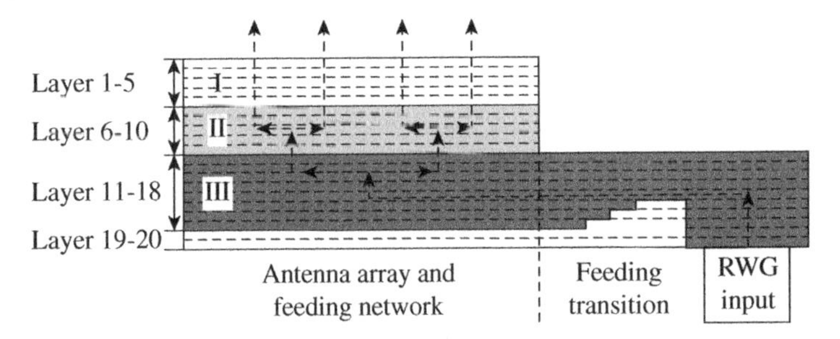

Figure 7.8 Side view of the 4 × 4 antenna array with a feed network [32].

the feeding network. The LTCC technology is used because of the advantages including multilayer configuration, flexible metallization, easy realization of blind vias, and low fabrication tolerance.

As shown in Figure 7.8, the antenna consists of 20 LTCC substrate layers and can be divided into three regions [32]: region I includes layers 1–5 for radiation apertures, region II includes layers 6–10 for feeding cavities, and region III includes layers 11–20 for the feeding network and the transition from input waveguide to the feeding network. The trace of power flow is also illustrated in Figure 7.8. The detailed design processes are discussed here, and the related parameters are not shown in this text but can be found in [32].

7.3.1 Wideband AR Element

The antenna element consists of three components: a rectangular patch, a feeding slot on top of the SIW, and a cavity at the end of the SIW. The feeding slot is parallel to the x-axis and makes an angle of α with respect to the patch. All these structures are fabricated with a 10-layer LTCC substrate of Ferro A6-M. Figure 7.9 shows the structure of the antenna element. The patch is printed on layer 1 and the SIW feeding structure is across five layers beneath the patch and positioned in layers 6–10.

Figure 7.10 shows the element performance over the required bandwidth, i.e., covering 57–64 GHz, for −10-dB reflection coefficient, 3-dB axial ratio, and 3-dBic gain bandwidths. Different from the reviewed techniques for wideband AR performance, in this design, the AR bandwidth is enhanced by using two different CP modes, which can be generated within this structure. A dipole mode is excited in the rectangular patch fed by the slot. It has been studied previously that such an antenna structure consisting of a dipole and a slot is capable for CP radiation, as discussed in the part of slot-coupled dipole antenna. Another mode is introduced by the rotated patch. A current ring mode is excited along the edges of the rectangular patch whose perimeter is one wavelength.

The snapshots of the time-varying current distributions on the patch in a period are depicted in Figure 7.11. The current periodically lies along the short sides and the long sides of the patch, which generates an additional CP radiation. These two modes work at different frequencies close to each other, thus broadening the bandwidth to 7.16 GHz, covering the required bandwidth for 60-GHz WPAN. The parameters of the antenna element have varying effects on the impedance matching and AR bandwidth. The size of the patch has a greater influence on the center frequency and the AR bandwidth. The parameters of the cavity, however, only contribute to the impedance matching, indicating the function as an impedance transformer of the cavity between the SIW and the feeding slot. The size of the feeding slot affects both the CP performance and the impedance matching; the effect on the impedance matching is larger.

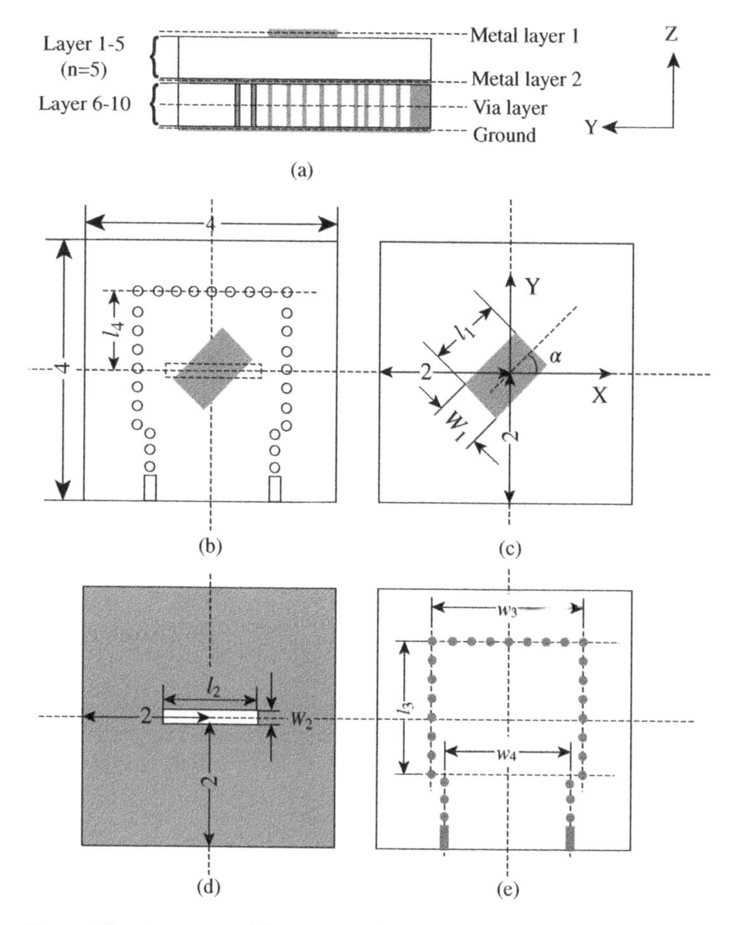

Figure 7.9 Geometry of the proposed antenna element (unit: mm). (a) Side view, (b) top view, (c) metal layer 1, (d) metal layer 2, and (e) via layer.

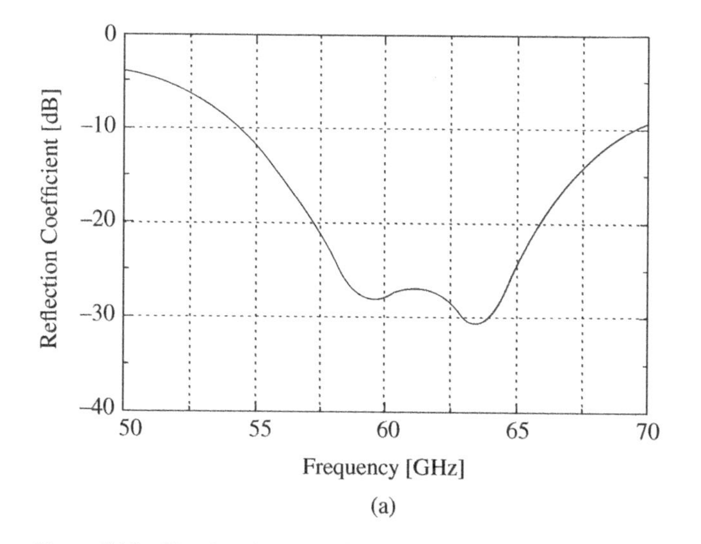

Figure 7.10 Simulated results of the element in Figure 7.9. (a) Reflection coefficient and (b) gain and AR.

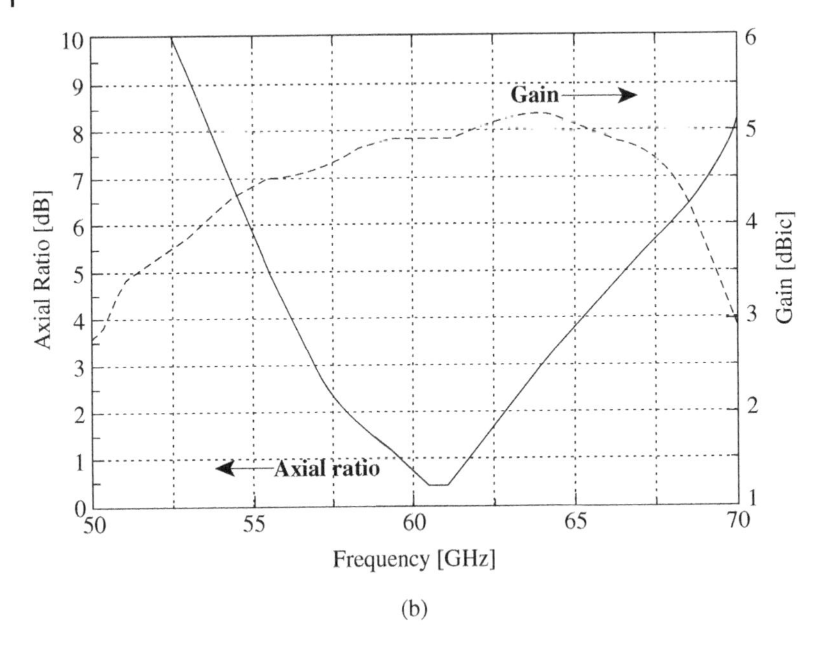

Figure 7.10 (*Continued*)

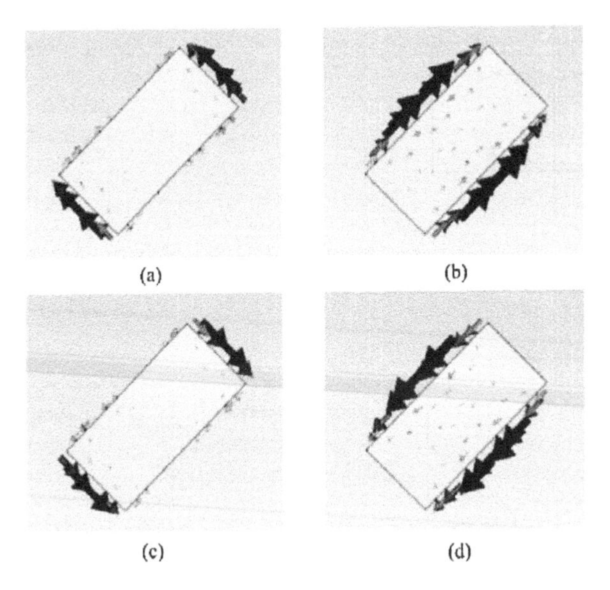

Figure 7.11 Snapshots of the time varying current distributions on the patch in a period. (a) $t = 0$, (b) $t = T/2$, (c) $t = 3T/2$, and (d) $t = T$.

7.3.2 Isolation Consideration

For the antenna array arrangement, the mutual coupling between the adjacent elements should be eliminated to maintain the low correlation. As shown in Figure 7.12a, two elements are positioned side by side at a distance of s. The surface wave with the electric field perpendicular to the metal can be easily excited and propagates due to the high relative permittivity of Ferro A6-M substrate. In this design, the surface wave propagating along the metal layer causes strong mutual coupling between the two elements and reduce the AR bandwidth of the antenna arrays. As shown in Figure 7.12b, the mutual coupling is near −20 dB with different distances and will decrease the AR bandwidth obviously, approximately 3 GHz, compared with the one over 8 GHz

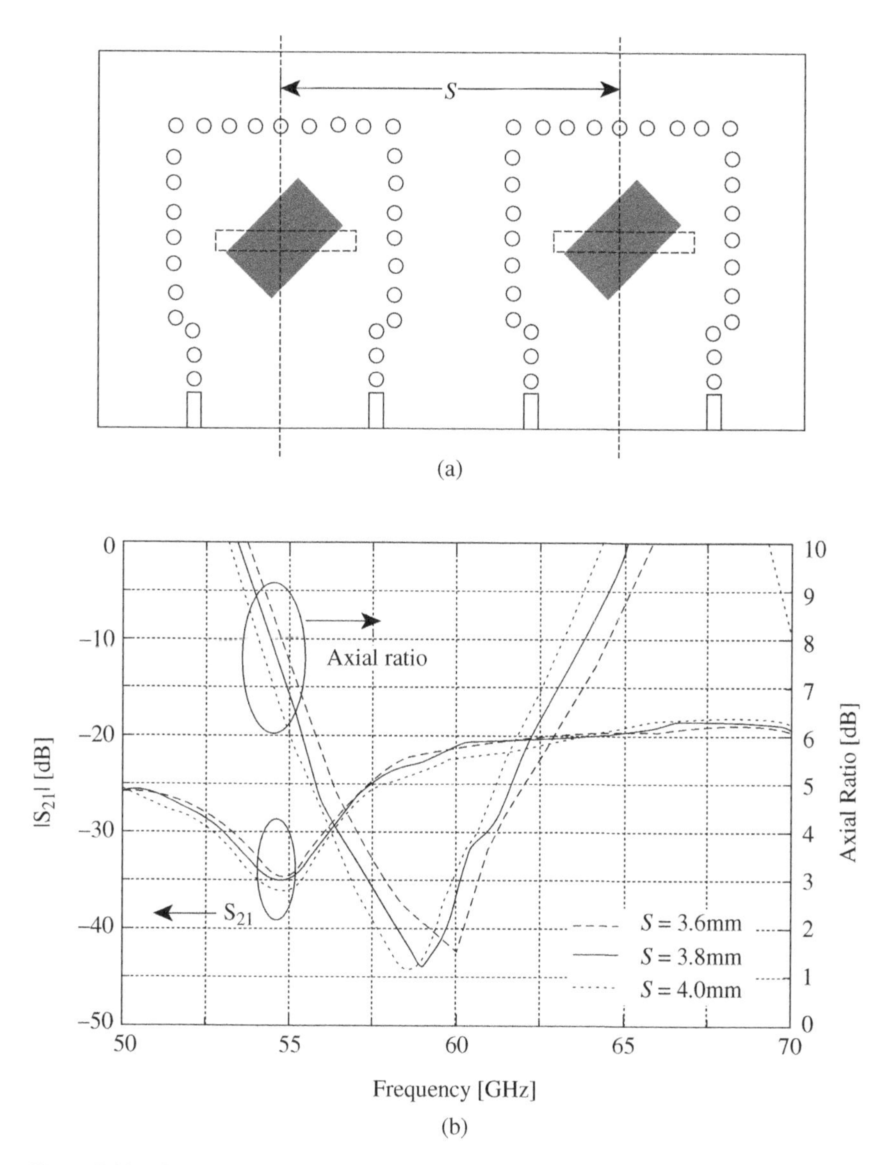

Figure 7.12 Geometry and the simulated results of the two-element array. (a) Geometry from the top view and (b) simulated mutual coupling and AR.

in Figure 7.10b. For this reason, extra structures to block the surface wave should be introduced to maintain the AR bandwidth by reducing the mutual coupling between adjacent elements. Such structures will not deteriorate the impedance matching.

To maintain the CP performance when an antenna array is formed, approaches to suppress the surface wave are needed. In this example, we use two rows of metal topped vias between the two elements to block the surface wave, as shown in Figure 7.13a. The rows of the vias form a fence surrounding each element and thus the surface wave is blocked.

In order to validate the design, simulations are launched on a two-element array, as shown in Figure 7.13a, and also for single element antenna with the via fences. As illustrated in Figure 7.13b, an improved isolation of more than 25 dB is observed. The AR bandwidth is much broader than that in Figure 7.12b and even broader than the single element structure.

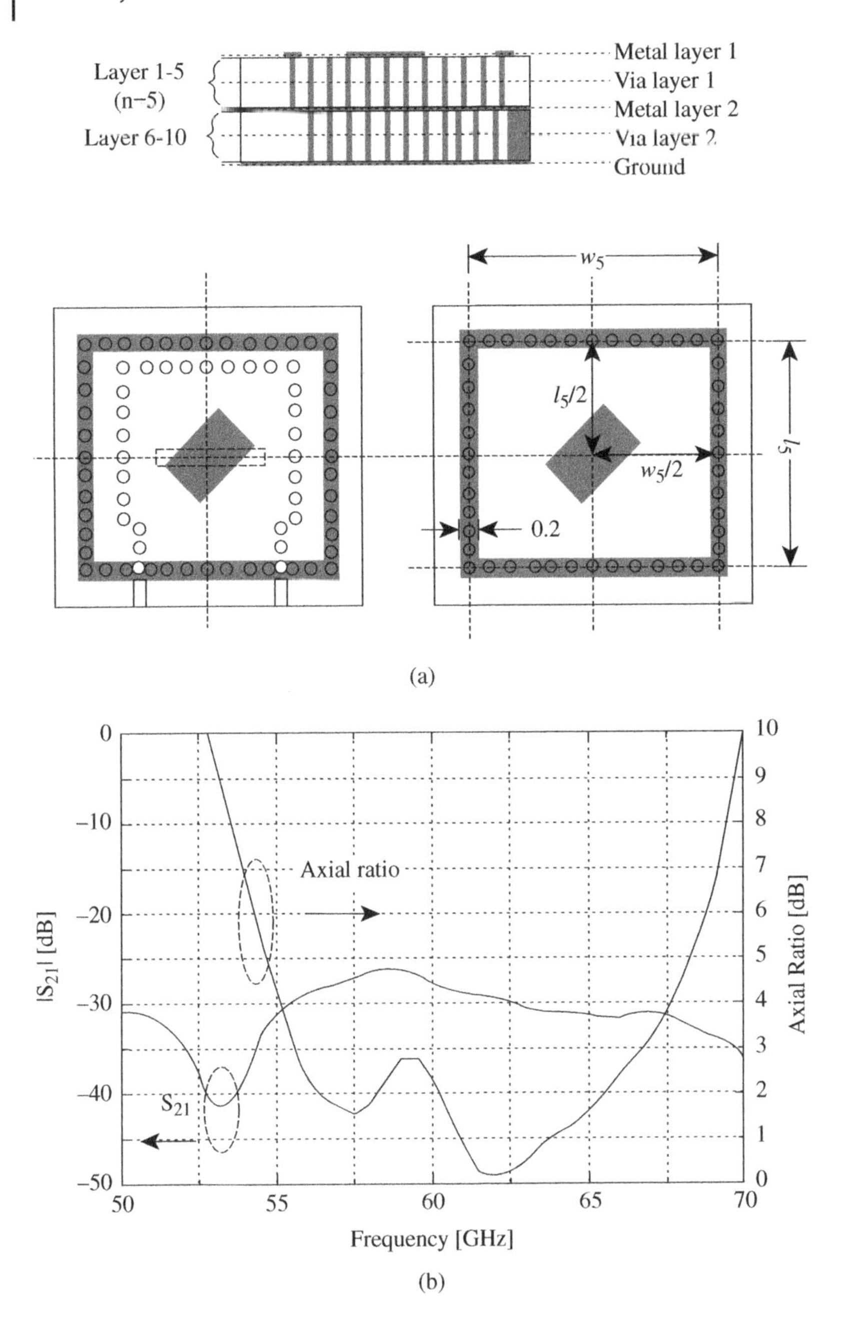

Figure 7.13 Geometry and simulated results of the two-element array with a metal-topped via fence. (a) Geometry from the top view and the side view and (b) simulated mutual coupling and AR.

The reflection coefficient of a single element with a via-fence is shown in Figure 7.14a, maintaining the wideband property from Figure 7.10a. The related AR bandwidth and gain against frequency of a single element with via-fence are shown in Figure 7.14b. The AR bandwidth recovers from the deterioration in Figure 7.12b. What's more, the via-fence serves as a loading cavity and offers additional freedoms in the antenna parameter tuning. By optimizing the length and the width of

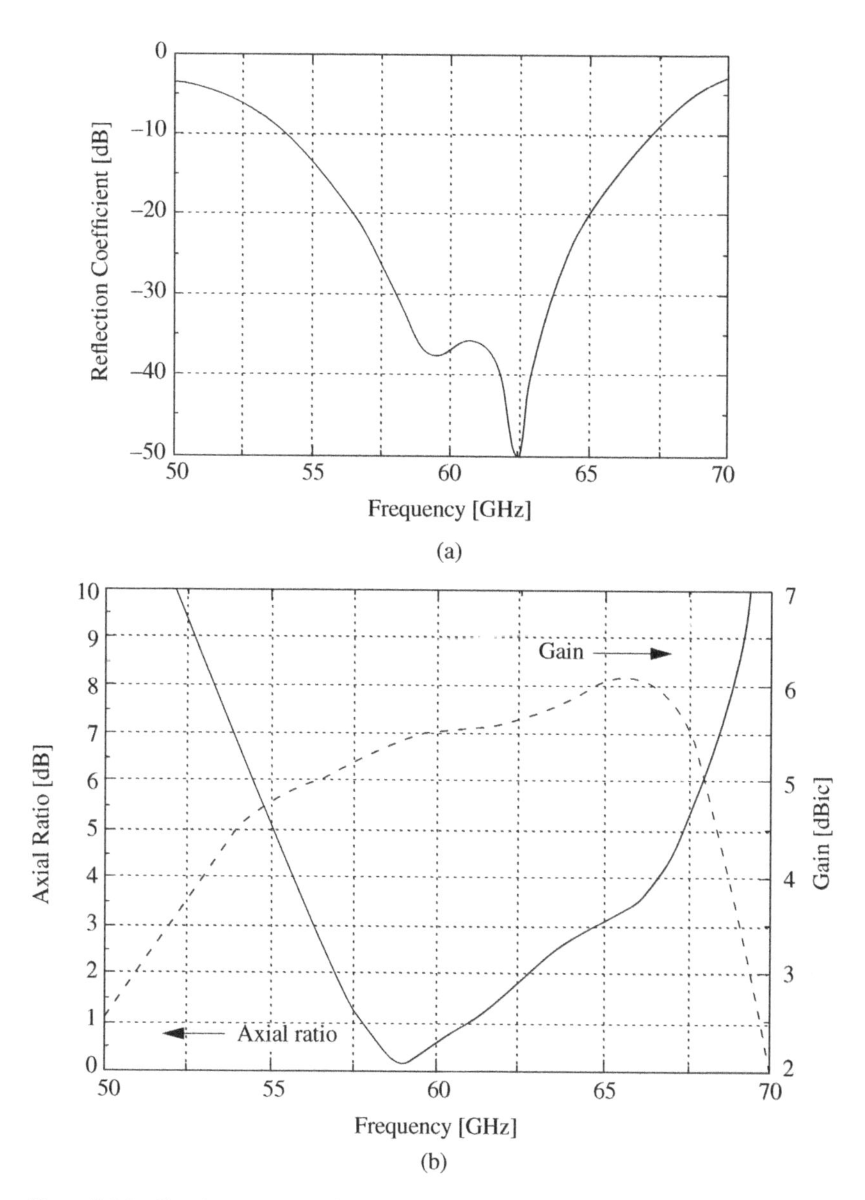

Figure 7.14 Simulated results of the single element with the via-fence. (a) Reflection coefficient and (b) AR and gain.

the via-fence, the optimized AR bandwidth and gain can be achieved with the suppression of the surface wave.

7.3.3 Experiment Results and Discussion

Based on the via-fence structure, a 4 × 4 CP array is designed, and fabricated on a 20-layered LTCC substrate. There are 16 elements placed at a distance of 3.8 mm between each other. The rectangular patches are printed onto the top and the via-fences are in layers 1–5. Layers 6–10 include eight feeding cavities. Two feeding slots of each cavity are placed on top of the cavity to feed the

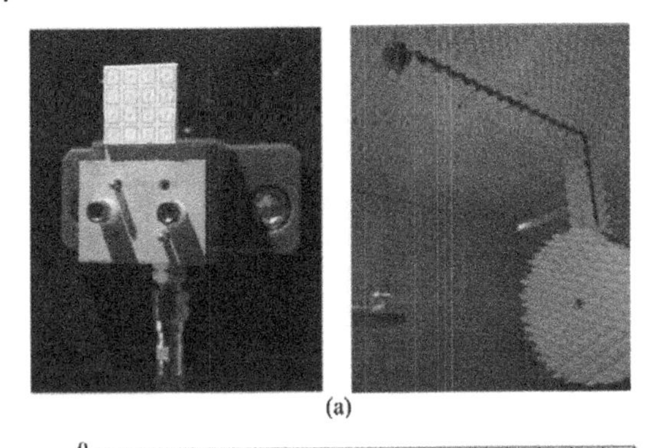

Figure 7.15 (a) Photograph of the fabricated antenna prototype and the measurement system and (b) measured and simulated reflection coefficients.

antenna elements and a coupling slot is cut onto the bottom of the cavity to feed the cavity. The 16 feeding slots are positioned on top of layer 6, and 8 coupling slots on top of layer 11. The feeding network includes 10 layers, layers 11–20. The part below the radiating apertures and the feeding cavities contains a 1-to-8 power divider for feeding in layers 11–18 and a stepped feeding transition is designed in layers 11–20. The SIW-rectangular waveguide (RWG) transition serves as an impedance transformer between RWG and SIW in layers 11–18 for measurement. The detailed structures and dimensions of all the regions can be found in [32].

The antenna array is fabricated as shown in the left panel of Figure 7.15a with a total antenna area of $15.4 \times 15.4\,\text{mm}^2$. The performance of the proposed antenna array is measured using the mmW measurement system, which is shown in right panel of Figure 7.15a. In Figure 7.15b, the simulated and measured reflection coefficients are shown and compared. A reflection coefficient lower than −6 dB is measured from 56.65 to 65.75 GHz, indicating an impedance bandwidth of 9.1 GHz approximately. It means that most of the feeding power can be transferred to the antenna. Compared with the simulated result, the measured reflection coefficient is a little larger, which is possibly due to the defects or inaccuracies in LTCC fabrication.

Such inaccuracies are mainly caused by the difference of thicknesses between the simulation and the fabricated prototype. Compared with the predesigned thickness of 2.02 mm at every point of the antenna array, the actual thicknesses are smaller and vary at different positions across the antenna array. Figure 7.16a shows exact thicknesses at different positions. The reason is due to the shrinking effect of the LTCC process. The shrinking in the size of the antenna array leads to a higher resonant frequency of the CP modes than the ones introduced before. The results for the measurement verify this shrinking effect and explain the results. Figure 7.16b shows the simulated and measured AR bandwidth. The 3-dB AR band ranges from 60.2 to 67 GHz, which shifts to a higher frequency band

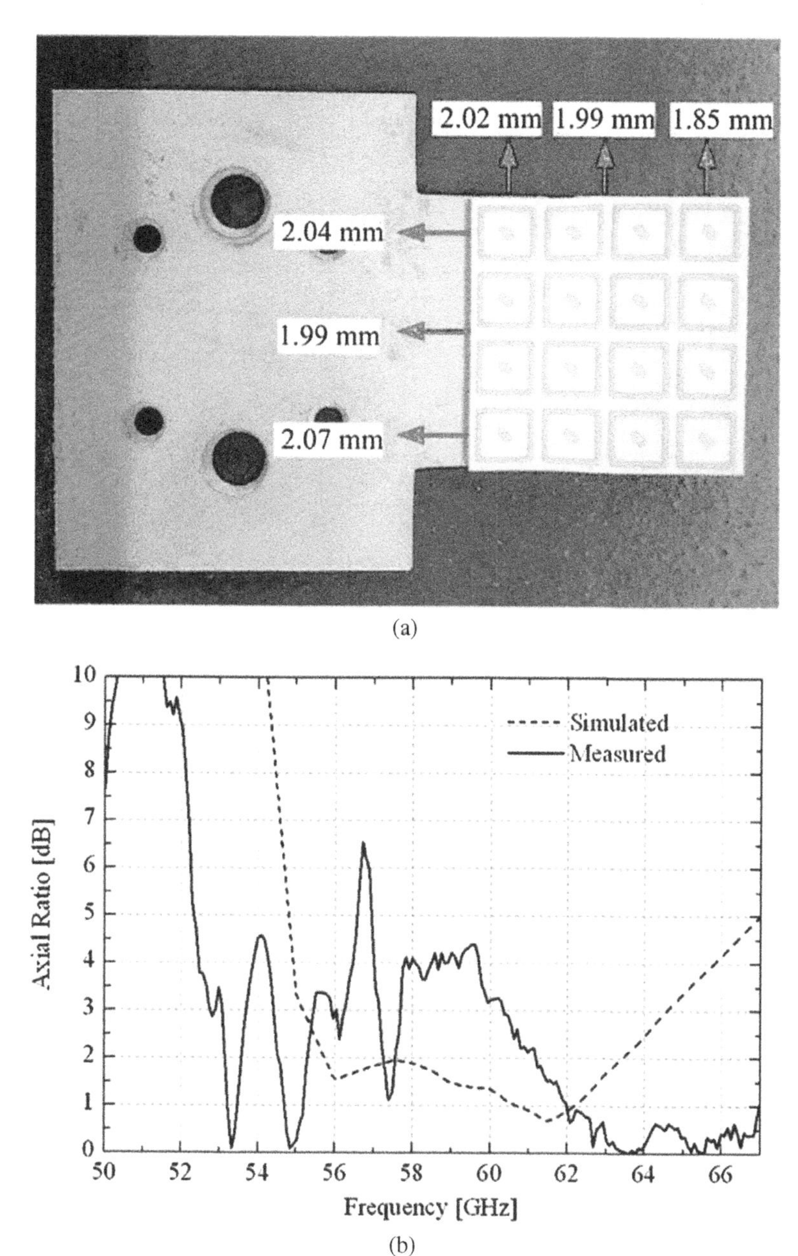

Figure 7.16 (a) The thicknesses at different positions and (b) measured and simulated AR.

than the simulated one while the AR bandwidth maintains to be more than 7 GHz. In Figure 7.17, normalized radiation patterns in both the *xz*-plane and the *yz*-plane are measured and compared to the simulated ones at 60 GHz. The measured results are in agreement with the simulations. The side lobe level (SLL) comes to be lower than −10 dB.

The CP gain at each angle is measured, and the peak gain is shown in Figure 7.18a. The gain is measured to be 15 dBic at 60 GHz, where a drop of 2.5 dB can be observed. At 60 GHz, the simulated efficiency is 89% but reduces to 49% in the measurement. This result is mainly caused by two factors. The first one is the impedance mismatching caused by the shrinking effect of the thicknesses

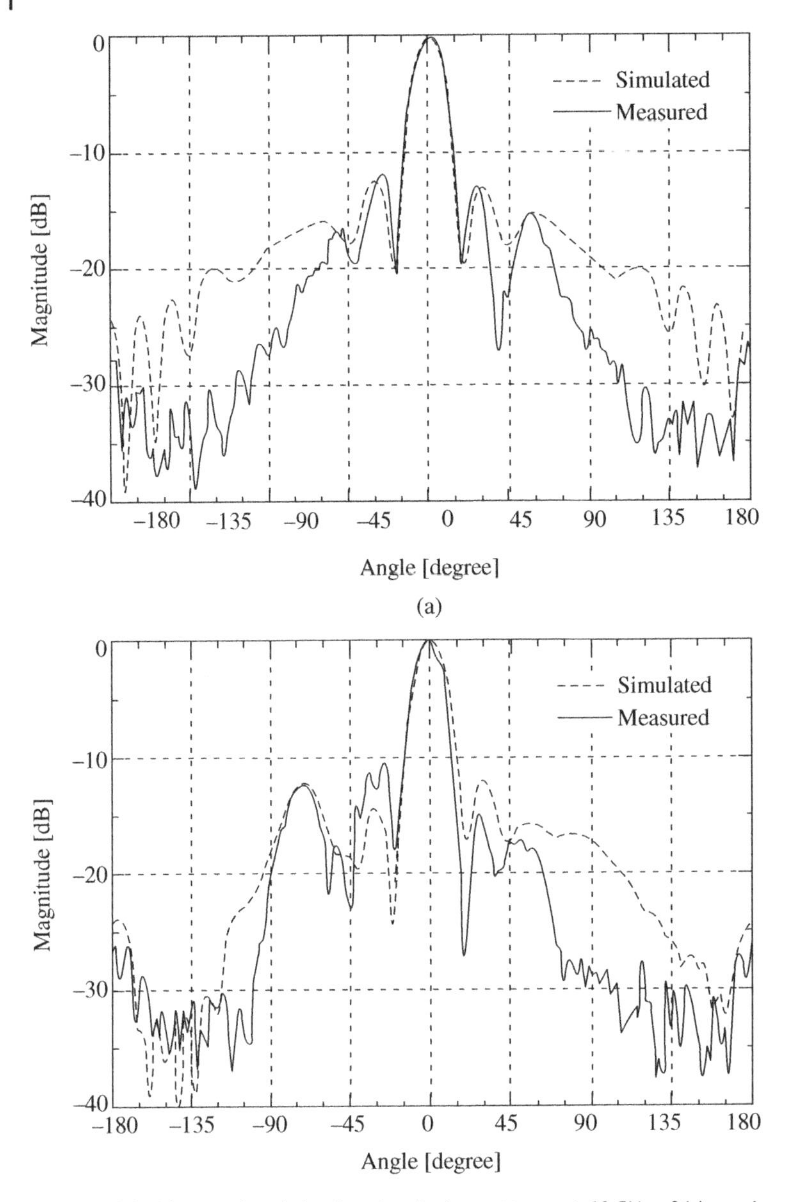

Figure 7.17 Measured and simulated radiation patterns at 60 GHz of (a) *xz*-plane and (b) *yz*-plane.

discussed above. The other one is due to the dielectric loss of the LTCC substrate and metals. In order to understand the loss issue quantitively, Figure 7.18b shows the gains of several cases with different dielectrics and conductors. From the simulated results, we can see that the dissipation loss is from both dielectric of LTCC and copper metal since the difference is minor by using copper or PEC as the metal layer. However, compared with other processes in mmW fabrication, the overall loss of LTCC is quite small, which is an advantage. Figure 7.18b also confirms this conclusion. We can see that the dielectric loss and the finite conductivity of the metals introduce only an 0.5 dB reduction to the gain of the antenna. From this example, we believe that LTCC process and SIW structures are good candidates in mmW antenna array design with required performance.

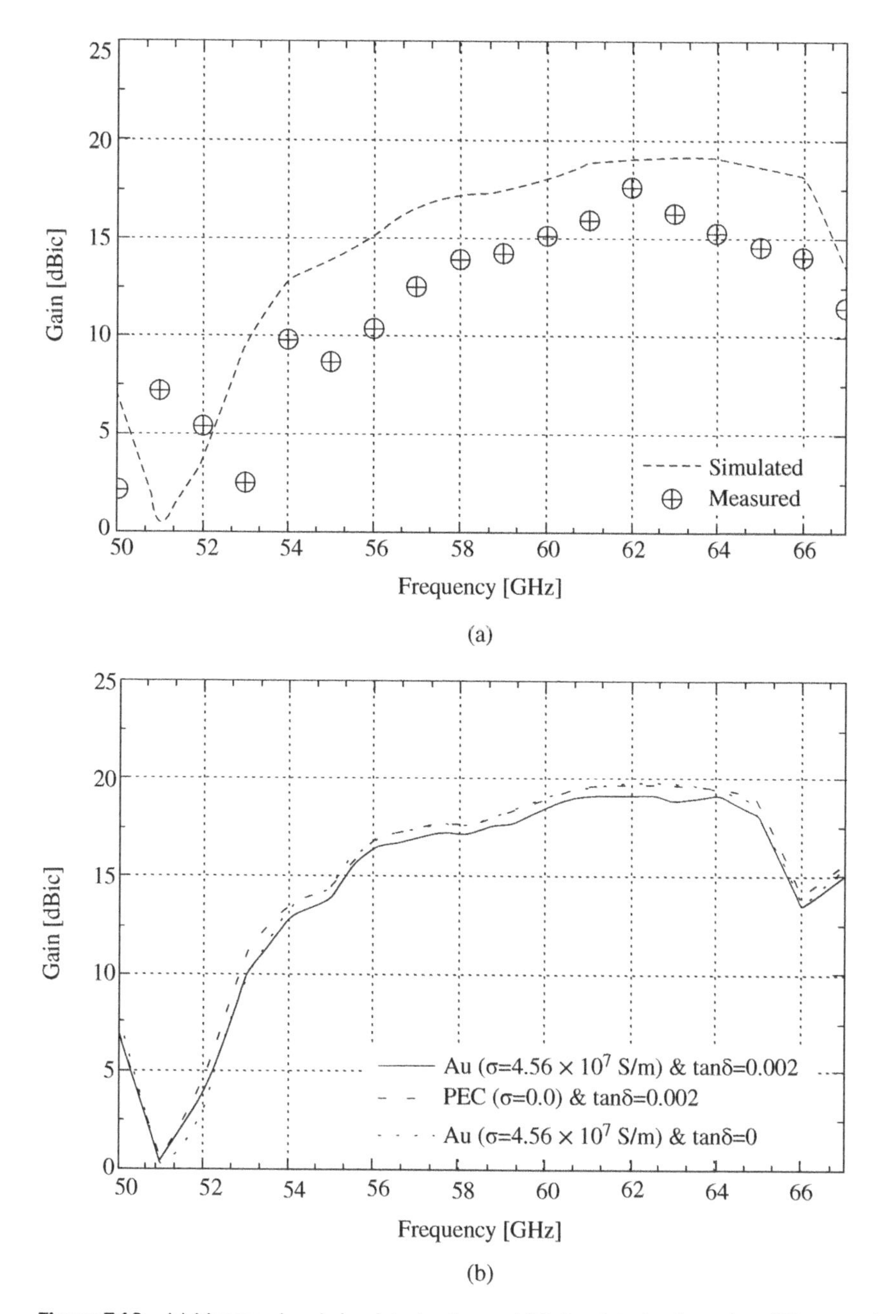

Figure 7.18 (a) Measured and simulated gains and (b) simulated gains using different substrates and conductors.

7.4 Summary

In this chapter, a brief review for mmW CP antenna array and a design example at 60 GHz using LTCC process have been introduced for future 5G and WPAN applications. For the array design, we should consider the strategies of antenna element selection and wideband AR. If we aim to design a wideband CP antenna array at 60 GHz, there are three feasible solutions in the existing literatures: (i) microstrip element with SRFN; (ii) helix element with in-phase feeding network; and (iii) cavity loading. Besides that, multiple CP mode coupling to enhance the AR bandwidth is also useful in mmW antenna array design. As mentioned in the example using LTCC, acceptable CP performance has been achieved based on the dual CP mode coupling and verified by experiment results.

References

1 Zwick, T., Liu, D., and Gaucher, B. (2006). Broadband planar superstrate antenna for integrated millimeterwave transceivers. *IEEE Trans. Antennas Propag.* 54 (10): 2790–2796.
2 Promotion Group (2016). IMT-2020(5G)PG white paper. Beijing.
3 Stutzman, W.L. and Thiele, G.A. (2012). *Antenna Theory and Design*, 3e. New York: Wiley.
4 Balanis, C.A. (2005). *Antenna Theory: Analysis and Design*, 3e. New York: Wiley.
5 Herth, E., Rolland, N., and Lasri, T. (2010). Circularly polarized millimeter-wave antenna using 0-level packaging. *IEEE Antennas Wireless Propag. Lett.* 9: 934–937.
6 Qian, K. and Tang, X. (2011). Compact LTCC dual-band circularly polarized perturbed hexagonal microstrip antenna. *IEEE Antennas Wireless Propag. Lett.* 10: 1212–1215.
7 Guntupalli, A.B. and Wu, K. (2014). 60-GHz circularly polarized antenna array made in low-cost fabrication process. *IEEE Antennas Wireless Propag. Lett.* 13: 864–867.
8 Park, S.-J. and Park, S.-O. (2017). LHCP and RHCP substrate integrated waveguide antenna arrays for millimeter-wave applications. *IEEE Antennas Wireless Propag. Lett.* 16: 601–604.
9 Xu, J., Hong, W., Jiang, Z.H. et al. (2017). A Q-band low-profile dual circularly polarized array antenna incorporating linearly polarized substrate integrated waveguide-fed patch subarrays. *IEEE Trans. Antennas Propag.* 65 (10): 5200–5210.
10 Zhao, Y. and Luk, K.M. (2018). Dual circular-polarized SIW-fed high-gain scalable antenna array for 60 GHz applications. *IEEE Trans. Antennas Propag.* 66 (3): 1288–1298.
11 Sun, M., Zhang, Y.Q., Guo, Y.X. et al. (2011). Integration of circular polarized array and LNA in LTCC as a 60-GHz active receiving antenna. *IEEE Trans. Antennas Propag.* 59 (8): 3083–3089.
12 Shen, T.M., Kao, T.Y.J., Huang, T.Y. et al. (2012). Antenna design of 60-GHz micro-radar system-in-package for noncontact vital sign detection. *IEEE Antennas Wireless Propag. Lett.* 11: 1702–1705.
13 Sun, H., Guo, Y.X., and Wang, Z. (2013). 60-GHz circularly polarized U-slot patch antenna array on LTCC. *IEEE Trans. Antennas Propag.* 61 (1): 430–435.
14 Du, M., Dong, Y., Xu, J., and Ding, X. (2017). 35-GHz wideband circularly polarized patch array on LTCC. *IEEE Trans. Antennas Propag.* 65 (6): 3235–3240.
15 Liu, C., Guo, Y.X., Bao, X., and Xiao, S.-Q. (2012). 60-GHz LTCC integrated circularly polarized helical antenna array. *IEEE Trans. Antennas Propag.* 66 (3): 1329–1335.
16 Fakharzadeh, M. and Mohajer, M. (2014). An integrated wide-band circularly polarized antenna for millimeter-wave applications. *IEEE Trans. Antennas Propag.* 62 (2): 925–929.
17 Wu, Q., Hirokawa, J., Yin, J. et al. (2017). Millimeter-wave planar broadband circularly polarized antenna array using stacked curl elements. *IEEE Trans. Antennas Propag.* 65 (12): 7052–7062.
18 Zhu, Q., Ng, K.B., and Chan, C.H. (2017). Printed circularly polarized spiral antenna array for millimeter-wave applications. *IEEE Trans. Antennas Propag.* 65 (2): 636–643.
19 Bisharat, D.J., Liao, S., and Xue, Q. (2015). Circularly polarized planar aperture antenna for millimeter-wave applications. *IEEE Trans. Antennas Propag.* 63 (12): 5316–5324.

20 Bisharat, D.J., Liao, S., and Xue, Q. (2016). High gain and low cost differentially fed circularly polarized planar aperture antenna for broadband millimeter-wave applications. *IEEE Trans. Antennas Propag.* 64 (1): 33–42.
21 Zhu, J., Liao, S., Yang, Y. et al. (2018). 60 GHz dual-circularly polarized planar aperture antenna and array. *IEEE Trans. Antennas Propag.* 66 (2): 1014–1019.
22 Wu, Q., Yin, J., Yu, C. et al. (2017). Low-profile millimeter-wave SIW cavity-backed dual-band circularly polarized antenna. *IEEE Trans. Antennas Propag.* 65 (12): 7310–7315.
23 Bai, X., Qu, S.W., Yang, S. et al. (2016). Millimeter-wave circularly polarized tapered-elliptical cavity antenna with wide axial-ratio beamwidth. *IEEE Trans. Antennas Propag.* 64 (2): 811–814.
24 Bai, X. and Qu, S.W. (2016). Wideband millimeter-wave elliptical cavity-backed antenna for circularly polarized radiation. *IEEE Antennas Wireless Propag. Lett.* 15: 572–575.
25 Asaadi, M. and Sebak, A. (2017). High-gain low-profile circularly polarized slotted SIW cavity antenna for MMW applications. *IEEE Antennas Wireless Propag. Lett.* 16: 752–755.
26 Li, Y. and Luk, K.M. (2016). A 60-GHz wideband circularly polarized aperture-coupled magneto-electric dipole antenna array. *IEEE Trans. Antennas Propag.* 64 (4): 1325–1333.
27 Ruan, X., Qu, S.W., Zhu, Q. et al. (2017). A complementary circularly polarized antenna for 60-GHz applications. *IEEE Antennas Wireless Propag. Lett.* 16: 1373–1376.
28 Perro, A., Denidni, T.A., and Sebak, A.R. (2010). Circularly polarized microstrip/elliptical dielectric ring resonator antenna for millimeter-wave applications. *IEEE Antennas Wireless Propag. Lett.* 9: 783–786.
29 Lin, S.K. and Lin, Y.C. (2011). A compact sequential-phase feed using uniform transmission lines for circularly polarized sequential-rotation arrays. *IEEE Trans. Antennas Propag.* 59 (7): 2721–2724.
30 Li, Y., Zhang, Z., and Feng, Z. (2013). A sequential-phase feed using a circularly polarized shorted loop structure. *IEEE Trans. Antennas Propag.* 61 (3): 1443–1447.
31 Deng, C., Li, Y., Zhang, Z., and Feng, Z. (2014). A wideband sequential-phase fed circularly polarized patch array. *IEEE Trans. Antennas Propag.* 62 (7): 3890–3893.
32 Li, Y., Chen, Z.N., Qing, X. et al. (2012). Axial ratio bandwidth enhancement of 60-GHz substrate integrated waveguide-fed circularly polarized LTCC antenna array. *IEEE Trans. Antennas Propag.* 60 (10): 4619–4627.

8

Gain Enhancement of LTCC Microstrip Patch Antenna by Suppressing Surface Waves

Zhi Ning Chen[1] *and Xianming Qing*[2]

[1] *Department of Electrical and Computer Engineering, National University of Singapore, Singapore 117583, Republic of Singapore*
[2] *Signal Processing, RF, and Optical Department, Institute for Infocomm Research, Singapore 138632, Republic of Singapore*

8.1 Introduction

8.1.1 Surface Waves in Microstrip Patch Antennas

Surface waves: The simplest scenario is when electromagnetic waves are propagating in free space or homogeneous media. The electromagnetic waves experience the reflection and refraction with the changes in terms of the distribution of their amplitudes and phases, the velocity (speeds and directions) of their traveling, and/or even the polarization of propagation when they encounter a change in a medium. One more phenomenon of the electromagnetic waves propagating on the interface between two dielectric media having different dielectric constants or permittivity or along a refractive index gradient has been discovered; such propagating waves have been named "electromagnetic surface waves." On the interface, the electromagnetic waves are guided along the refractive index gradient or the interface. This phenomenon has been one of the most important types of electromagnetic waves propagation from both the physical and engineering points of views across the spectral range of electromagnetic waves from long waves to light waves. Within microwave bands, this phenomenon has long been physically discussed and applied in the design and modeling of radio frequency paths as well as wave transmission systems, for instance transmission lines and even single-wire transmission lines such as Goubau lines [1], and antennas.

The basic characteristic of the electromagnetic surface wave is that the electromagnetic energy propagates along the interface. If the propagation speed of the electromagnetic waves normal to the interface is slower than the speed of light, the waves become evanescent along the direction normal to the interface without any electromagnetic energy propagating away from the interface. Therefore, the field components of the surface wave diminish against the distance from the interface. The electromagnetic waves propagate along the interface and the form of energy is kept unchanged unless they are lossy surface waves [2].

In the 1950s, S. A. Schelkunoff extensively discussed the phenomenon, definition, and characteristics of the electromagnetic surface waves [3]. According to Schelkunoff [3], Lord Rayleigh discovered that a source with finite dimensions can excite two types of waves, namely, the space waves and the surface waves in a semi-infinite elastic medium. The former spreads in all directions while the latter along with the boundary of the medium. The readers can find more information about the conditions of the generation of the surface waves and the classification of waves that look like a surface wave in the reference [3].

Substrate-Integrated Millimeter-Wave Antennas for Next-Generation Communication and Radar Systems, First Edition.
Edited by Zhi Ning Chen and Xianming Qing.

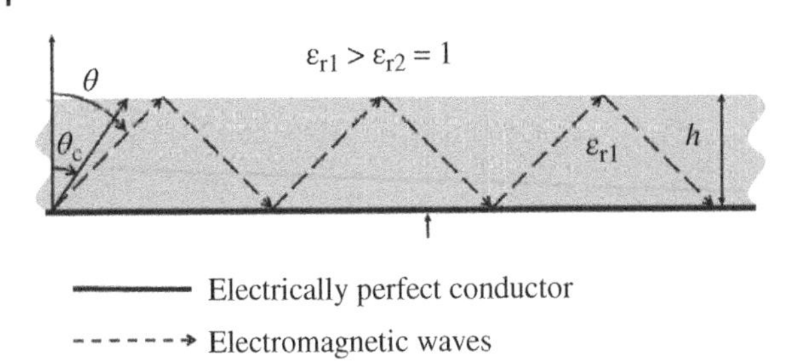

Figure 8.1 Full reflection between the boundary of two media with different dielectric constants.

Figure 8.1 shows the infinite dielectric layer with relative permittivity ε_{r1} covering a perfect electrical conductor (PEC) surface. The medium surrounding the dielectric is vacuum or free space having the relative permittivity of ε_{r2}=1. We can also consider this case as the dielectric slab grounded by a PEC. This configuration is associated with a typical microstrip patch antenna when a radiating patch is printed onto the dielectric substrate and well excited. With such a structure, when the wave incident angle from below is larger than the critical angle of $\theta_c=\sin^{-1}(1/\sqrt{\varepsilon_{r1}})$, the waves are propagating in the dielectric layer only.

Furthermore, according to R. F. Harrington, the surface waves at both transverse electric (TE) and transverse magnetic (TM) modes can be excited in the grounded dielectric layer [4]. The cutoff frequency of the modes is given by

$$f_c = \frac{nc}{4h\sqrt{\varepsilon_{r1}-1}}, \tag{8.1}$$

where c is the speed of light in free space, and n = 0, 1, 2, ... for TM_n and TE_{2n+1} surface modes. From equation (8.1), it is noted that the TM_0 surface mode having a zero cutoff frequency is always excited for any layer thickness h. Therefore, we can draw the following important conclusions from our discussion:

1. the surface waves are trapped by the dielectric layer grounded by a PEC surface;
2. the dielectric layer traps more electromagnetic wave energy and the waves evanesce more rapidly away from the dielectric surface for larger h, the thickness of the dielectric coating layer, as shown in Figure 8.1; and
3. the higher the dielectric constant, the more energy the dielectric layer traps.

The surface waves, at least the TM_0 mode, are always excited when two media are with different dielectric constants. The electrically thicker dielectric layer with either physically larger thickness or higher operating frequencies, or both, traps more energy with more TM and TE surface modes. The trapped surface wave energy is propagating along the surface till the reflection by the discontinuity of the medium and/or the disappearance due to dissipative medium or leaky waves.

Surface waves in microstrip patch antennas: In a microstrip patch antenna design as shown in Figure 8.2, the dielectric substrate is with a thickness of about $h = 0.01\lambda_0$ (λ_0 is the operating wavelength in free space) and typical value of (32 ~ 64 mil, or 0.813 ~ 1.626 mm); a relative dielectric constant lower than 5 (such as 4.0 ~ 4.8 for FR4) at lower microwave bands (such as operating frequencies lower than 10 GHz). As a result, the surface waves generated in microstrip structures with electrically thin grounded dielectric substrate can be ignored [5].

At the mmW bands of 30–300 GHz, the operating wavelengths are smaller than those at the frequencies below 30 GHz, which makes the grounded dielectric with typical thickness of 32 ~ 64 mil electrically thicker. Such dielectric layer very likely generates the surface waves when the electromagnetic waves are excited in the PCB. Therefore, the effects caused by the surface waves on the performance of the microstrip patch antennas operating at mmW bands cannot be ignored in their design [6]. Different from the ohmic losses caused by lossy dielectric and imperfect conductors,

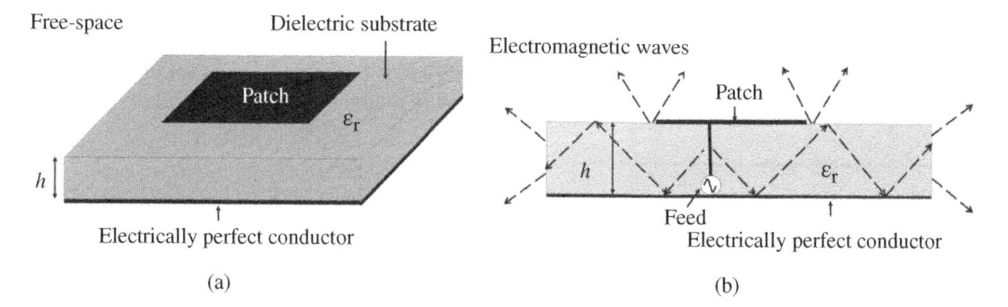

Figure 8.2 (a) The sketch of a probe-fed microstrip patch antenna and (b) electromagnetic waves in a microstrip patch antenna of arbitrary shape.

wherein the energy is converted to heat, the loss from the surface waves involves in the radiation, usually distorting the radiation from the radiator and inevitably reducing the gain of the antenna at desired directions [7]

8.1.2 Surface Waves Effects on Microstrip Patch Antenna

The effects of the surface waves on antenna performance may include the reduction of radiation efficiency and the distortion of the radiation patterns. When a patch antenna printed on an infinitely sized substrate, the energy distributed to the surface waves is trapped in the substrate, and the surface waves don't contribute to the radiation of the antenna. However, the existence of the surface waves reduces the antenna radiation efficiency since part of the feeding energy is trapped rather than radiated. When the patch antenna is printed on a finite-sized substrate, the situation may even worsen. The surface waves radiate into space when they reach the edges of the dielectric substrate, which may cause higher sidelobe levels, higher cross-polarization levels, and/or pattern distortion with ripples. Moreover, the diffracted surface waves may be coupled into the circuits on the dielectric substrate or/and behind the ground plane. This may cause more losses and even electromagnetic compatibility (EMC) issues.

8.2 State-of-the-Art Methods for Suppressing Surface Waves in Microstrip Patch Antennas

The suppression of the undesired surface waves is important and challenging for microstrip patch antenna designs, in particular for the operation at mmW bands with electrically thick dielectric substrate (large thickness and/or high permittivity). There have been many solutions to reduce the effects caused by surface waves on the performance of patch antennas. This chapter focuses on the techniques based on a theoretical analysis model to suppress the TM_0 mode surface wave and the partial removal of the dielectric substrate to suppress all surface wave modes.

Suppression of TM_0 mode surface wave: As mentioned above, the thickness and permittivity of the dielectric substrate of a microstrip patch antenna determine the excitation of surface waves at either TM or TE or both modes. In particular, the TM_0 surface wave always exists even if the substrate is electrically very thin because only the TM_0 surface wave is above the cutoff frequency. Thus, it is a fundamental challenge to suppress the excitation of the dominant TM_0 surface wave.

Jackson et al. presented two solutions to suppress the TM_0 surface wave in a circular patch antenna [8]. The circular patch antenna layout is designed to have a radius equal to a proposed critical value in order to resonate at the design frequency as shown in Figure 8.3. According to the

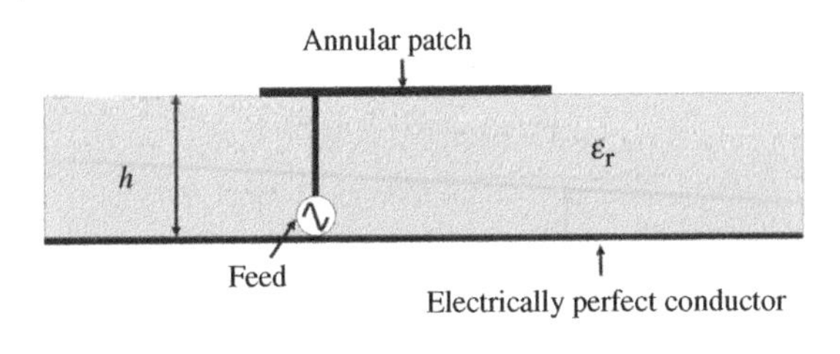

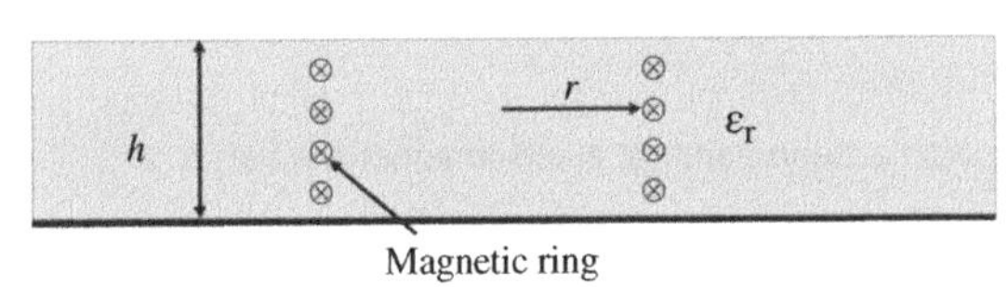

Figure 8.3 The geometry of an annular patch antenna and its magnetic current model for the exterior fields based on the equivalent theory and cavity model of a patch antenna.

equivalence principle and the cavity model, a circular microstrip patch can be modeled as a ring of magnetic current, so a circular patch having the same critical radius as the magnetic current ring will not excite the TM_0 surface wave if the radius of the circular patch is selected to meet the condition [8]:

$$J_1'\left(\beta_{TM_0} r\right) = 0, \tag{8.2}$$

where, β_{TM_0} is the propagation constant of the TM_0 mode surface wave, r is the radius of the magnetic ring/patch, and $J_1'(\cdot)$ the first-order derivative of the Bessel function of the first kind. The selection of r is determined by the dielectric constant and thickness of the dielectric substrate.

Once the radius is set, the radius of the patch is fixed. However, the resonant frequency of the antenna is also determined by the radius of the patch. To tune the resonance to the designed frequency, two methods were introduced in Ref. [8] as shown in Figure 8.4. The one shown in Figure 8.4a is presented to raise the resonance frequency by removing the dielectric right behind the patch. This removal does not affect the modeling mentioned above because according to the equivalent theory, the fields within the magnetic ring are zero. The increase in the cavity radius, a, raises up the resonance frequency of the patch. The other one as shown in Figure 8.4b is to raise the resonance frequency by decreasing the effective radius of the circular patch. The annular ring antenna is shorted by pins. As a result, these two patch designs excite only very little surface wave power. With the reduced diffraction of surface waves, smoother radiation patterns are achieved when the circular patch antenna is mounted on a finite-size ground plane.

Suspended Patch Antennas: Suspended patch antennas are well known with suppressed surface waves and achieved broad impedance operating bandwidths [9–11]. Such antennas are

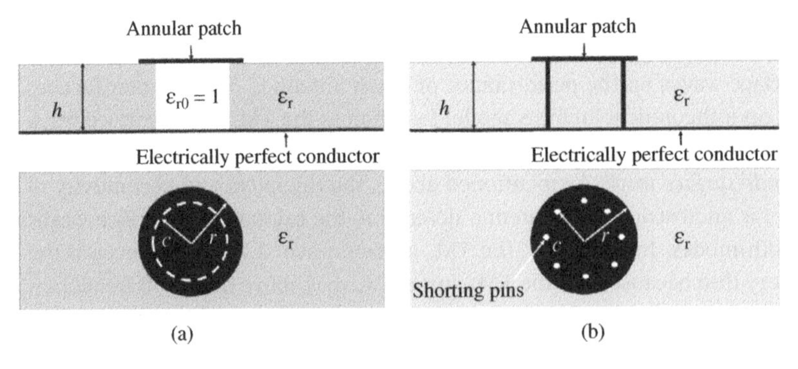

Figure 8.4 (a) The annular patch having a central air core of radius of c and (b) The annular patch shorted by grounded pins positioned at a radius of c.

easily implemented by replacing the substrate of the patch antenna shown in Figure 8.2a with a much lower permittivity dielectric such as air, for which $\varepsilon_r \cong 1$. The patch is usually supported by a spacing material such as Styrofoam ($\varepsilon_r \cong 1.07$) or plastic posts. Such a type of the suspended patch antennas have a larger physical size and are not desirable to configure large-scale antenna arrays.

An alternative is to perforate the substrate, i.e., drill holes in the substrate, hence synthesizing a lower dielectric constant substrate [12–14]. For ease of fabrication, the latter concept of synthesizing a lower dielectric constant by partially removing the substrate surrounding the patch is more practical. The method has also been applied by referring to it as trenches; no detailed information is provided in [15–18]. Furthermore, a cavity beneath the microstrip patch is also used to suppress the generation of surface waves as shown in Figure 8.5 [19–21].

Furthermore, by introducing PBG or EBG structures, the surface waves, in particular operating at TM_0 modes, in a microstrip antenna can be suppressed [22, 23]. The EBG structures can be formed by drilling thin holes on the dielectric substrate or arraying thin dielectric cylinders surrounding the patch. As a result, the radiation patterns and gain of the patch antennas can be improved. Two-dimensional periodic patch structures also can block the surface waves from propagating in microstrip patch antennas at a certain frequency band [24, 25]. Such structures have been used to enhance the isolation between patch antennas by suppressing the surface waves propagating between the patch antennas as shown in Figure 8.6a [21]. Functioning as soft-surfaces, the EBG structures have been used in microstrip patch antenna design on LTCC [26–28] as shown in Figure 8.6b. The cells of the soft surfaces can be of different types that are usually resonant with electrically small size.

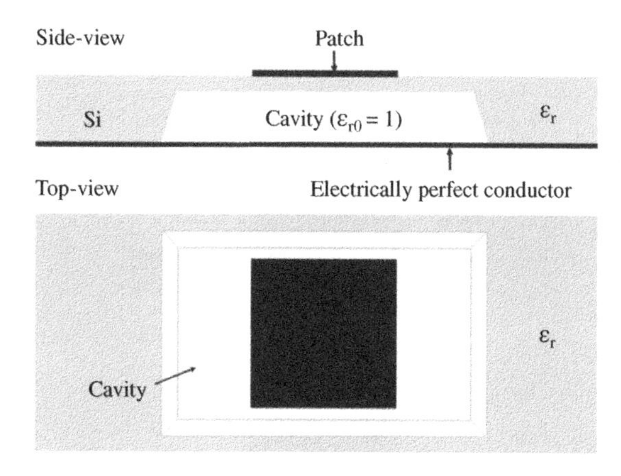

Figure 8.5 The microstrip patch antenna backed by a cavity.

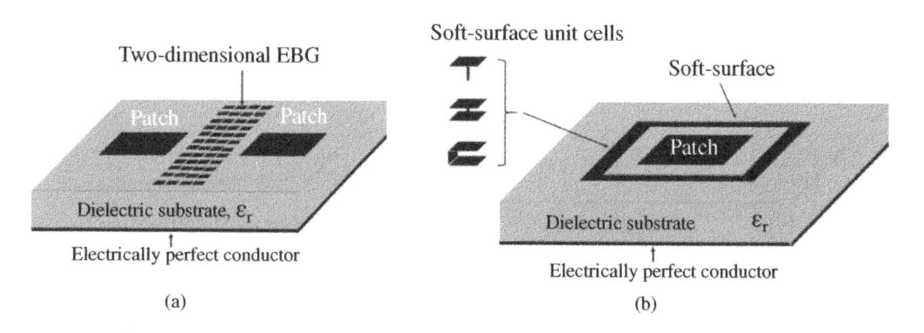

Figure 8.6 (a) The EBG structure used for improving isolation between the patch antennas and (b) patch antenna surrounding with soft-surface.

8.3 Microstrip Patch Antennas with Partial Substrate Removal

It is clear that synthesizing or removing the dielectric substrate of microstrip patch antennas improves the radiation patterns and enhances the gain. This section introduces two designs based on such techniques. One shows the detailed information about the mechanism of microstrip patch antenna based on the partial substrate removal for surface wave suppression. The other one demonstrates the detailed procedure and practical design considerations of the technique of surface wave suppression using partial substrate removal on LTCC at 60 GHz mmW band.

8.3.1 Technique of Partial Substrate Removal

Aperture-coupled microstrip patch with partial substrate removal: Consider a conventional aperture-coupled microstrip patch antenna operating at 2.4 GHz on a dielectric substrate of Roger 6006 with ε_r of 6.15 and loss tangent of 0.0027. The geometry of the aperture-coupled microstrip patch is shown in Figure 8.7a. The patch of size $l \times w$ is centered on top of the upper substrate with thicknesses of h_1. The feeding strip on the bottom of the lower substrate with thicknesses of h_2 is designed with 50 Ω to excite the patch through a narrow rectangular coupled slot positioned right underneath the center of the patch. Figure 8.7b shows the simulated gain and impedance matching against frequency. The result shows that the design achieves the impedance bandwidth of 10% over the bandwidth of 2.33–2.57 GHz for $|S_{11}| < -10$ dB. More detailed dimensions can be found in Ref. [29].

The substrate surrounding the patch is then removed to form an annular open air cavity with a width g and depth h_1 as shown in Figure 8.7c. The dimensions of the antenna are accordingly changed to keep the antenna resonating at 2.4 GHz when the effect of varying g on the gain of the

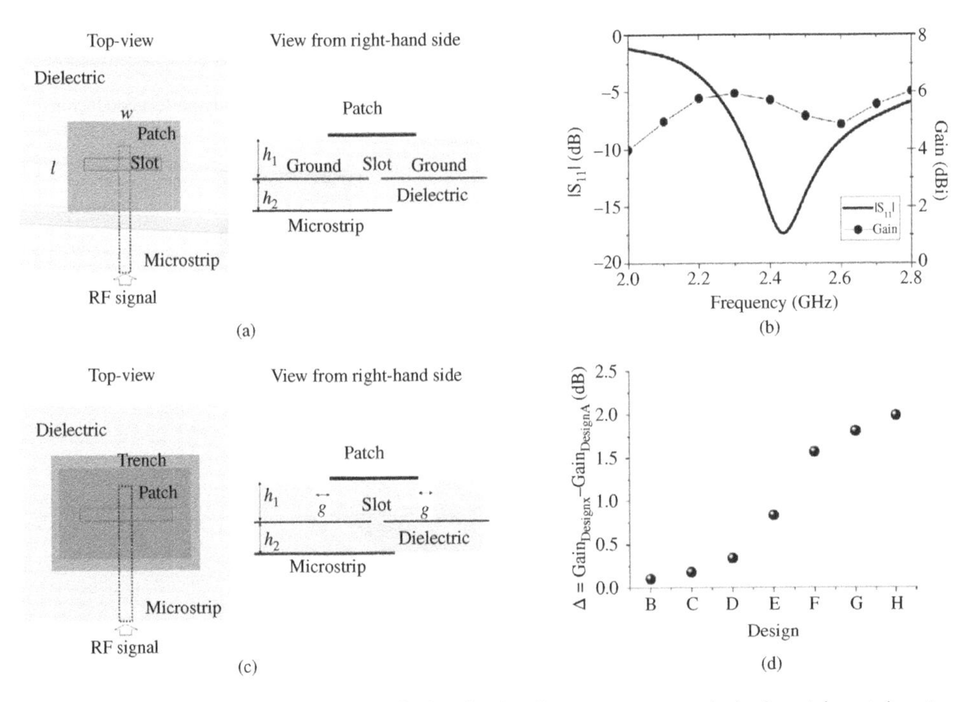

Figure 8.7 (a) Geometry, (b) simulated $|S_{11}|$ and gain of an aperture-coupled microstrip patch antenna against frequency, (c) aperture-coupled microstrip patch antenna with substrate removal, and (d) gain improvement of Designs B–H compared to Design A at 2.4 GHz [29].

antenna is studied as shown in Figure 8.7d. There are eight designs with varying the cavity width g. Design A with $g = 0$ mm is the original design while dielectric substrate in Design H is fully removed except the substrate beneath the patch. For Designs B–G, the width g gradually increases from 1.25 to 44 mm.

Compared with the gain of Design A at 2.4 GHz, Figure 8.7d shows that the gain of all Designs B–H increases when the width g of the patch antenna increases. In particular, in Designs E–H, with $g > 12$ mm, the gain improvement increases from 0.7 to 2.0 dB. It should be noted that when the width increases, the effective dielectric constant decreases accordingly. To keep the antenna operating at 2.4 GHz, the patch size inevitably increases although the ground plane size is kept unchanged. Therefore, all the factors including the increase in the dimensions of the patch, the reduction of dielectric loss due to reduced dielectric substrate, and suppression of surface waves cause the gain increase.

Aperture-coupled microstrip patch with varying configurations of partial substrate removal: Besides varying the open air cavity size as mentioned above, Figure 8.8 shows the effects of removing substrate in different configurations on the impedance matching, resonant frequency, and gain of antennas. Six configurations, namely antennas 1–6, are taken into consideration as below; note that the dimensions of the patch antennas were changed accordingly to keep their resonance at 2.4 GHz.

- **Antenna 1**: the same reference design as design A in the last section.
- **Antenna 2**: the same as design H in last section.
- **Antenna 3**: modified antenna 1 by fully removing the dielectric substrate along the radiating edges of the patch.
- **Antenna 4**: modified antenna 1 by fully removing the dielectric substrate along the non-radiating edges of the patch.
- **Antenna 5**: the same design as antenna 1 with lower permittivity.
- **Antenna 6**: modified antenna 1 by fully removing the dielectric substrate right below the patch.

To validate the fact that the gain improvement is mainly attributed to the surface waves suppression instead of the enlarged patch size, the Antenna 5 is with the same patch size as Antenna 2 by changing the dielectric constant of the substrate from $\varepsilon_r = 6.15$ to 3.6 whereas the antenna still operates at 2.4 GHz. Antenna 6 has an embedded air cavity right below the patch where the cavity size is the same as the patch. All the substrates have the same thickness h_1 as the previous designs.

The simulated and measured $|S_{11}|$ in Figure 8.8a shows that all the designs achieve good impedance matching at 2.4 GHz. The impedance matching bandwidth for 10-dB return loss for Antennas 1–6 is 8.3%, 5%, 5.1%, 6%, 11.2%, and 12%, respectively. Note that the higher surface wave loss usually is one of the possible factors for wider impedance bandwidth.

Figure 8.8b displays the normalized electric field distributions on the plane of the patch at 2.4 GHz. It is observed that compared with Antenna 1, Antennas 2 and 3 have much weaker radiation from the edges of the ground plane/substrate whereas Antennas 5 and 6 illustrate strong radiation from their edges. Compared with Antennas 2 and 3, Antenna 4 has stronger radiation from its substrate edges and ground plane along the dielectric substrate. It suggests that a majority of surface waves generated in microstrip patch antennas stem from the radiating edges of the patch. Compared with Antenna 1, Antenna 5 has the weaker radiation from the edges of its dielectric substrate and ground plane. This observation proves the fact again that the dielectric substrate causes the surface waves and the higher the dielectric constant, the stronger the surface waves.

Considering the generation of surface waves in the six antennas, the gain of the patch antennas inevitably changes. Antennas 1–6 achieve gains of 5.7, 7.6, 7.5, 6.6, and 6.3 dBi, respectively.

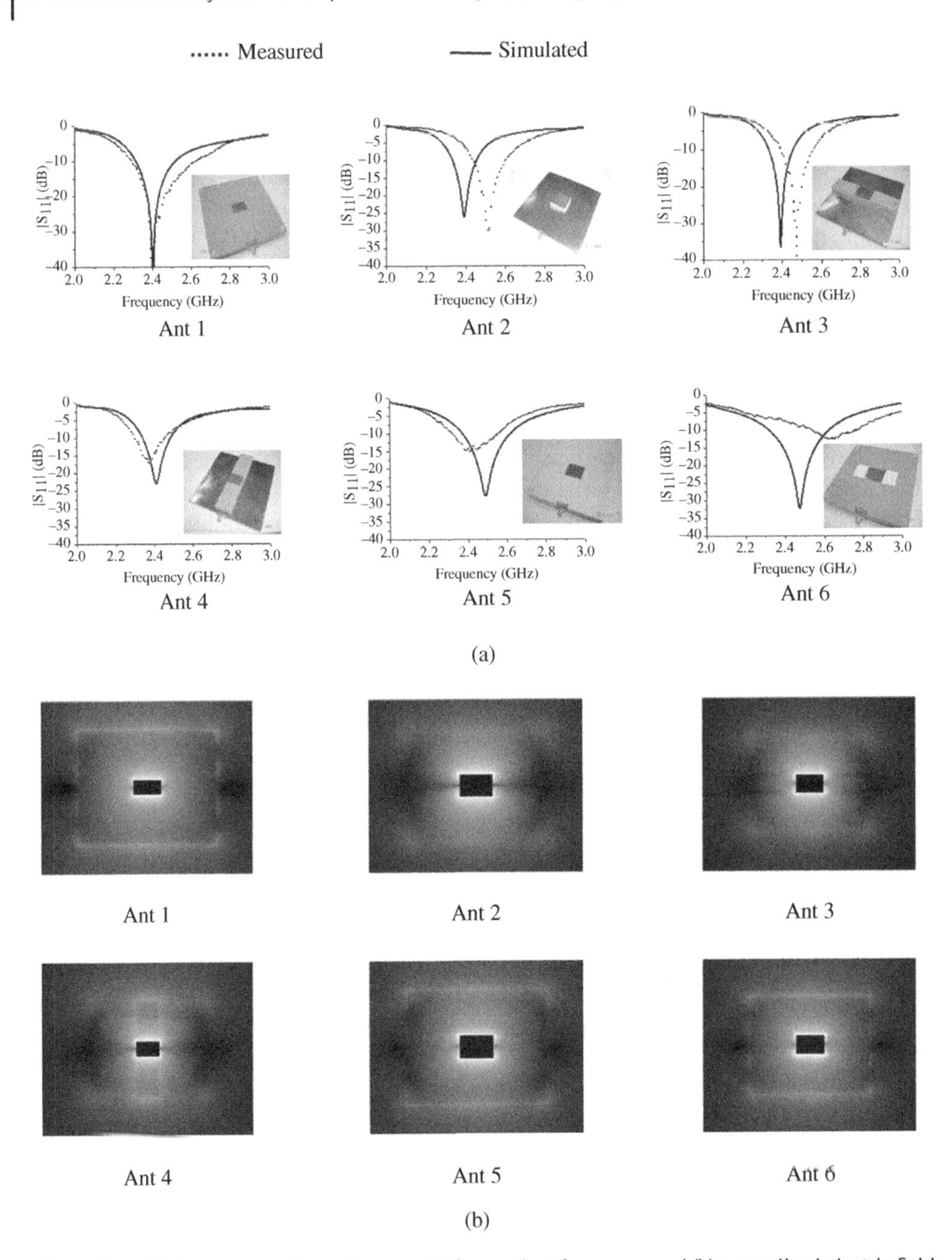

Figure 8.8 (a) Geometry and impedance matching against frequency and (b) normalized electric field distributions at 2.4 GHz of the aperture-coupled microstrip patch antennas with different substrate removal configurations.

The results well agree with the observation from Figure 8.8b and verify the fact that Antennas 2 and 3 with less surface waves can focus more radiation to the boresight direction compared to the other antennas with more surface waves so that both achieve the highest gain. The gain improvement in this study reaches 1.9 dB. This also suggests that the gain increase of Designs F–H in the last section is mainly attributable to the suppression of surface waves in the designs.

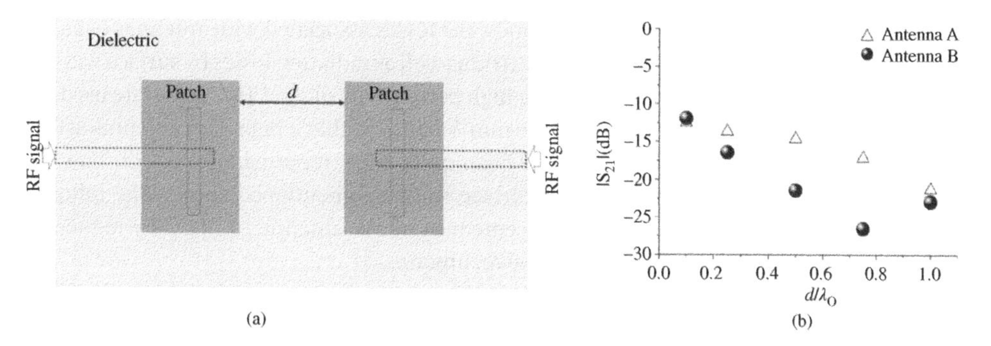

Figure 8.9 (a) Two aperture-coupled patch antennas placed opposite to each other with separation *d* and (b) comparison of isolation at 2.4 GHz for the pairs of Antennas A and B with varying separation *d*.

The inter-antenna mutual coupling or isolation: With the suppression of surface waves of microstrip patch antennas, a lower mutual coupling, or higher isolation, between the microstrip patch antennas can be expected. Figure 8.9 shows the configurations of the two antennas. The mutual-coupling between the two adjacent microstrip patch antennas positioned along their *E*-planes is examined. The antenna pairs A and B are with/without substrate removal. The antennas are placed opposite to each other as shown in Figure 8.9a at a distance of *d* away from each other. The separation *d* is chosen at 0.1, 0.25, 0.5, 0.75, and $1\lambda_0$.

Figure 8.9b shows the isolation for two pairs of antennas, A and B, for a varying *d* at 2.4 GHz, respectively. The pair of antennas B achieves better isolation $|S_{21}|$ or $|S_{12}|$ for $d = 0.25 \sim 1.0\lambda_0$ and a maximum for $d = 0.75\lambda_0$ with isolation difference of 9.4 dB. The isolation then becomes almost similar as $d \sim \lambda_0$ when the surface waves strength starts to deteriorate. Another point to note is that the pairs of Antennas A and B have the same isolation when spacing *d* is around $0.1\lambda_0$ as space waves also directly contribute to and dominate the mutual coupling at a small separation.

In summary, the technique of partial substrate removal is capable of significantly suppressing the surface waves of microstrip patch antennas. Therefore, the technique can be used to improve the gain of microstrip patch antennas and isolation between the closely positioned antennas.

8.3.2 60-GHz LTCC Antenna with Partial Substrate Removal

Millimeter-wave antennas on LTCC: With higher operating frequency at mmW bands, the antennas have unique design challenges in terms of material selection, fabrication tolerance, and measurement methodology of the antennas. In particular, the losses caused by materials, fabrication tolerance, connection, and so on, become much more critical in antenna engineering compared with the designs at lower microwave bands. The conventional PCB process is not good enough for the mmW antennas with higher dielectric loss and lower fabrication tolerance. Therefore, the fabrication of mmW antennas using LTCC are getting increasing attention. Besides lower substrate loss and higher fabrication tolerance, the LTCC process also enjoys flexibility in realizing an arbitrary number of layers and ease of integration with circuit components through stacked vias and cavity-buried components as mentioned in Chapter 1. Furthermore, the reliable and flexible fabrication process offers much more design freedom, in particular, cross-layer vias and open and embedded cavities, in mmW antenna and circuit design than the conventional PCB process.

Since the 2000s, there have been many types of mmW antennas fabricated on LTCC, in particular at operating frequencies around and higher than 60 GHz. The antennas on LTCC include almost all types of antennas such as patch antennas, slot antennas, Yagi antennas, and linear tapered slot antennas for wireless communication, radar, and imaging systems [19–21, 26–28, 30–35].

Surface wave loss in LTCC antennas: At mmW bands, the losses associated with antennas usually include ohmic losses caused by conductor and dielectric as well as radiation losses by surface waves. In particular, the larger electrical thickness and the high permittivity of the LTCC substrate used in the antenna design result in significant losses at the mmW bands so that a high-gain antenna/array on LTCC becomes a design challenge. There have been quite a few reported methods on how to suppress the losses, especially losses caused by surface waves as mentioned above. The following design example demonstrates the gain enhancement of mmW antenna in LTCC by reducing surface-wave loss with the partial substrate removal technique.

60-GHz patch antenna on LTCC: The technique of partial substrate removal instead of an embedded cavity is much more suitable for antennas on LTCC, more easily processed and more mechanically stable without fear of collapse, and easier to identify fabrication deformation under a microscope. Note that fully removing the substrate surrounding the patch would lead to a substrate "island" being formed. Due to limitations in the LTCC process, it is not possible to fabricate such a configuration. Alternatively, a more realistic approach is to partially remove the substrate around the two radiating sides of the patch only based on the findings in the last Subsection, 8.3.1.

Consider a conventional aperture-coupled patch antenna operating at 60 GHz as shown in Figure 8.6. There are two designs with and without the partially removed substrate. Different from the design at 2.4 GHz band, the substrate here is LTCC Ferro A6-M with ε_r of 5.9 and a loss tangent of 0.001. Each LTCC layer has a thickness of 0.09652 mm. The antenna is designed from a five-layer LTCC substrate with an overall size of l = 4 mm, w = 4 mm, h_1 = 0.38608 mm, and h_2 = 0.09652 mm. The patch is fed by a 50-Ω microstrip line via a slot that is positioned on the bottom of the substrate. The feeding slot is cut onto the ground plane that is one layer above the feeding strip. More detailed dimensions of the antenna designs can be found in Ref. [36].

Figure 8.10 shows the three designs including the conventional aperture-coupled patch (Design A) with full substrate as a reference design, the aperture-coupled patch supported by substrate right beneath the patch (Design B), and the aperture-coupled patch with a substrate stub along its non-radiating edges only (Design C). Design B is unable to be fabricated using the LTCC process but just simulated for comparison.

Figure 8.11 compares the simulated impedance matching and the gain response of the three designs against frequency. Figure 8.11a shows the impedance bandwidth of Design A of 18% and Designs C and D of about 7%. Figure 8.11b shows that Designs B and C achieve a gain enhancement of 2.3–2.8 dB compared with Design A over the frequency band of 57–64 GHz. The comparison suggests that Design A achieves a wider operating bandwidth but suffers from a lower gain because of the higher surface wave loss generated in the thicker higher-permittivity substrate. Therefore, the technique is a feasible way to suppress the surface waves for improving the gain of the mmW antennas based on the LTCC process.

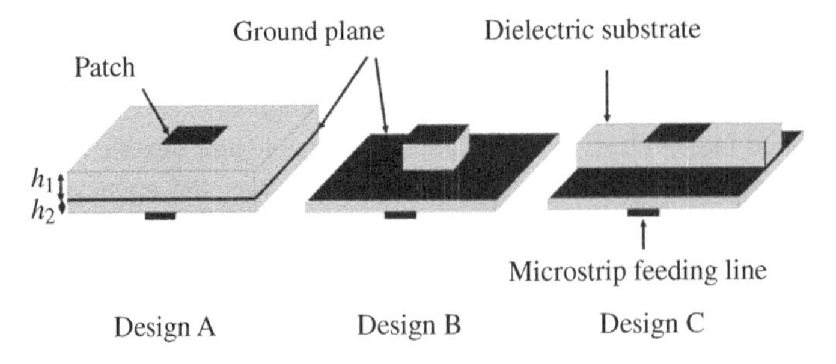

Figure 8.10 Three aperture-coupled patch antennas.

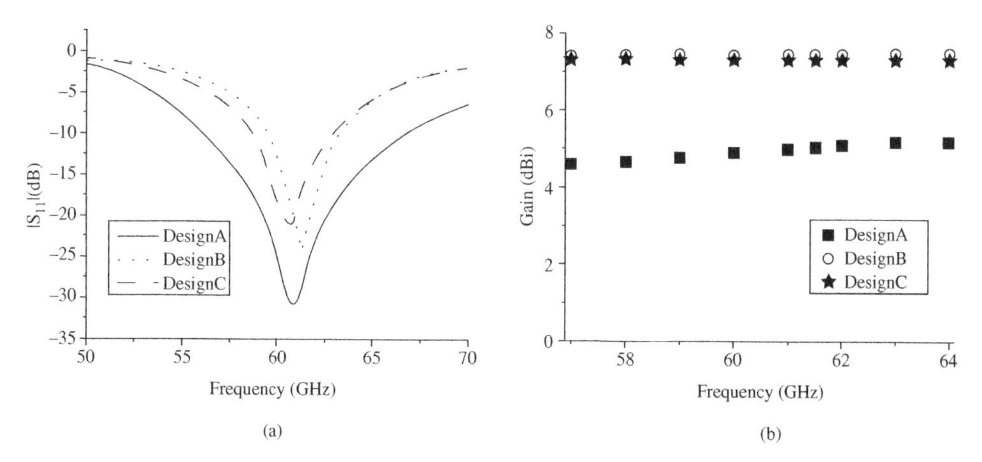

Figure 8.11 Comparison of the simulated performance. (a) $|S_{11}|$ and (b) gain of Designs A, B, and C.

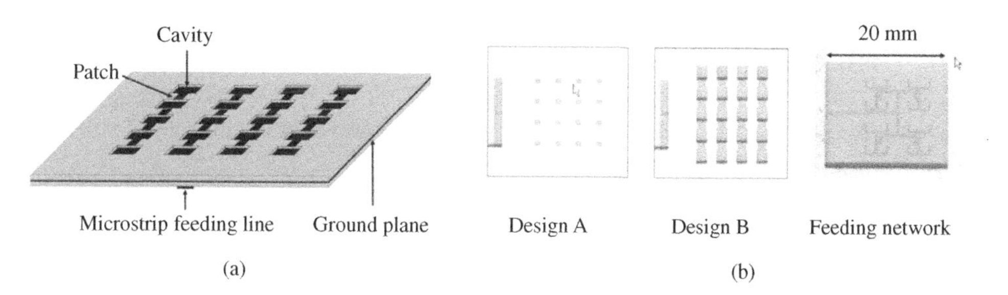

Figure 8.12 (a) A 4 × 4 array design with open air cavities by the radiating edges of the patches and (b) prototypes of reference array and array with open air cavities and feeding networks on the bottom of LTCC substrate.

60-GHz patch antenna arrays on LTCC: Designs A and C are chosen as the antenna elements for the 4 × 4 array designs, respectively; the latter is shown in Figure 8.12. The inter-element spacing of the antenna array is set at $0.6\lambda_0$ (λ_0 is the free space wavelength at 60.0 GHz). Due to the limitations of the LTCC process, 20 open air cavities with ground plane as their bottoms are fabricated on the LTCC substrate because the substrate in between the rows could not be removed entirely along the radiating edges of the patches as shown in Figure 8.12. Ref. [36] demonstrated three types of feeding structures used in the array, namely the cavity with a feeding microstrip line, the cavity covered by an extra LTCC layer with a feeding microstrip line, and the cavity with a feeding stripline.

Performance comparison: The antenna arrays A and B achieve the $|S_{11}| \leq -10$ dB bandwidths of 17% and 13% and gains of 15.3 and 17.4 dBi, respectively. It is concluded that the gain of the patch antenna on LTCC is greatly enhanced using a partial substrate removal technique because of the significant suppression of the severe surface waves in the dielectric substrate. More importantly, this technique is readily implemented using the LTCC process.

8.4 Summary

The generation of surface waves in microstrip patch antennas is not desirable. In mmW antenna designs, the existence of surface waves is the main cause for non-ohmic loss, lowering the antenna gain and isolation between the microstrip patch antennas. Partial substrate removal is a feasible and effective mean to suppress the severe surface waves in the patch antenna design using the LTCC process. Such implementation can significantly improve the gain of the microstrip patch antennas, in particular, at mmW bands.

References

1 Goubau, G. (1950). Surface waves and their application to transmission lines. *J. Applied Physics. Amer. Inst. Phys.* 21: 119.

2 Collin, R.E. (1990). Surface waveguides. In: *Field Theory of Guided Waves*, Chapter 11. New York: Wiley-IEEE Press.

3 Schelkunoff, S.A. (1959). Anatomy of "surface waves". *IRE Trans. Antennas Propaga.* 7 (5): S133–S139.

4 Harrington, R.F. (2001). *Time-Harmonic Electromagnetic Fields*. Chapter 4–8. New York: IEEE Press.

5 Garg, R., Bhartia, P., Bahl, I.J., and Ittipiboon, A. (2001). *Microstrip Antenna Design Handbook*. Chapter 1. Boston: Artech House.

6 Pozar, D.M. (1983). Considerations for millimeter wave printed antennas. *IEEE Trans. Antennas Propaga.* 31 (5): 740–747.

7 Bhattacharyya, A.K. (1990). Characteristics of space and surface-waves in a multilayered structure. *IEEE Trans. Antennas Propaga.* 38 (9): 1231–1238.

8 Jackson, D.R., Williams, J.T., Bhattacharyya, A.K. et al. (1993). Microstrip patch designs that do not excite surface waves. *IEEE Trans. Antennas Propaga.* 41 (8): 1026–1037.

9 Chen, Z.N. (2000). Broadband probe-fed plate antenna. In: *30th European Microwave Conference*, 1–5.

10 Chen, Z.N. and Chia, M.Y.W. (2005). *Broadband Planar Antennas: Design and Applications*. London: Wiley.

11 Lee, H.S., Kim, J.G., Hong, S., and Yoon, J.B. (2005). Micromachined CPW-fed suspended patch antenna for 77 GHz automotive radar applications. *Eur. Microwave Conf.* 3: 4–6.

12 Gauthier, G.P., Courtay, A., and Rebeiz, G.M. (1997). Microstrip antennas on synthesized low dielectric-constant substrates. *IEEE Trans. Antennas Propoga.* 45 (8): 1310–1314.

13 Colburn, J.S. and Rahmat-Samii, Y. (1999). Patch antennas on externally perforated high dielectric constant substrates. *IEEE Trans. Antennas Propaga.* 47 (12): 1785–1794.

14 Yook, J.G. and Katehi, L.P.B. (2001). Micromachined microstrip patch antenna with controlled mutual coupling and surface waves. *IEEE Trans. Antennas Propaga.* 49 (9): 1282–1289.

15 Solis, R.A.R., Melina, A., and Lopez, N. (2000). Microstrip patch encircled by a trench. *IEEE Int. Sym. Antennas Propaga. Soc.* 3: 1620–1623.

16 Chen, Q., Fusco, V.F., Zheng, M., and Hall, P.S. (1998). Micromachined silicon antennas. *Int. Conf. Microwave Millimeter-wave Tech. Proc.*: 289–292.

17 Chen, Q., Fusco, V.F., Zhen, M., and Hall, P.S. (1999). Trenched silicon microstrip antenna arrays with ground plane effects. In: *29th European Microwave Conference*, vol. 3, 263–266.

18 Chen, Q., Fusco, V.F., Zheng, M., and Hall, P.S. (1999). Silicon active slot loop antenna with micromachined trenches. *IEE Nat. Conf. Antennas Propaga.*: 253–255.

19 Panther, A., Petosa, A., Stubbs, M.G., and Kautio, K. (2005). A wideband array of stacked patch antennas using embedded air cavities in LTCC. *IEEE Microwave Wirel. Compon. Lett.* 15 (12): 916–918.

20 Lamminen, A.E.I., Saily, J., and Vimpari, A.R. (2008). 60-GHz patch antennas and arrays on LTCC with embedded-cavity substrate. *IEEE Trans. Antennas Propaga.* 56 (9): 2865–2874.

21 Byun, W., Kim, B.S., Kim, K.S. et al. (2009). 60GHz 2×4 LTCC cavity backed array antenna. *IEEE Int. Sym. Antennas Propaga.*: 1–4.

22 Gonzalo, R., de Maagt, P., and Sorolla, M. (1999). Enhanced patch-antenna performance by suppressing surface waves using photonic-bandgap substrates. *IEEE Trans. Microwave Theory Tech.* 47 (11): 2131–2138.

23 Boutayeb, H. and Denidni, T.A. (2007). Gain enhancement of a microstrip patch antenna using a cylindrical electromagnetic crystal substrate. *IEEE Trans. Antennas Propoga.* 55 (11): 3140–3145.

24 Yang, F. and Rahmat-Samii, Y. (2001). Mutual coupling reduction of microstrip antennas using electromagnetic band-gap structure. *IEEE Int. Sym. Antennas Propaga. Soc.* 2: 478–481.

25 Llombart, N., Neto, A., Gerini, G., and de Maagt, P. (2005). Planar circularly symmetric EBG structures for reducing surface waves in printed antennas. *IEEE Trans. Antennas Propaga.* 53 (10): 3210–3218.

26 Li, R.L., DeJean, G., Tentzeris, M.M. et al. (2005). Radiation-pattern improvement of patch antennas on a large-size substrate using a compact soft-surface structure and its realization on LTCC multilayer technology. *IEEE Trans. Antennas Propaga.* 53 (1): 200–208.

27 Byun, W., Eun, K.C., Kim, B.S. et al. (2005). Design of 8×8 stacked patch array antenna on LTCC substrate operating at 40GHz band. *Asia Pacific Microwave Conf.*: 1–4.

28 Lamminen, A.E.I., Vimpari, A.R., and Saily, J. (2009). UC-EBG on LTCC for 60 GHz frequency band antenna applications. *IEEE Trans. Antennas Propagat* 57 (10): 2904–2912.

29 Yeap, S.B. and Chen, Z.N. (2010). Microstrip patch antennas with enhanced gain by partial substrate removal. *IEEE Trans. Antennas Propaga.* 58 (9): 2811–2816.

30 Seki, T., Honma, N., Nishikawa, K., and Tsunekawa, K. (2005). A 60GHz multilayer parasitic microstrip array antenna on LTCC substrate for system-on-package. *IEEE Microwave Wirel. Compon. Lett.* 15 (5): 339–341.

31 Vimpari, A., Lamiminen, A., and Saily, J. (2006). Design and measurements of 60 GHz probe-fed patch antennas on LTCC substrates. In: *Proceedings of the 36th European Microwave Conference*, 854–857.

32 Lee, J.H., Kidera, N., Pinel, S. et al. (2006). V-band integrated filter and antenna for LTCC front-end modules. *IEEE Int. Microwave. Symp.*: 978–981.

33 Jung, D.Y., Chan, W., Eun, K.C., and Park, C.S. (2007). 60-GHz system-on-package transmitter integrating sub-harmonic frequency amplitude shift-keying modulator. *IEEE Trans. Microwave Theory Tech.* 55 (8): 1786–1793.

34 Aguirre, J., Pao, H.Y., Lin, H.S. et al. (2008). An LTCC 94 GHz antenna array. *IEEE Int. Sym. Antennas Propaga.*: 1–4.

35 Xu, J., Chen, Z.N., Qing, X., and Hong, W. (2011). Bandwidth enhancement for a 60 GHz substrate integrated waveguide fed cavity array antenna on LTCC. *IEEE Trans. Antennas Propaga.* 59 (3): 826–832.

36 Yeap, S.B., Chen, Z.N., and Qing, X. (2011). Gain-enhanced 60-GHz LTCC antenna array with open air cavities. *IEEE Trans. Antennas Propaga.* 59 (9): 3470–3473.

9

Substrate Integrated Antennas for Millimeter Wave Automotive Radars

Xianming Qing[1] *and Zhi Ning Chen*[2]

[1] *Signal Processing, RF, and Optical Department, Institute for Infocomm Research, Singapore 138632, Republic of Singapore*
[2] *Department of Electrical and Computer Engineering, National University of Singapore, Singapore 117583, Republic of Singapore*

9.1 Introduction

Making driving safer and more convenient has been one of the key promises for every new car generation during the past four decades [1–4]. The radar-based sensors have been identified as a key technology to further decrease the amount and severity of traffic accidents, as they are robust against weather conditions and other environmental influences. Automotive radars have been on the market since 1999, in both 24 GHz and 77 GHz bands first and then with a new frequency band of 77–81 GHz intended for medium and short-range sensors. By 2003, most major car manufacturers offered radar systems in their upper-class segments. Since then, several new generations of automotive radar sensors have become available from a number of manufacturers. A rapidly growing number of radar-based sensors are being integrated into new vehicles to allow driver assistance functions, such as adaptive cruise control (ACC), emergency braking (EB), parking assistant system (PAS), lane change assist (LCA), front/rear cross-traffic alert (F/RCTA), rear-collision warning (RCW), and blind spot detection (BSD) for comfort and safety [5–11]. The evolution of automotive radar from its inception to the present has been thoroughly discussed in [2, 3]. Today's radar-based advanced driver assistance systems (ADASs) have become available even for middle class cars. With highly integrated and inexpensive mmW circuits implemented in silicon, compact automotive radar safety systems have become a popular feature [12, 13].

9.1.1 Automotive Radar Classification

No super radar sensor features all the needed functions for ADASs. Instead, a specific radar sensor is needed to address different requirements such as the range or field of view (FOV) properly. In the automotive industry, the radar sensors can be broadly categorized into three groups as follows:

- Long-range radar (LRR): for applications with the need of a narrow-beam forward-looking view such as ACC;
- Medium-range radar (MRR): for applications with a medium distance and speed profile, such as CTA; and
- Short-range radar (SRR): for applications for sensing in direct proximity of the vehicle, such as PAS.

Substrate-Integrated Millimeter-Wave Antennas for Next-Generation Communication and Radar Systems, First Edition.
Edited by Zhi Ning Chen and Xianming Qing.

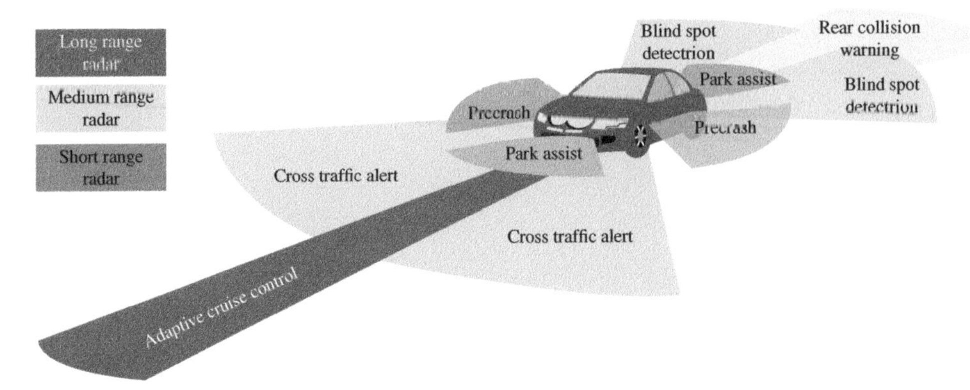

Figure 9.1 Radar subsystems in ADASs [4].

Table 9.1 The classification of automotive radars based on range detection capability.

Radar type	Long-range radar	Medium-range radar	Short-range radar
Range (m)	10–250	10–100	0.15–30
Distance resolution (m)	0.5	0.5	0.1
Distance accuracy (m)	0.1	0.1	0.02
Speed resolution (m/s)	0.6	0.6	0.6
Speed accuracy (m/s)	0.1	0.1	0.1
Angular accuracy (°)	0.1	0.5	1
Azimuthal field of view (°)	±5	±5	±10
Elevation field of view (°)	±15	±40	±80
Applications	ACC, EB	F/RCTA, BSD, RCW	PAS, PC

Furthermore, the radar sensors can be grouped into corner radars and front radars as well. Corner radars usually implemented at both the rear and front corners of the car are typically for short-range detection while the front radars are typically for medium and long-range detection.

Figure 9.1 depicts the radar subsystems in ADASs. Each subsystem has unique functions and specific requirements in terms of radar range and angular measurement capability as exhibited in Table 9.1.

9.1.2 Frequency Bands for Automotive Radars

Currently, two different frequency bands are predominantly used for automotive radar sensors, namely 24 GHz and 77 GHz bands. Other frequencies, for instance, below 10 GHz or above 100 GHz, have also been investigated; however, currently they do not play a practical role.

The 24 GHz band includes an industrial, scientific, and medical (ISM) band from 24.0 to 24.25 GHz with a bandwidth of 250 MHz, which is often called the narrowband (NB). The 24 GHz band also includes an ultra-wideband (UWB) from 21.65 to 26.65 GHz with a bandwidth of 5 GHz. For an SRR, the 24 GHz NB and UWB bands have been used in automotive sensors. For basic blind-spot detection, the 24 GHz NB can be used while for the cases including the ultra SRR applications, the need for high-range resolution dictates the use of the 24 GHz UWB band. However,

Table 9.2 Frequency bands available for automotive radars over the world.

Country	24 GHz Narrow Band (ISM 24.05–24.25 GHz, NB)	24 GHz Ultra-wideband (21.65–26.65 GHz UWB)	60 GHz Band (60–61 GHz)	77 GHz Band (76–77 GHz)	79 GHz Band (77–81 GHz)
Europe	√	√		√	√
USA	√	√		√	√
China	√			√	
Japan			√	√	√
Russia	√	√		√	√
Korea	√	√		√	√

due to the spectrum regulations and standards developed by the European Telecommunications Standards Institute (ETSI) and the Federal Communications Commission (FCC), the 24 GHz UWB band is being phased out and will not be available after January 1, 2022 [14–18].

The first 77 GHz band for automotive radar is from 76 to 77 GHz, which is available nearly worldwide. The second band is the directly neighboring band from 77 to 81 GHz, which has been introduced to replace the 24 GHz UWB. Table 9.2 shows the current frequency regulatory status in different countries.

9.1.3 Comparison of 24 GHz and 77 GHz Bands

There is an ongoing competition in the choice between the two frequency bands for automotive radars. In the early days, a 24-GHz radar sensor was often used for short and mid-range detection, while the 77-GHz radar sensor found its way into long-range detection. As the technology improves with lower cost and better performance at higher frequencies, there is a tendency to replace 24-GHz radar sensors with 77-GHz radar sensors, although the shipment of 24-GHz radar sensors, particularly 24-GHz side-looking short range radar sensors still prevails in the mmW radar market. However, 77-GHz radar sensors have nearly caught up with the 24-GHz radar sensors in market size by 2020.

The limited bandwidth of the 24 GHz band, coupled with the need for higher performance in emerging radar applications, makes the 24 GHz band less attractive for new automotive radar sensor implementations. This is especially true if considering the significant interest in the automotive industry for advanced applications such as automated parking and a 360° view.

At the 77 GHz band, there is a 76–77 GHz band available for vehicular LRR applications. This band has the benefit of higher allowed equivalent isotropic radiated power (EIRP), which is desired for LRR applications such as adaptive cruise control. Recently, both regulators and industry have paid much attention to the new 77–81 GHz SRR band because of the availability of wide bandwidth up to 4 GHz, in particular for the applications requiring high range resolution. Very likely, 77-GHz automotive radar sensors will replace the majority of 24-GHz ones in the near future.

The merits of 77-GHz automotive radar sensors over the 24-GHz ones are summarized below [19]:

- Higher range resolution and range accuracy:
  A 77-GHz radar sensor is able to achieve 20 times better performance in terms of range resolution and accuracy than those of 24-GHz ones. The achievable range resolution can be 4 cm, which is useful for automotive park-assist applications because higher range resolution offers better separation of objects and helps the sensor achieving a shorter minimum distance.

- Improved speed resolution and accuracy:
  The speed resolution and accuracy are inversely proportional to the operating frequency. Therefore, the speed resolution and accuracy of a 77-GHz radar sensor will be three times of those at 24 GHz. The enhanced speed resolution and accuracy is desired for automotive park-assist applications where an accurate maneuvering of the vehicle at slow speeds is needed.
- Smaller form factor:
  For a specific antenna FOV and gain, the aperture of the antenna at 77 GHz is about one ninth of that at 24 GHz. The smaller antenna enables a small sensor, which is particularly useful for automotive applications wherein the radar sensors need to be mounted in the tight spots behind the bumper or other spots around the car such as door and trunks.

9.1.4 Antenna System Considerations for Automotive Radar Sensors

The antenna system is one of the main differentiating factors of an automotive radar sensor. Besides the antenna element performance such as gain, impedance matching, bandwidth, sidelobe level, and cross polarization level, there are system requirements such as the FOV and the number of channels to be considered [13, 20].

In general, a wider FOV is desired for the automotive radar sensor while it is limited by the narrow beamwidth of the antenna with higher gain. Thus, much effort has been put to the design of the antenna systems in combination with modified overall sensor arrangements, including multi-beam antennas, switched beam antennas, and analog/digital beam forming antenna arrays.

9.1.4.1 Lens Antenna and Reflector Antenna

Dielectric lens antenna [21] and reflector antenna [8, 22] have been used for LRR sensors where a narrow FOV is required. If more than one antenna feed is used, an offset between the different feeds allows the generation of simultaneous or sequential multiple beams looking at different directions as shown in Figures 9.2 and 9.3 respectively. If each antenna feed is connected to an individual receiver channel, parallel data acquisition can be achieved, allowing processing of multiple antenna beams simultaneously. Monopulsing can be used to gather the angular information of a detected object using multiple antenna beams [9].

9.1.4.2 Planar Antennas

The most popular type of planar antennas is based on a microstrip process. Single-patch antennas can be used as feeding elements for a lens antenna, or arrays of microstrip patches can be used directly as automotive antennas as shown in Figure 9.4 [10–12]. For large-scale antenna arrays, feeding network losses may pose a limit for antenna size.

The antenna array can be connected to a feeding network to generate a fixed beam. The antenna elements or sub-arrays can also be connected to an analog beamformer or digital beamformer to provide electronically beam scanning as shown in Figure 9.5. The analog beamforming using phased array technology is exhibited in Figure 9.5a, where the beamforming is generated directly in the mmW frontend [23, 24]. By controlling the phase weights of each phase shifter electronically, the direction and beamwidth of the antenna can be adjusted to a desired value almost instantaneously. This property provides multi-mode ability via beam shape versatility. By tapering the amplitude of the antenna elements, the sidelobe levels can be controlled as well. Until the last decade, the use of the phased array concept for the automotive radars was almost impossible due to the high cost of mmW phase shifters. However, improved silicon semiconductor technology allows the integration of multiple channels of phase shifters and variable amplifiers into one single chip, which enables the possibility of utilizing analog beamforming techniques for automotive radar sensor, in particular, at the 77 GHz band.

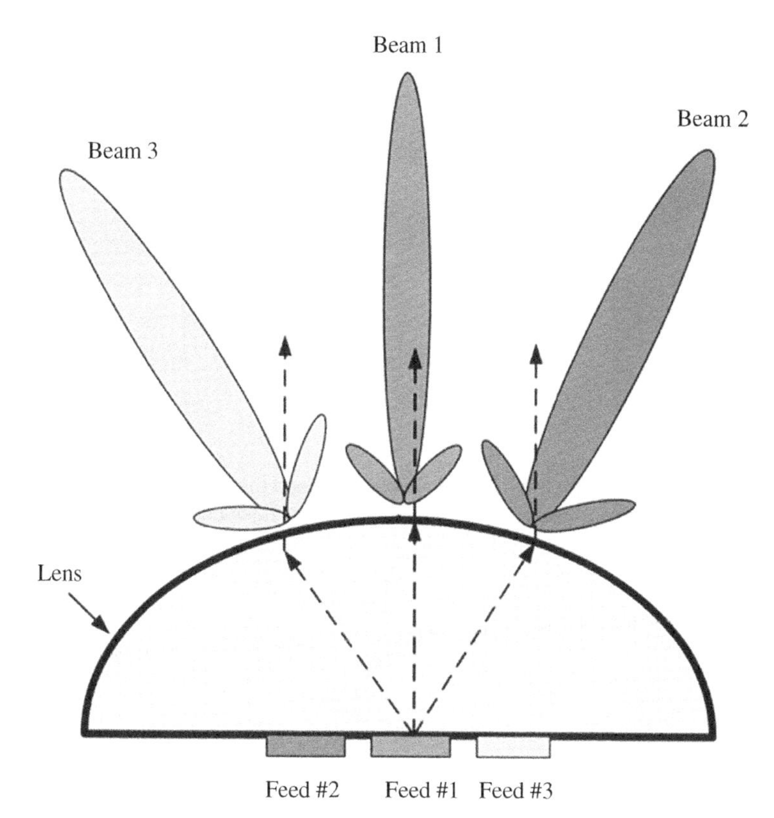

Figure 9.2 Dielectric lens antenna with switched beams.

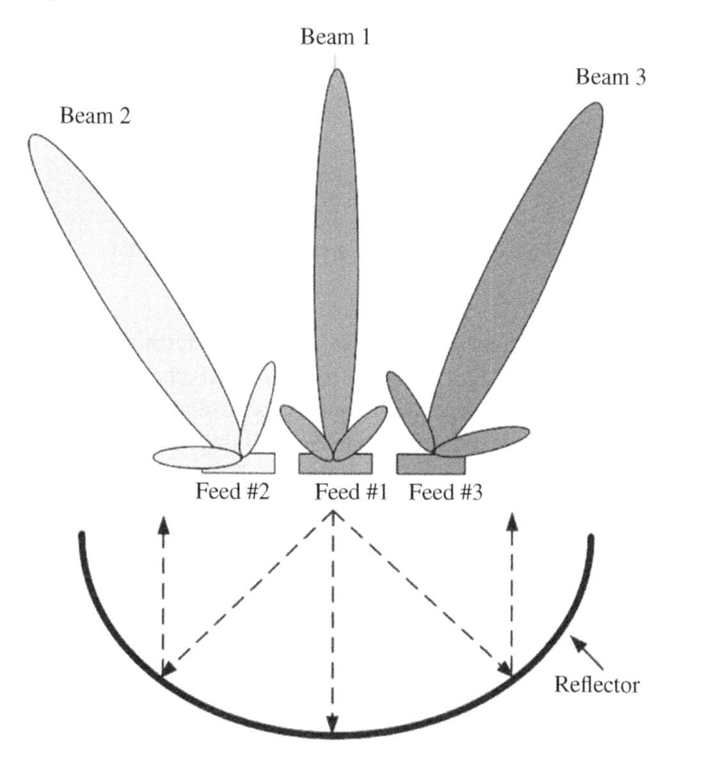

Figure 9.3 Reflector antenna with switched beams.

Figure 9.4 (a) Automotive radar sensor using microstrip antenna elements to feed the lens and (b) sensor with three transmit and three receive patch antenna arrays (courtesy of Bosch and Toyota Laboratories).

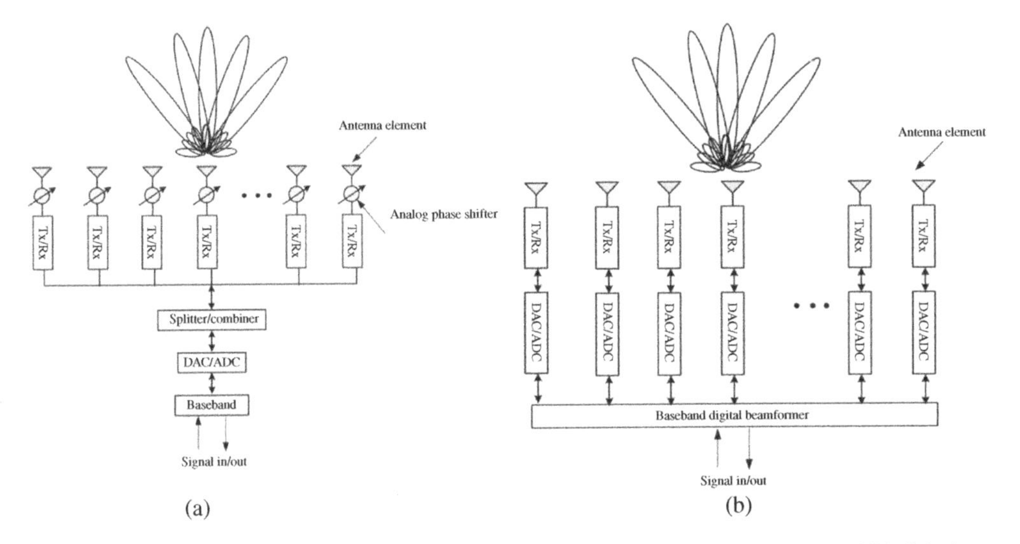

Figure 9.5 Schematic diagram of beamforming techniques. (a) Analog beamforming and (b) digital beamforming.

Compared with the analog beamforming, the beamsteering of the digital beamforming shown in Figure 9.5b is carried out in the digital baseband [25]. There is a separate RF chain for each antenna element. The beam is then formed by matrix-type operations in the baseband where the amplitude and phase weighting are applied. Digital beamforming has the advantage that the digital data streams can be manipulated and combined in many possible ways to get many different output signals in parallel. The signals from all directions can be detected simultaneously, and the signals can be integrated for a longer time when studying far-off objects and simultaneously integrated for a shorter time to study fast-moving close objects, and so on.

9.1.5 Fabrication and Packaging Considerations

The automotive radar sensors have to operate in different harsh environments under the conditions of rain, snow, or ice with a varying temperature range from −40 to 85 °C or even higher, and to withstand shock and vibrations as well. Therefore, materials, fabrication processes, and packaging have to be selected carefully. Substrate materials need to be suitable for the high frequency at 77 GHz, their physical and electrical properties should be reasonably constant over the temperature range, and they should not absorb moisture. On the other hand, the standard microwave substrates may be too expensive, so a compromise has to be made in this respect. For a low-cost point of view, the fabrication using standard PCB processes is desired while it is a big challenge to achieve the accuracies down to 20 μm for the 77 GHz operating frequency range. This also poses a great challenge toward tolerance-optimized antenna design. Metal surfaces, in addition, need be protected against corrosion by suitable plating. The antenna and the radar sensor as a whole, finally, have to be protected by a suitable package. The antenna radome must be transparent to the electromagnetic wave in the respective frequency range, leading to an optimized thickness for the used plastic materials in the order of multiples of half a guide wavelength within the material. Moreover, the radome should be optimized for all other angles instead of just for boresight.

9.2 State-of-the-Art Antennas for 24-GHz and 77-GHz Automotive Radars

Compared with the reflector and lens antennas, the substrate integrated planar antenna and array are more preferable for automotive radar sensors. The microstrip patch antenna is a desired choice for automotive radar sensors due to its low cost, low profile, simple integration with systems, and easy implementation in an array structure [26, 27]. The main concerns of the microstrip antennas come from the undesired radiation as well as surface wave coupling and loss. Alternatively, the SIW [28–30] or post-wall waveguide fed antennas feature the same advantages of microstrip antennas because they can be designed and implemented using the same PCB fabrication technology. Furthermore, as a waveguide-like structure, SIW does not suffer from the unintentional radiation and surface wave loss, which alleviates the limitation of using a thin substrate for antenna design. These merits offer the SIW-based antennas the possibility of achieving higher efficiency.

In this section, selected state-of-the-art substrate integrated planar antennas for automotive radar sensors are briefly reviewed.

9.2.1 Selected State-of-the-Art Antennas for 24-GHz Automotive Radars

9.2.1.1 Shorted Parasitic Rhombic Patch Antenna Array with Lower Cross-Polarization Levels

The antenna for rear and side detection system (RASD) typically requires a low cross-polarization level since the 45° slant polarization is utilized to reduce the interference from the car traveling in an adjacent lane. The signals from other cars are orthogonally polarized by using 45° slant polarization [31–33]. To reduce the high cross-polarization level of a conventional 45° slant polarized microstrip antenna, Shin et al. presented a shorted parasitic rhombic array antenna as shown in Figure 9.6. The parasitic rhombic antenna element is made from the rhombic patch antenna by slitting both sides of the rhombic patch. These parasitic elements increase the currents related to the co-polarization while decreasing the current related to the cross-polarization. The grounded pins at the center of the parasitic element are inserted to reduce the current related to the cross-polarization. The six-element array shown in Figure 9.6 is designed on a substrate with a dielectric constant of $\varepsilon_r = 3.48$, $\tan\delta = 0.0031$, and a thickness of 0.254 mm.

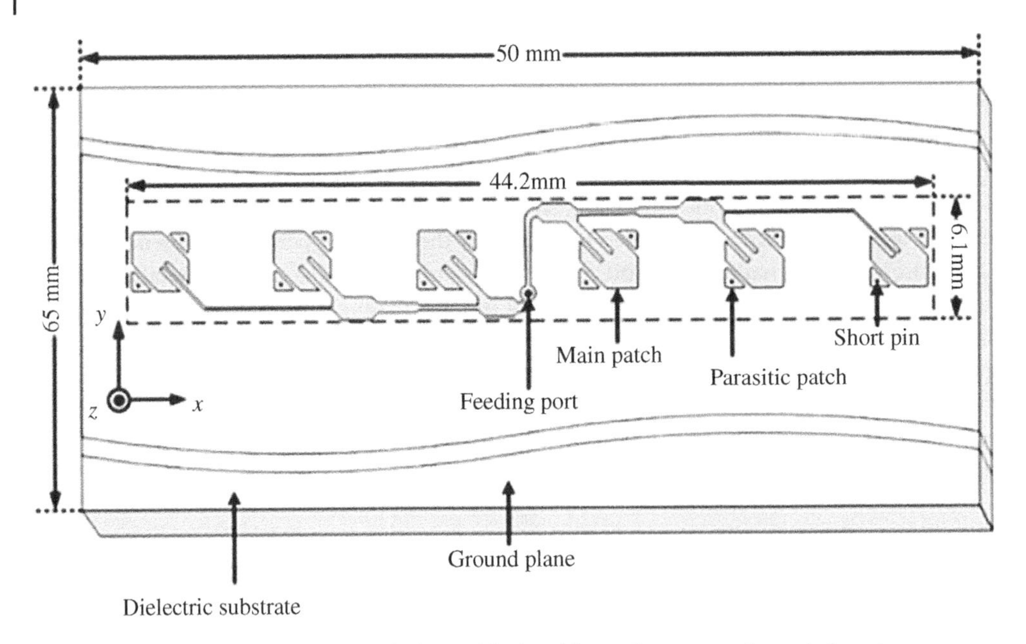

Figure 9.6 Array antenna with grounded parasitic rhombic patch antenna elements for a lower cross-polarization level [31].

The radiators occupy an area of $44.2 \times 6.1\ \text{mm}^2$, and the size of the substrate is $50 \times 65\ \text{mm}^2$. The antenna under test achieves a bandwidth of 660 MHz from 23.87 to 24.53 GHz, a gain of 11.04 dBi, a cross-polarization level of −20.73 dB, a sidelobe level of −20.59 dB, a front-to-back ratio of −29.10 dB, and a beamwidth of 16.2° in the *xz*-plane and 100.8° in the *yz*-plane.

9.2.1.2 Compact Two-Layer Rotman Lens-Fed Microstrip Antenna Array

The Rotman lens is a constrained lens in which a wave is guided along constrained paths upon design equations. It can generate multiple beams with phase relationships that are determined from the path length of the wave passing through the lens. A compact Rotman lens is a desired beamformer for implementing the multi-beam radar sensor on the surface of a vehicle. In [34], a compact two-layer Rotman lens-fed microstrip antenna array at 24 GHz is reported. As shown in Figure 9.7, the lens-fed antenna has the form of two layers, which is a new approach for reducing the size of the Rotman lens. The lens consists of a top metal layer, a dielectric, a common ground, a dielectric, and a bottom metal layer, in sequential order. The layout of the lens body is placed on the bottom layer, and the antennas are placed on the top layer. Both of them are electrically connected through slot transitions. This two-layer structure reduces not only the total size of the lens but also the loss of the delay lines because the lines can be designed to be as short and straight as possible.

Figure 9.8 illustrates the antenna prototype with seven array ports, five beam ports, and six dummy ports. It was fabricated on an RO3003 substrate with ε_r of 3.0, tanδ of 0.0013, and thickness of 0.508 mm. Both the focal angle and the corresponding scanning angle are ±30°, and the spacing between the antennas is 0.6 times wavelength at 24 GHz. The diameter of the lens is approximately 27 mm, and the overall size of the lens including the lens body, ports, and transitions is $75 \times 80\ \text{mm}^2$. The lens is fed by 50-Ω microstrip lines.

The series-fed four-element patch array was designed based on an RO3003 substrate, where the patch is with a width of 4.4 mm and length is 3.4 mm as resonating at 24 GHz. The patches are connected in series by a narrow microstrip line with an input impedance of 360 Ω. The distance between the patches is 3.0 mm, and the patches are positioned periodically with a spacing

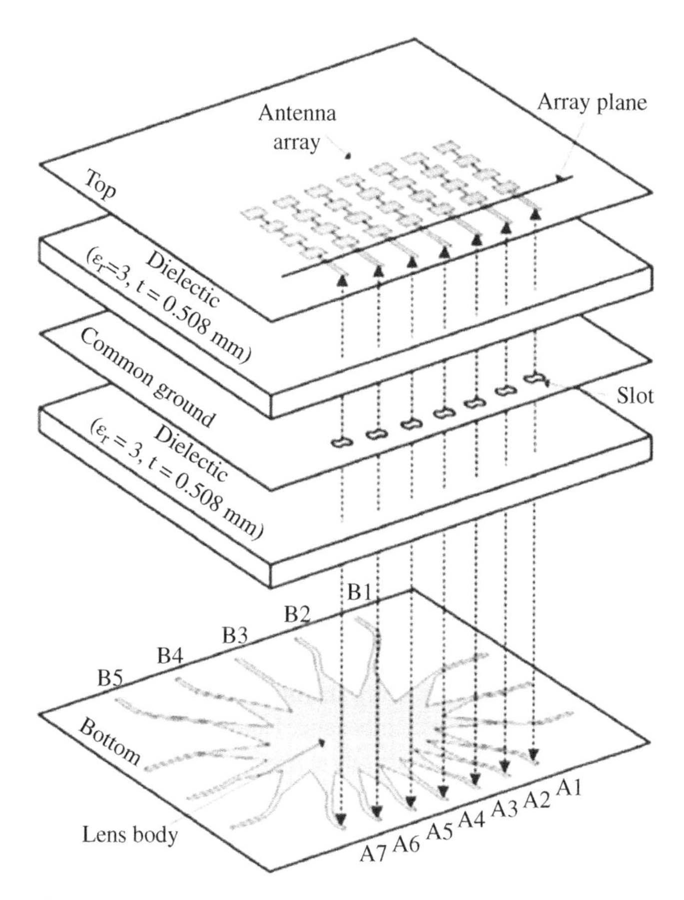

Figure 9.7 Geometry of the two-layer Rotman lens-fed antenna array that consists of a top metal layer, a dielectric substrate, a common ground plane, a dielectric substrate, and a bottom metal layer [34].

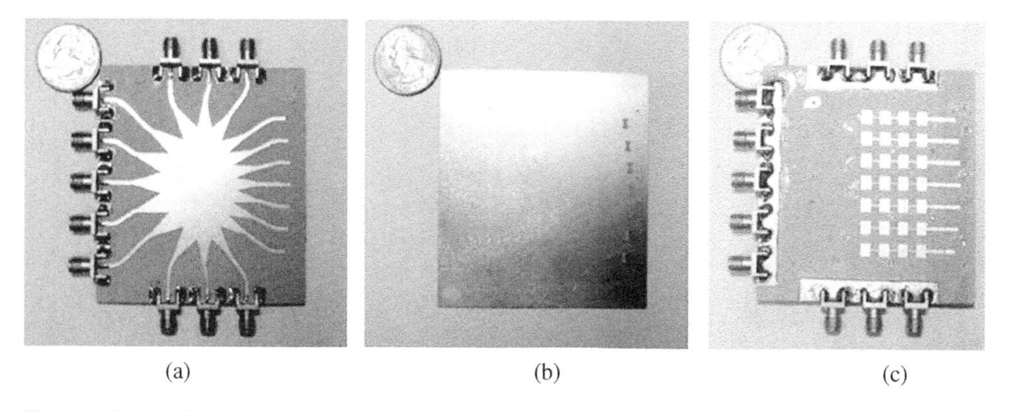

Figure 9.8 The fabricated two-layer Rotman lens-fed antenna. (a) The bottom layer, (b) the ground plane, and (c) the top layer [34].

of 6.4 mm, or 0.512 wavelength at 24 GHz. The antenna exhibits good impedance matching with return loss of greater than 10 dB from 23 to 25 GHz. The measured beam patterns show that the beam directions are −28.1°, 14.9° 0°, 15.5°, and 28.6° with beamwidths of 13.4°, 13.2°, 12.8°, 13.5°, and 13.0°, respectively.

9.2.1.3 SIW Parasitic Antenna Array Without Feeding Network

In [35], an SIW parasitic compact antenna array without any feeding network is reported as shown in Figure 9.9. In contrast to a conventional antenna array where the antenna elements are excited by an elaborate feeding network consisting of multiple stages of power dividers, only one antenna element in the antenna array is excited and the adjacent array elements are nonradiatively coupled via inductive windows, along both the x-axis and the y-axis. As such, this coupling technique results in an extremely compact design, facilitating integration in close proximity with active devices. Moreover, the insertion loss and the parasitic radiation from the feeding network are significantly reduced, improving the overall radiation efficiency. The amplitude distribution can be optimized by tuning the amount of coupling between the adjacent array elements, effectively eradicating the sidelobes in the H-plane and minimizing them in the E-plane. By tuning the amount of coupling between the adjacent elements via the widths of the inductive windows, an optimal amplitude distribution of the E-field across different slots is achieved. As such, a directive radiation pattern along the broadside without sidelobes is obtained in the H-plane. The sidelobes in the E-plane are minimized by tuning the height of the radiating slots.

As shown in Figure 9.9, the single-fed 3×3 parasitic SIW array antenna prototype is with an overall size of $18.00 \times 19.60 \times 0.56\ \text{mm}^3$. It achieves a measured impedance bandwidth of 315 MHz, a maximum array gain of 10.3 dBi at 24.15 GHz, and cross-polarization level of more than 30 dB. The 3-dB beamwidth is 43° and 46° in the H-plane and E-plane, respectively.

9.2.1.4 SIW Pillbox Antenna Integrating Monopulse Amplitude-Comparison Technique

TekkouK et al. reported an amplitude-comparison monopulse slotted SIW antenna array with a pillbox parabolic quasi-optical beamforming network at the 24 GHz band [36]. As shown in Figure 9.10, the beamformer is illuminated by a cluster of horn pairs located in the focal plane of

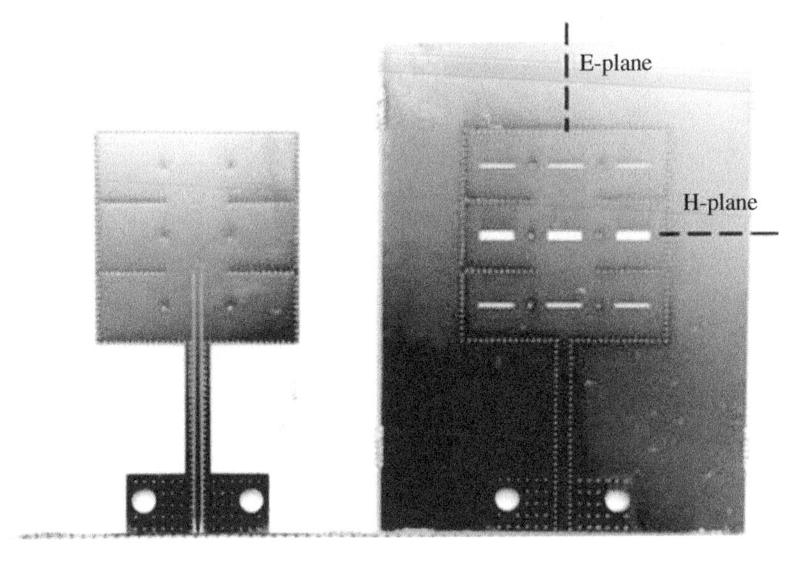

Figure 9.9 SIW parasitic antenna array [35].

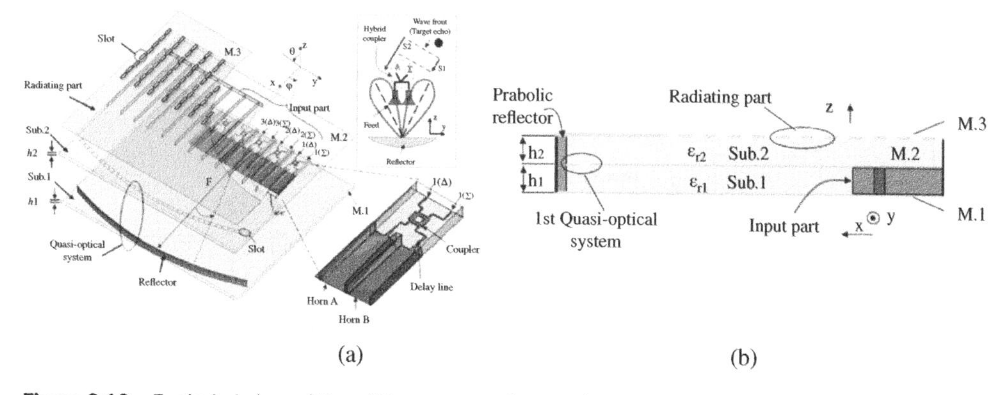

Figure 9.10 Exploded view of the pillbox antenna integrating monopulse amplitude-comparison technique. Inset: principle of the amplitude comparison monopulse technique for the central horn pair. (a) 3-D view and (b) cross-sectional view [36].

the pillbox reflector. Each horn pair is composed of two H-plane sectorial horns connected to a coupler. In-phase and out-of-phase operations of each horn pair allow radiation of the Σ and Δ beams along the antenna scanning E-plane. The scanning and tracking operations are realized along the same scanning plane. These functions are ensured by switching between input ports of a cluster of horn pairs located in the focal plane of a pillbox transition without any mechanical orientation of the antenna system. To ensure a proper monopulse operation with a deep null of radiation for the Δ patterns, each horn pair has been designed and optimized separately. Furthermore, the antenna can also be applicable in multi-tracking systems where a single antenna can track multiple targets simultaneously by rapidly switching from one input port to another.

The antenna prototype as exhibited in Figure 9.10 demonstrates an antenna FOV of $\pm 26°$, with a null depth better than -20 dB for all the measured Δ beams at 24.15 GHz, the antenna gain varies between 21.6 and 20.5 dBi for the Σ beams, and the measured input reflection bandwidth is about 2.4% for VSWR <2.

9.2.2 Selected State-of-the-Art Antennas for 77-GHz Automotive Radars

9.2.2.1 SIW Slot Array for Both Medium- and Long-Range Automotive Radar Sensor

LRR, MRR, and SRR have different requirements on the FOV such that LRR demands a high-gain and a narrow FOV, while MRR needs a relatively lower gain and a wider FOV. When the MRR and LRR are required to be realized in one module, one possible method is to design two sets of antennas, including a high-gain one for LRR and another low-gain one for MRR, and then switch between them by an mmW controlling circuit. The drawback of such solution is the poor power efficiency because a significant portion of the mmW power is wasted due to switch losses. In addition, a larger area is required because of the use of multiple antennas, and the complexity of signal processing at the baseband also becomes more complicated. Therefore, it is desired for an antenna to support both the medium- and long-range radar sensors (MLRRs) simultaneously [37, 38].

The ideal radiation pattern (horizontal plane) of an antenna to support MLRR sensors is shown in Figure 9.11, wherein the pattern has a "flat-shoulder shape." The θ_{LRR} is defined as a 3-dB beamwidth (BW) of the main beam of antenna, $G_p = G_{t1} - G_{t2} = 40\lg(k)$ $(k = R_{LRR}/R_{MRR})$ is the gain difference between peak and flat shoulder, and θ_{MRR} is defined as the $-G_p$-dB BW of the radiation pattern. For example, assume $R_{LRR} = 200$ m and $R_{MRR} = 100$ m, therefore $k = 2$ and $G_p = 12$ dB. If the θ_{LRR} and θ_{MRR} are set to be 15° and 80°, respectively, the most important feature of the -12-dB BW in the horizontal plane is required to be approximately $\pm 40°$. The ripple level in the flat-shoulder

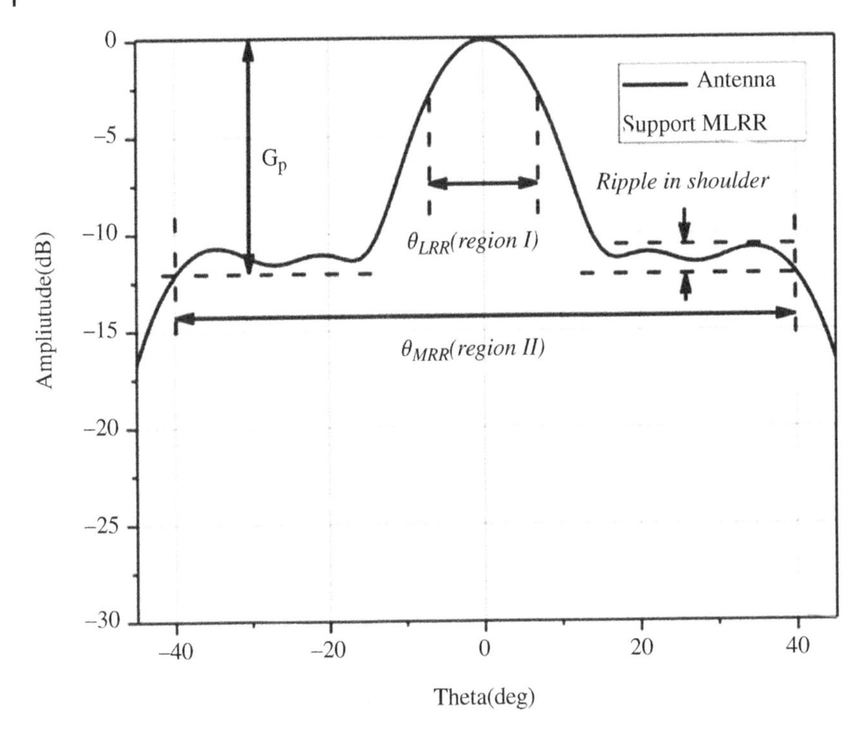

Figure 9.11 Ideal radiation pattern of an antenna supporting MLRR (horizontal plane) [37].

area should be as small as possible. Usually, a 2~3-dB ripple is acceptable when considering its relatively small influence on the detection range and the manufacturing inaccuracy of the antenna.

Yu et al. developed an SIW slot antenna array with a flat-shoulder shaped radiation for MLRR sensors at the 77 GHz band. The antenna prototype on a piece of Rogers 5880 (ε_r = 2.2 and thickness h = 0.508 mm) is shown in Figure 9.12. Due to the requirement of θ_{LRR}, the antenna array is composed of six identical SIW linear slot subarrays that comprise 16 SIW slot elements each. Each linear subarray was designed using the Elliott method [39, 40], and a structure named "the first-order inductive window" [41] was utilized to improve the impedance matching. From 76.4 to 77.8 GHz, the antenna array achieves the measured 3-dB BW in the E-plane (xz-plane or horizontal plane) of about ±7°–7.5° with a peak gain of 21.7 dBi, which is applicable for LRR sensors. The maximal G_p (shoulder level) and ripples in the flat-shoulder area are around 12 and 3.3 dB over the impedance bandwidth, respectively, which is suitable for MLRR applications when $R_{LRR} \approx 2R_{MRR}$.

9.2.2.2 16 × 16 Phased Array Antenna/Receiver Packaged Using Bond-Wire Technique

The increasing requirements in functionality and performance of automotive radar sensors equalize the cost reduction achieved by a higher degree of integration of the main parts such as the microcontroller, the baseband circuitry, and the radar frontend. The GaAs technology is now replaced by SiGe, BiCMOS, or CMOS processes, which enables the highly integrated radar frontends with transmitter [42–45], receiver [46], or even transceivers at 77 GHz [47, 48], with a different amount of receive (Rx), transmit (Tx), or Tx and Rx channels. These monolithic microwave integrated circuits (MMICs) are mounted in a cavity on the top or bottom side of the PCB and are contacted with bonding wires. In [42], Ku et al. first demonstrated a complete phased array system at the 77 GHz band with pattern scanning capabilities up to ±50°. The work also shows that complex

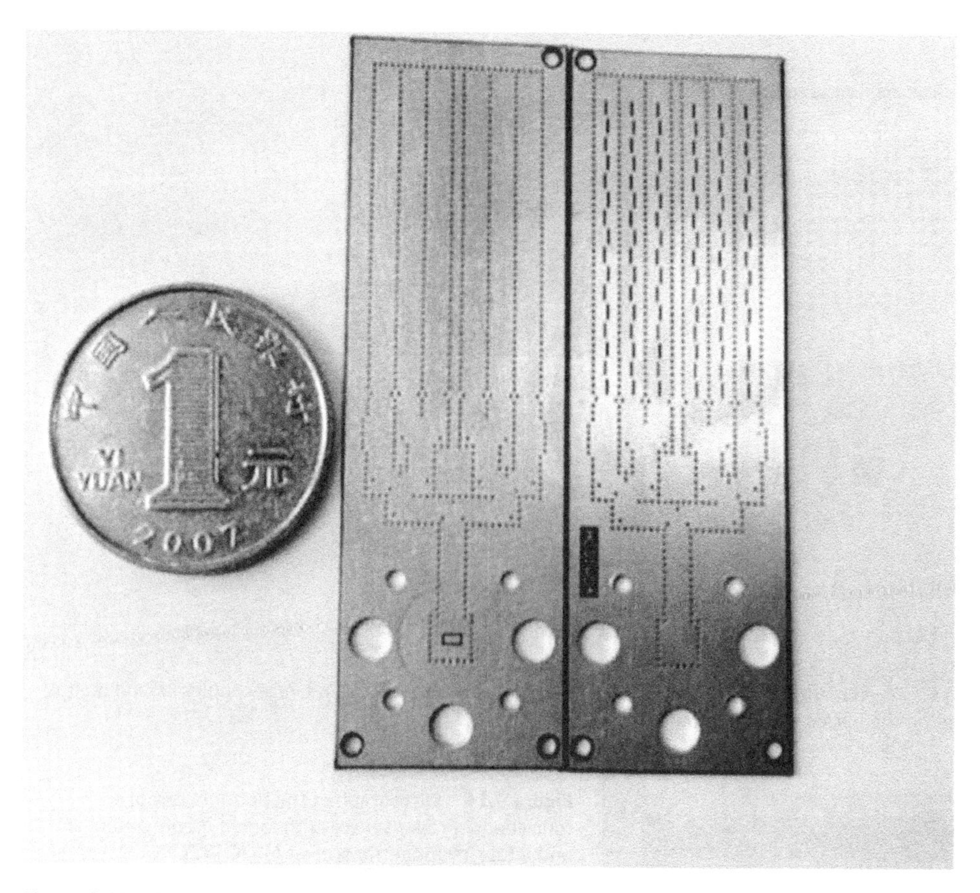

Figure 9.12 Photograph of the SIW slot antenna array for MLRR sensors [37].

mmW phased array chips can be packaged using traditional bond-wire techniques and thus are suitable for low-cost high-volume applications. This is in contrast to the traditional assumption that complex phased array chips must be packaged using flip-chip techniques to achieve acceptable performance at mmW frequencies.

Figure 9.13 shows the 16×16 phased array antenna with single SiGe chip receiver. Each subarray with a series-fed 16-element microstrip patch array on 0.125 mm thick RO3003 board with different microstrip antenna widths so as to create a varying shunt impedance and a taper in the elevation plane. The design is based on a standing wave half-wave spacing and therefore does not scan with frequencies. The simulated mutual coupling over 77–81 GHz between two adjacent antenna arrays is less than −25 dB at a spacing of 0.6 times wavelength. The simulated directivity of the vertical antenna element is 18.3 dBi at 77 GHz with a gain of 17.0 dBi. The elevation pattern has a 3-dB beamwidth of 6° over 77–80 GHz with 17-dB sidelobe. The 16×16 element antenna array occupies an area of $38.5 \times 38.5\ \text{mm}^2$ and achieves directivity up to 29.3 dBi considering the tapering in the vertical and horizontal and a gain of 28.0 dBi.

9.2.2.3 Antenna/Module in Package

Connecting the antennas and the MMICs via a wire bonding technique is applicable for 77 GHz radar sensor integration while it is still costly. Packaging the MMIC in a radio frequency (RF) suitable case [49, 50] is one of the solutions for cost reduction. Nevertheless, the connection between the antenna and the RF frontend has to be implemented on the PCB, which is usually expensive

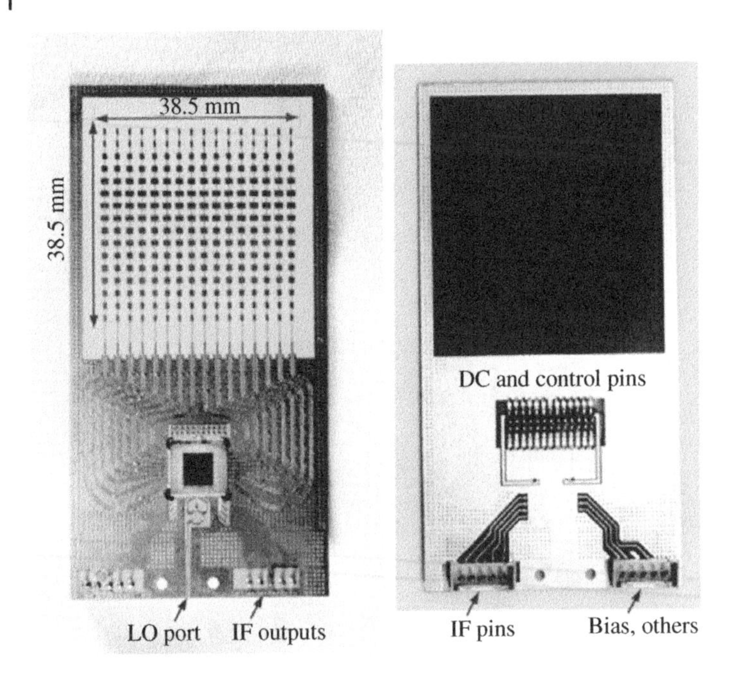

Figure 9.13 16-element phased array antenna including antennas and LO rat-race coupler. IF and digital control are on the back side of the board [42].

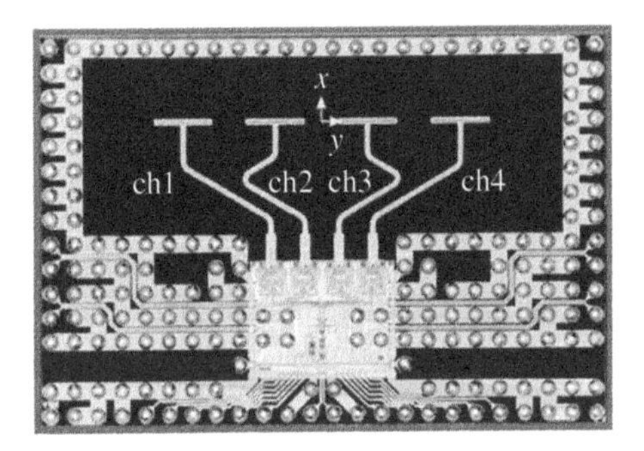

Figure 9.14 Micrograph of the bottom view of a four-channel radar sensor with folded dipole antennas and a four-channel transceiver MMIC [52].

due to the required special electrical properties. Instead, integration of the antennas into a package is preferred while more challenging.

The embedded wafer level ball grid array (eWLB) technology makes the encapsulation of the frontend chip with an antenna in a package feasible [51, 52]. Figure 9.14 demonstrates a four-channel radar sensor with horizontally oriented folded dipole antennas integrated in an eWLB package, where antenna elements are separated at a distance of half a wavelength, or 1.96 mm at 76.5 GHz. A single folded dipole antenna is able to achieve the maximum gain of 6.2 dBi, 3-dB beamwidth of $\theta_E = 57°$, and sidelobe level SLL_E of =9.6 dB in the E-plane over the frequency range from 76 to 81 GHz. Considering the four-channel operation, the mutual coupling shows some effect on the antenna radiation properties. The gain pattern of the antenna array with the channel 1–2 and channel 3–4 behaves like the outer channels 1 and 4, respectively. The pattern of the array with channel 2–3 shows a good symmetry, and the gain pattern of the full antenna array with channel 1–2–3–4 is symmetric to the z-axis with a maximum gain of 8.2 dBi and an HPBW of $\theta_e = 23°$.

9.3 Single-Layer SIW Slot Antenna Array for 24-GHz Automotive Radars

For LRR sensors, the antenna is required to be with a narrow beamwidth, typically <6°, and low SLLs, typically <−20 dB in a horizontal plane. In addition, the main beam should be centered horizontally and vertically without beam squinting across the bandwidth [53]. In [28], an end-fed SIW slot antenna has been reported to achieve a narrow beam and the SLLs as low as −30-dB in both the E- and H-planes, while the antenna array suffers from beam squinting against the operating frequency. In general, a center-fed antenna array is more desired [54–56] to overcome the beam squinting problem. However, for a conventional single-layer center-fed waveguide-based slot antenna array configuration, the center portion is always occupied by the feeding structure, and therefore there is a large aperture blockage or a slot-free area, so it is a big challenge to achieve low SLLs in the H-plane or horizontal plane. Some studies have been reported to reduce the aperture blockage effect and lower the SLLs. A post-wall waveguide feeding network for a center-fed antenna array has been reported [54]. The large blockage area results in a high SLL of −7.8 dB but is suppressed to −11.1 dB by using a tapered amplitude distribution. In [55], the SLL of −9.5 dB associated with the aperture blockage is improved to −14.7 dB by applying a genetic algorithm to control the slot excitation distribution. In [56], the first SLL is reduced from −10 dB to −13 dB using the E- to H-plane cross-junction power dividers.

In this section, a compact co-planar waveguide (CPW) center-fed SIW slot antenna array for a 24 GHz automotive radar sensor is exemplified. Implemented with a single-layer substrate and a normal, low-cost PCB process, the antenna array with 32 × 4 slot elements has an overall size of 195 × 40 × 0.79 mm^3 and achieves a gain of >22.8 dBi, efficiency of >67%, return loss of >10 dB, sidelobe level (SLL) of < −21 dB, and a fixed boresight beam of <4.6° in an H-plane or horizontal plane over 24.05–24.25 GHz.

9.3.1 Antenna Configuration

Figure 9.15a shows the top view of the SIW slot array antenna design. There are a total of 4 × 32 slots on the broad wall of the SIWs. In each row, two 16-element linear arrays are placed end to end at a distance of d_1 between the two edge slots. The central portion of the antenna is occupied by the feeding structure and is thus slot free, which is the bottleneck of sidelobe suppression. The spacing between the adjacent slots in the H-plane, d_h, is $\lambda_g/2$ (λ_g is the guided-wavelength of the SIW at the center frequency of 24.15 GHz). The spacing between the adjacent slots in the E-plane,

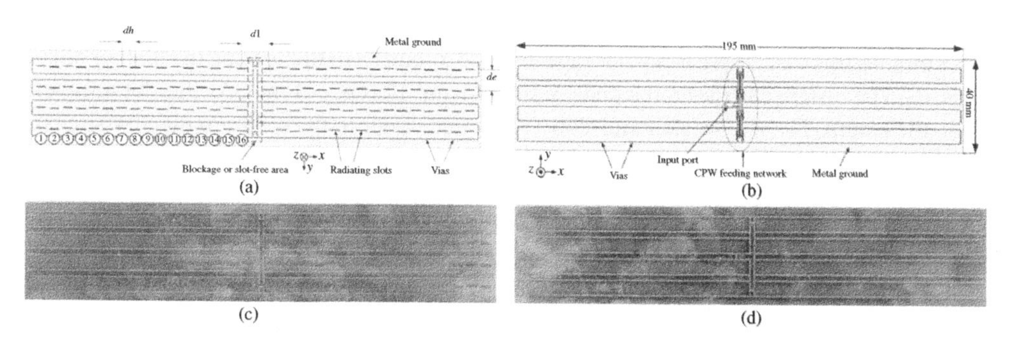

Figure 9.15 Top and bottom view of the CPW fed SIW slot array antenna. (a) Top view the antenna design, (b) bottom view of the antenna design, (c) top view of the antenna prototype, and (d) bottom view of the antenna prototype [53].

Table 9.3 Dimensions of the slot array (unit: mm).

l_1	4.76	l_2	4.6	l_3	4.52	l_4	4.62	l_5	4.52
l_6	4.58	l_7	4.48	l_8	4.54	l_9	4.52	l_{10}	4.5
l_{11}	4.5	l_{12}	4.4	l_{13}	4.32	l_{14}	4.22	l_{15}	4.14
l_{16}	4	x_1	0.34	x_2	0.33	x_3	0.32	x_4	0.31
x_5	0.29	x_6	0.27	x_7	0.25	x_8	0.23	x_9	0.21
x_{10}	0.2	x_{11}	0.18	x_{12}	0.15	x_{13}	0.2	x_{14}	0.2
x_{15}	0.2	x_{16}	0.2	d_1	11.6	d_e	8.7	d_h	5.8

d_e, is λ_w (λ_w is the guided-wavelength of the CPW at the center frequency of 24.15 GHz) so that the adjacent linear arrays are fed in phase.

Figure 9.15b exhibits the bottom view of the slot array antenna. The input of the array is connected to an external mini-SMP connector. The compact CPW feeding network is located in the center of the antenna with an eight-way parallel feeding configuration. The antenna is printed onto a 0.79 mm thick single-layer PCB of Rogers 5880 with $\varepsilon_r = 2.2$ and $\tan\delta = 0.0009$. The conductor used for metallization is copper, whose conductivity is 5.8×10^7 S/m, with a thickness of 0.02 mm. The detailed geometrical dimensions of the antenna prototype as shown in Figure 9.15c,d are tabulated in Table 9.3, in which x_i (i=1, …, 16) and l_i (i=1, …, 16) are the offsets and lengths of the slots.

9.3.2 Slot Array Design

The design of the slot array follows the method in [28]. First, the parameter is extracted for a single slot on the SIW with various offsets. Resonant length, resonant conductance, and admittance are obtained for the array synthesis. Next, Elliott's iterative procedure for a waveguide-fed slot array [40], including all mutual couplings, is applied for the SIW-fed linear array to calculate the initial slot parameters for a targeted amplitude distribution. Further fine-tuning by electromagnetic simulation finalizes the slot parameters for desired SLLs. In order to achieve the requirement of −20 dB SLL in the H-plane, a Taylor distribution for −26 dB is chosen. To achieve the HPBW<6° in the H-plane or horizontal plane, 32 slots are used.

In the simulation of the 32-element linear array as shown in Figure 9.16, two ports excite each half of the slot arrays simultaneously. The spacing between slots 16 and 17, d_1, is first set to be $\lambda_g/2 = 5.8$ mm. In this case, there is no blockage because d_1 equals to the spacing of all other adjacent slots, d_h. The distance between the ends of slots 16 and 17 is $d_s = 1$ mm. As the slot number decreases, the lengths become smaller and slots are located closer to the center of the SIW so that the radiation from the "edge" slots weakens. The offsets of slots 1–4 are set a little larger than that of slot 5 for a 2-dB better SLL.

There is a trade-off between the blockage area and the requirement of implementing the feeding network. Figure 9.17 shows the H-plane radiation patterns of the linear array with different d_1 at 24 GHz. As d_1 increases from $\lambda_g/2$ to λ_g, or 11.6 mm, the innermost SLL changes slightly, but the grating lobes in the range of $30° < |\theta| < 75°$ increase due to the enlarged blockage area. When d_1 becomes 16 mm, the peak and first SLL degrades to −19 dB. To meet the requirement of −20-dB SLL, the case of −24.5-dB SLL is selected with 4.5 dB margin considering fabrication tolerance. With $d_1 = 11.6$ mm, d_s is set as 6.84 mm, applicable to accommodate a compact feeding network.

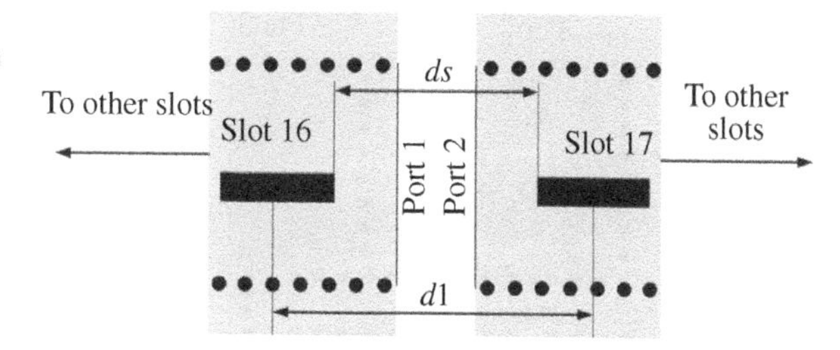

Figure 9.16 Simulation model of the two 1×16 linear arrays located end to end with different d_1.

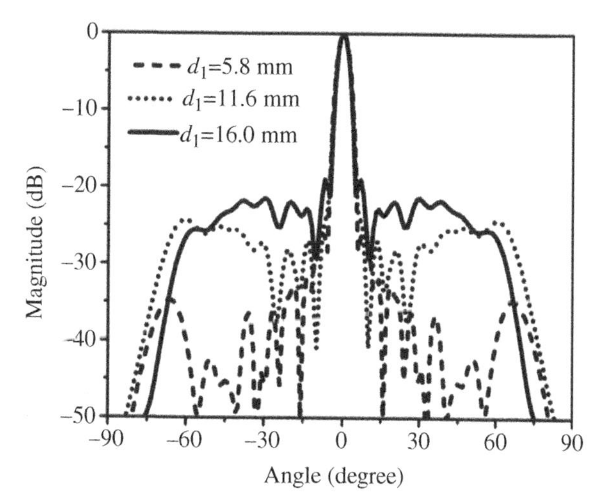

Figure 9.17 *H*-plane radiation patterns of the linear arrays with different d_1 at 24 GHz.

Aligning multiple linear arrays side by side to form the SIW planar array, the *H*-plane SLL of the whole planar array is almost unchanged. The reason is that SIW is a low-profile waveguide with about 8:1 width-to-height ratio, so the TE_{20}-mode internal mutual couplings inside the SIW are dominant among all kinds of internal and external mutual couplings [28]. After forming the SIW planar array, only external mutual couplings among slots at different branches of SIW are introduced. The effects of the additional external mutual couplings are much smaller than those mutual couplings caused by the TE_{20} mode, which have already been included in the SIW linear array design.

9.3.3 Feeding Network Design

The feeding network is illustrated in Figures 9.18 and 9.19. The *E*-plane spacing along the *y*-direction between the adjacent SIW is set to be the guided wavelength of the CPW at 24.15 GHz, λ_w so that the phase of S_{21} equals to the phase of S_{31}. Because of the structure symmetry, this configuration guarantees the phase balance at the eight outputs.

The input impedance of the half-wave slot dipole and the CPW with a length of l_c determine Z_1 as shown in Figure 9.18. If the characteristic impedance of the CPW is low, Z_1 is small and Z_2 (roughly $Z_1/4$) is thus even smaller. This small Z_2 requires an impedance transformer with a width exceeding the pre-assigned w_2 of 1.9 mm. Here, the characteristic impedance of the CPW is set to be 83 Ω. With the dimensions indicated in Figure 9.19a, Z_1 is $(211 - j30)\ \Omega$. The imaginary part of Z_1 is not necessarily zero and can be canceled out at the next stage.

$Z_2 = (50 + j37)\ \Omega$ is not exactly $Z_1/4$ because of the effects of the junctions. To match Z_2 to 100 Ω, a 46-Ω CPW line with an electrical length of 30° is needed. Because the width of the 46-Ω CPW line is 1.5 mm, it leads to a wide w_2 greater than pre-assigned 1.9 mm. Instead, a 55-Ω CPW line with an electrical length of 38° is used, which is only 0.8 mm wide, for the impedance transformer. Although

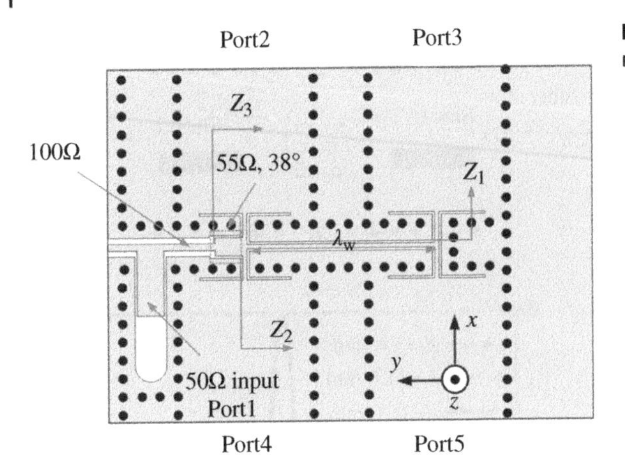

Figure 9.18 Input impedances at various reference planes of the feeding network.

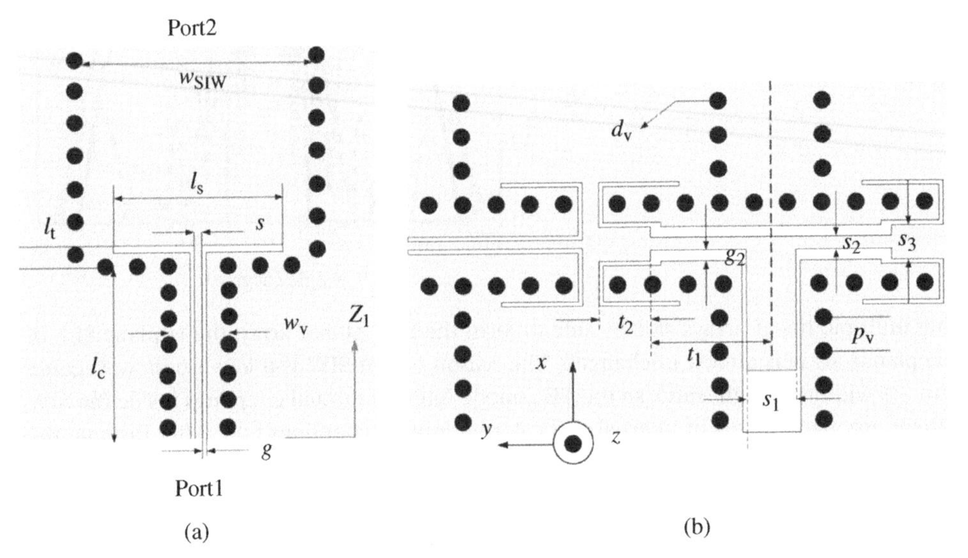

Figure 9.19 Detailed geometry of the feeding network. (a) CPW-SIW transition, $s = 0.2$, $g = 0.1$, $w_v = 1$, $l_t = 0.5$, $l_s = 4.3$, $w_{SIW} = 6.2$, $l_c = 0.95$, and (b) impedance transformer part. $s_1 = 1.2$, $s_2 = 0.3$, $s_3 = 0.8$, $g_2 = 0.25$, $t_1 = 2.9$, $t_2 = 1.25$, $p_v = 0.8$, $d_v = 0.4$. (unit: mm).

in this case Z_2 is matched to $Z_3 = (110 - j4)\ \Omega$, the overall impedance matching is acceptable. The rest of the feeding network includes two parallel 100-Ω CPW lines connecting to the 50-Ω input line. The detailed geometry of the rest of the feeding network is exhibited in Figure 9.19b.

Figure 9.20 exhibits the simulated S-parameters of the feeding network. Only four output ports, ports 2–5, are shown because the structure is symmetrical. At 24.15 GHz, $|S_{11}|$ is −18 dB. $|S_{21}|$, $|S_{31}|$, $|S_{41}|$, and $|S_{51}|$ are −9.5 dB, −9.53 dB, −9.64 dB, and −9.65 dB, respectively. In Figure 9.20b, the phase of S_{21}, S_{31}, S_{41}, and S_{51} are 173°, 171°, 171°, and 169°, respectively. Therefore, good impedance matching and amplitude/phase balance are achieved using the proposed feeding network.

9.3.4 Experiment Results

Figure 9.21 compares the measured and simulated $|S_{11}|$ of the antenna array prototype. The simulated $|S_{11}|$ is less than −10 dB in 23.8–24.2 GHz, and the measured $|S_{11}|$ is less than −10 dB in the range from 23.84 to 24.25 GHz.

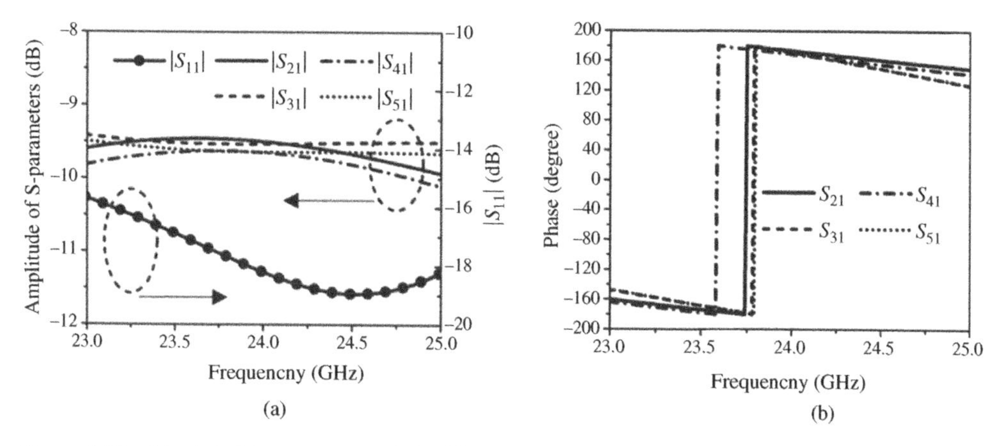

Figure 9.20 Simulated S-parameters of the CPW feeding network. (a) Amplitude and (b) phase.

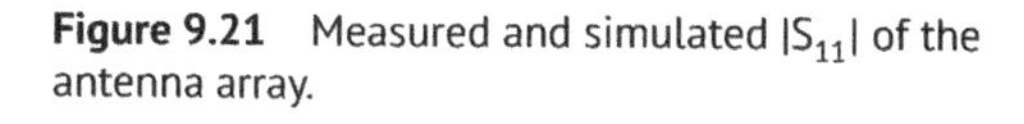

Figure 9.21 Measured and simulated $|S_{11}|$ of the antenna array.

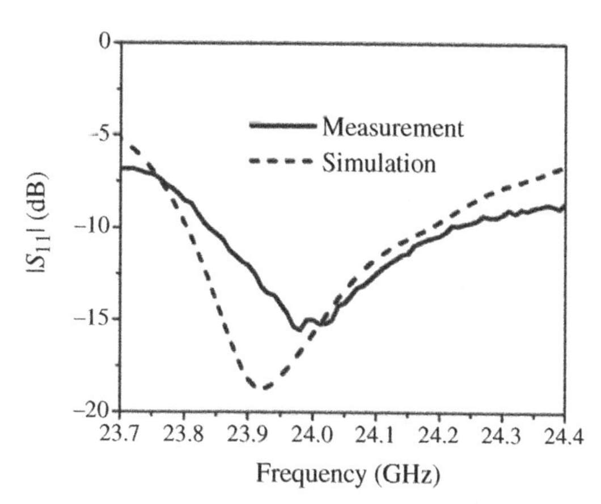

Figure 9.22 Measured and simulated boresight gains of the antenna array.

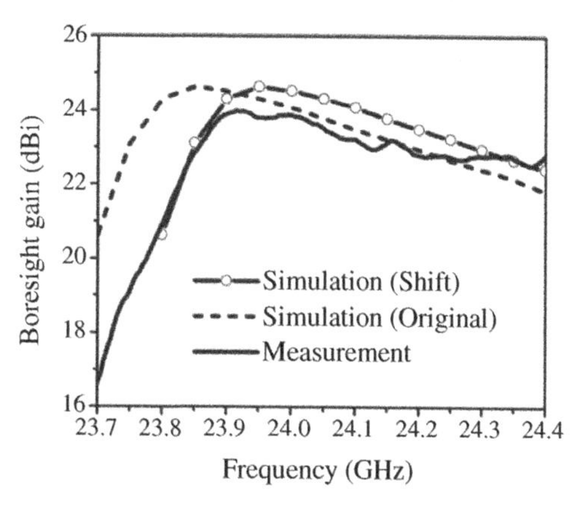

Figure 9.22 shows the measured and simulated boresight gain of the antenna array. The modified simulated gain shifting about 0.1 GHz upwards agrees well with the measured one. The difference between the measured and simulated gain is attributable to the actual antenna loss, which may be caused by lossy dielectric and higher than that in simulation. In the bandwidth of 23.86–24.12 GHz, the measured gain is higher than 23 dBi and the maximum gain is 24 dBi at 23.92 GHz.

Figure 9.23 shows the measured and simulated SLLs of the antenna array. In the bandwidth of 23.9–24.3 GHz, the simulated SLL is lower than −22.9 dB, and the lowest SLL is −25.4 dB at

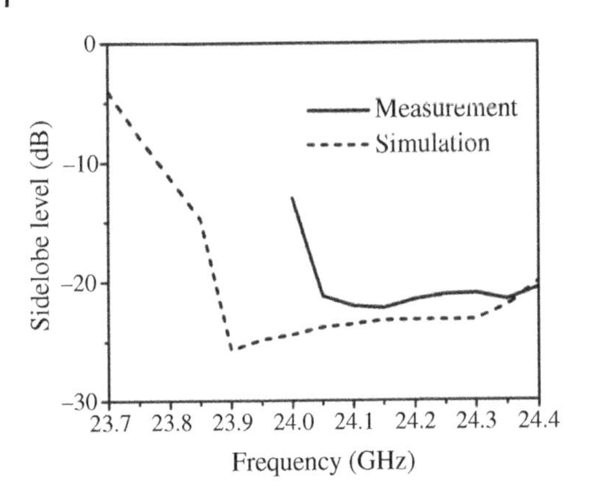

Figure 9.23 Measured and simulated H-plane side lobe levels of the antenna array.

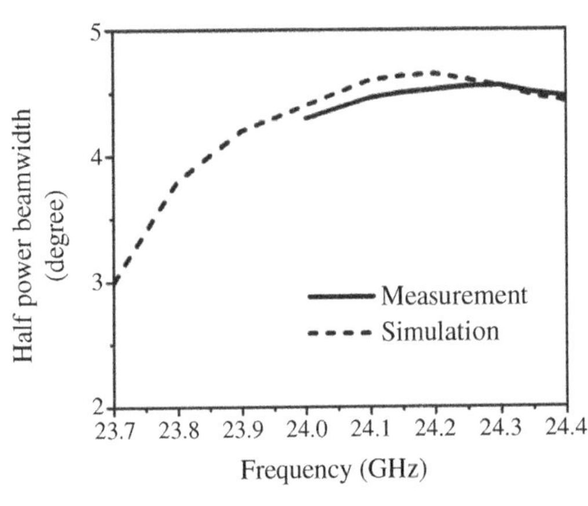

Figure 9.24 Measured and simulated H-plane HPBW of the antenna array.

23.9 GHz. The measured SLL is lower than −21 dB over the range from 24.05 to 24.40 GHz. An upward frequency shift is also observed between measurement and simulation. Compared with the simulated −24.5-dB SLL of the linear array at 24 GHz in Figure 9.17, the simulated SLL of the planar array without any tuning of the slots is −24.4 dB. The SLL of the planar array is almost unchanged compared with the linear array.

Figure 9.24 illustrates the measured and simulated H-plane HPBW of the antenna array. Over the range from 23.8 to 24.2 GHz, the simulated H-plane HPBWs keep less than 4.6°. In the bandwidth of 24.0–24.4 GHz, the measured H-plane HPBWs are less than 4.6°. In the E-plane, the measurement shows that the HPBWs are less than 20° across the band of 24.0–24.4 GHz.

Figures 9.25 and 9.26 show the comparison of the measured and simulated radiation patterns of the antenna array in the H- and E-plane, respectively. In addition, the main beam of the antenna keeps at boresight without any squinting as expected.

Figure 9.25 shows the H-plane radiation patterns over the frequency band of 24.05–24.25 GHz, the measured beamwidths agree well with the simulated ones. The measured inner-most SLLs increase because they are sensitive to fabrication tolerance and errors. In Figure 9.26, the measured beamwidths and peak SLLs in the E-plane are also very close to the simulated ones. The first sidelobes are not symmetrical in the measurement although with the E-plane symmetry in antenna configuration, the radiation pattern should be symmetrical as the simulation shows. Thus, the possible asymmetry in measurement and/or fabrication tolerance may lead to the asymmetry of the measured results.

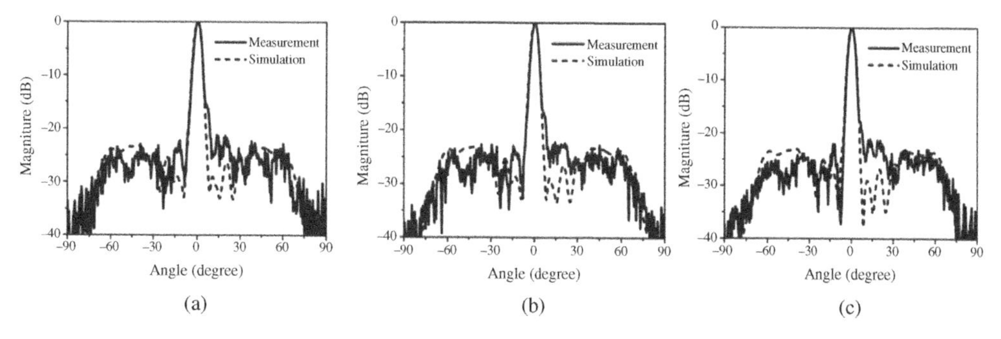

Figure 9.25 Measured and simulated *H*-plane radiation patterns of the antenna array. (a) Measurement at 24.05 GHz and simulation at 23.95 GHz, (b) measurement at 24.15 GHz and simulation at 24.05 GHz, and (c) measurement at 24.25 GHz and simulation at 24.15 GHz.

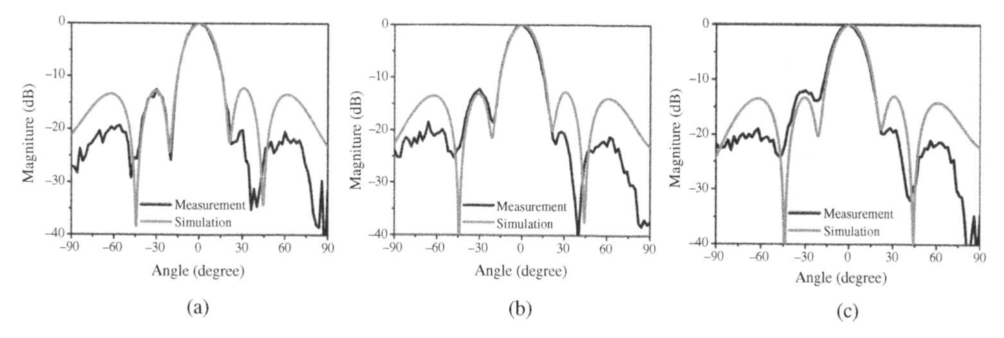

Figure 9.26 Measured and simulated *E*-plane radiation patterns of the antenna array. (a) Measurement at 24.05 GHz and simulation at 23.95 GHz, (b) measurement at 24.15 GHz and simulation at 24.05 GHz, and (c) measurement at 24.25 GHz and simulation at 24.15 GHz.

9.4 Transmit-Array Antenna for 77-GHz Automotive Radars

A lightweight, low-profile, and low-cost multi-beam antenna is always desired for automotive radar sensors [57]. Two types of free-space beamforming techniques have been studied for years, that is, the substrate-lens transmission type [58, 59] and transmit/reflect-array type [60–63]. Substrate lenses are bulky and heavy in nature while reflect-arrays may suffer from the shadowing effects caused by the blockage of their feed sources at certain positions in front of the reflector. The transmit-array is thus a strong candidate for low profile and high gain, as well as suitable for the close integration with primary feeds.

A transmit-array, also known as discrete lens, has been studied for years [60–67]. In general, the multi-layer transmit-array elements are required to provide specific phase compensation for an in-phase far-field radiation. In this section, a 77-GHz transmit-array on dual-layer PCB is demonstrated. Co-planar patch unit-cells are etched onto the opposite surfaces of the PCB and connected by through-vias. The unit-cells are arranged in concentric rings to form the transmit-array for 1-bit in-phase transmission. Combined with four-SIW slot antennas as the primary feeds, the transmit-array generates four beams with a coverage of ±15° in a horizontal plane and gain greater than 18.5 dBi at 76.5 GHz. The co-planar structure significantly simplifies the transmit-array design and eases the fabrication, in particular, at mmW frequencies.

9.4.1 Unit Cell

The unit-cell of the transmit-array is shown in Figure 9.27, where the co-planar patches are utilized as the receive- and transmit-elements. The co-planar patch has been well studied, with its resonant characteristics similar to a microstrip patch antenna rather than a slot loop antenna. The patch and ground of the co-planar patch are on the same layer of a PCB. This allows the receive- and transmit-patch arrays to be etched onto the opposite sides of the PCB as shown in Figure 9.27a. For the simplest 1-bit linear polarized transmit-array, two types of unit-cells, namely 0° unit-cell and 180° unit-cell, are designed. These unit-cells are arranged to synthesize a desired phase distribution across the array aperture with a 180° (1-bit) phase quantization. Each of the unit-cells has a size of $2\times2\,\text{mm}^2$ ($\lambda_0/2\times\lambda_0/2$ at the center frequency of 76.5 GHz). The patch facing the focal source is referred to as the receive-patch and the one facing free space is the transmit-patch. The patches are connected by feed via while the ground connected by ground vias at the edge of the unit-cell.

Multiple ground vias are used in the unit-cell to eliminate unwanted modes as well as to reduce surface wave losses. The ground vias are shown to be cut off at the boundary in the unit-cell simulation model so that the other halves are at the next unit-cell when combined in an array configuration. Figure 9.27b shows the 0° unit-cell, wherein the receive- and transmit-co-planar patches is identical with zero phase difference. Figure 9.27c,d show the receive- and transmit-co-planar patches of the 180° unit-cell, respectively. A slot ring is introduced around the feed via-pad of the transmit-co-planar patch to provide a reversed-phase feeding and generate 180° phase difference relative to the receive-patch.

The transmit-array is designed onto a piece of RO4003C of $\varepsilon_r = 3.38$, $\tan\delta = 0.0027$ with a thickness of $h = 0.203$ mm. The dimensions of the unit-cells are $W = L = 2$ mm, $W_1 = L_1 = 0.66$ mm, $g_1 = 0.22$ mm, $W_2 = 1.4$ mm, $L_2 = 1.075$ mm, and $g_2 = 0.08$ mm. The diameters of the feed via and the ground via are 0.1 and 0.15 mm, respectively. The simulated S-parameters of the co-planar patch unit-cells are shown in Figure 9.28. Under normal incidence, both co-planar unit-cells show very good impedance matching ($|S11| \leq -10$ dB) across the 76–77 GHz band, small insertion loss of 0.18 dB for the 0° unit-cell and 0.8 dB for the 180° unit-cell at 76.5 GHz, respectively. The 0° and 180° phase difference are achieved for the two unit-cells, respectively.

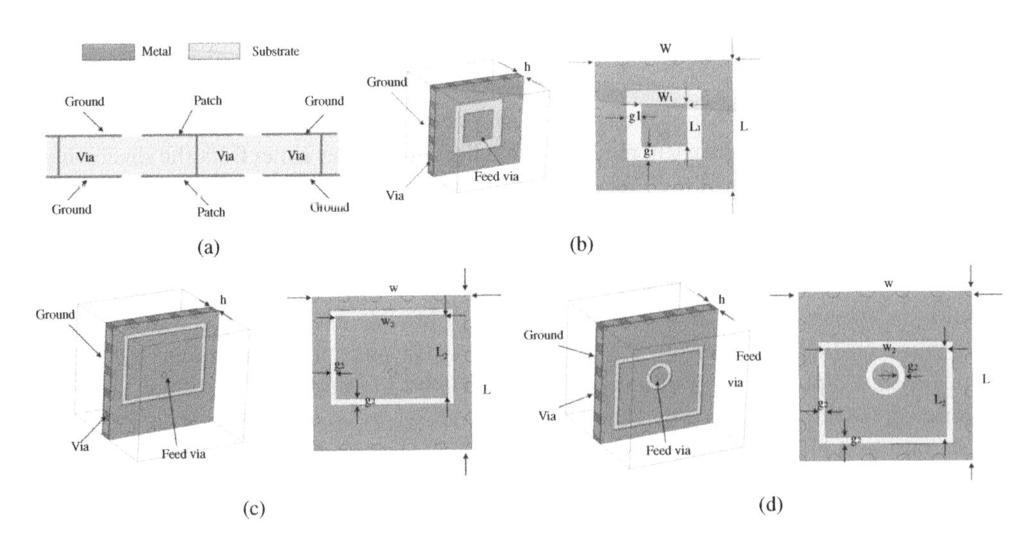

Figure 9.27 Configuration of the co-planar unit-cell. (a) Cross sectional view, (b) 0° unit-cell with identical patches on the opposite sides of the PCB, (c) receive-patch of the 180° unit-cell, and (d) transmit-patch of the 180° unit-cell.

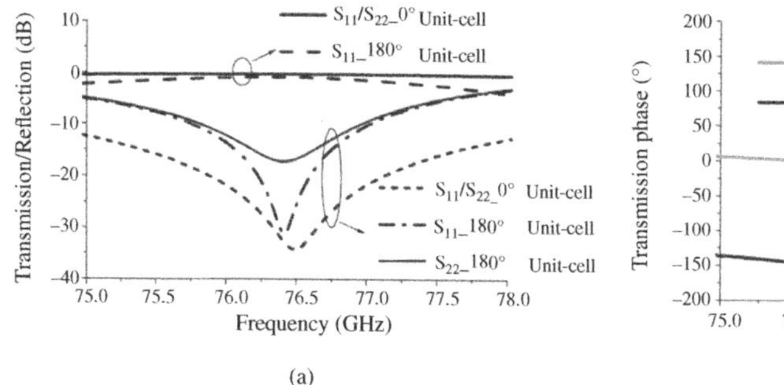

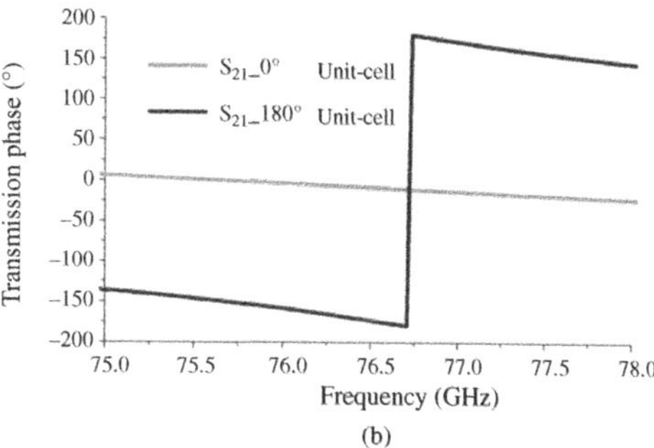

Figure 9.28 S-parameter of the 0° and 180° unit-cells. (a) Magnitude and (b) phase.

9.4.2 Four-Beam Transmit-Array

The 0° and the 180° unit-cells in the transmit-array are arranged in alternative concentric rings for linear polarization. The exact alternating configurations of the unit-cells depend on the diameter of the transmit-array (D) and the focal distance of the transmit-array to the primary feed (f). Figure 9.29 shows the configuration of the transmit-array with $D = 40.35$ mm or $\sim 10\lambda$ and $f = 20$ mm.

To develop a low-cost integrated 77-GHz automotive RF front-end package, the SIW slot antenna is feasible as the primary feed for integrating with a transceiver module. A 2×1 SIW slot antenna is excited through a SIW to 50λ grounded co-planar waveguide (GCPW) transition for measurement convenience at the focus of the transmit-array. Figure 9.30 shows the simulated pattern in the E-plane (horizontal plane) of the SIW slots with and without the transmit-array, respectively. The SIW slot is fabricated on RO4003C substrate with a height of 0.203 mm. The SIW width is 1.4 mm while the slot length and width, respectively are 1.65 mm and 0.15 mm. The SIW slots without transmit-array have the gain of 5.77 dBi with HPBW of 55.6° and 121° in the E- and H-plane, respectively. The transmit-array achieves the gain of 20.7 dBi with HPBW of 5.2° and 6.2° in the E- and H-plane, respectively.

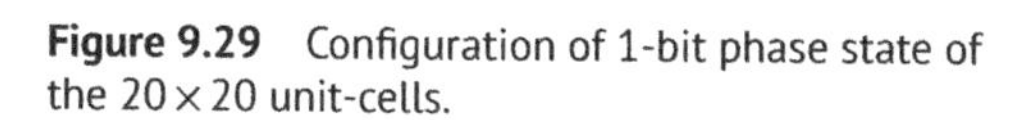

Figure 9.29 Configuration of 1-bit phase state of the 20×20 unit-cells.

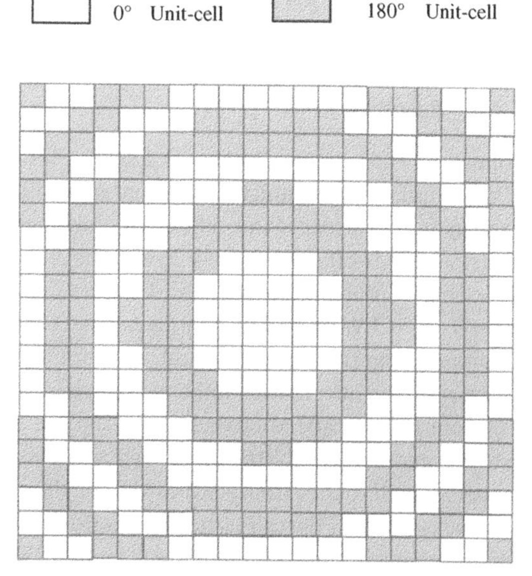

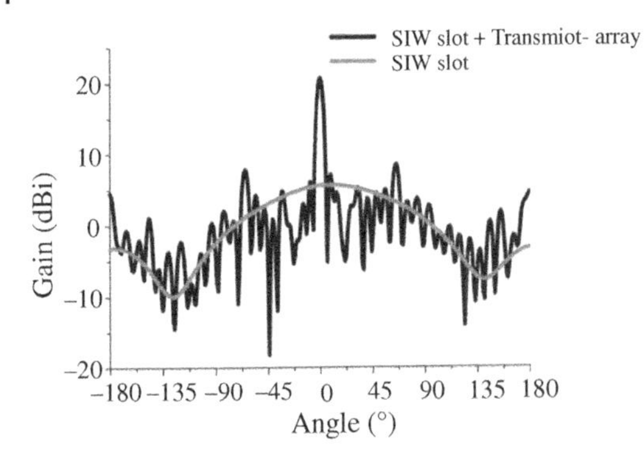

Figure 9.30 Simulated gain patterns of the SIW slot antenna with/without transmit-array at 76.5 GHz.

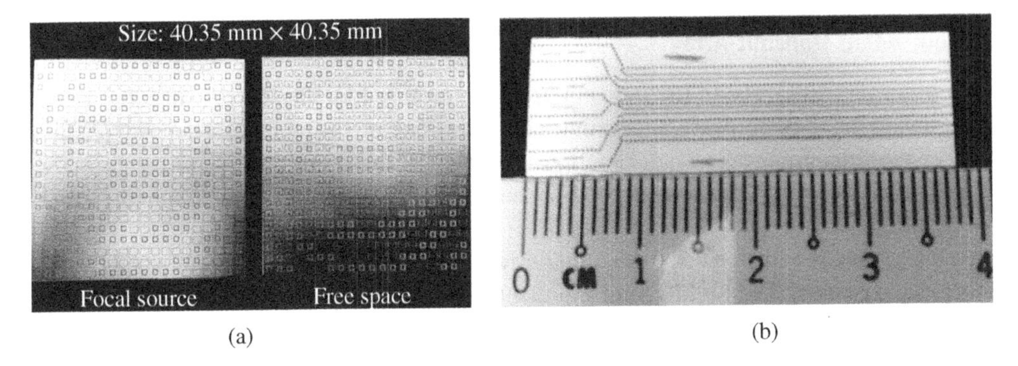

Figure 9.31 Transmit-array prototype at 76.5 GHz. (a) Receive- and transmit-patch arrays and (b) SIW slot primary feeds.

To demonstrate the functionality of beam scanning features, the transmit-array and four-SIW slot antennas were fabricated as shown in Figure 9.31. The four-SIW slots are displaced at 1.6 and 4.6 mm from the focus along the *E*-plane of the SIW slots, as well as routing and extending the GCPW for measurement purposes. The four SIW slots are labeled as P1 starting from the left-most SIW slot and correspondingly P2, P3, and P4, following the direction of the arrow shown in Figure 9.31b.

9.4.3 Results

For pattern and gain measurements, a special Teflon fixture was designed to hold and align both the SIW slot primary feeds and the co-planar transmit-array as shown in Figure 9.32a. The four SIW slot primary feeds are placed into a fitted shallow cavity and the transmit-array held in position by four poles, snapped at the four edges, 20 mm above the SIW slot antennas. The test fixture is then positioned on the probe station and the SIW slot primary feeds are connected to on-wafer probe through an SIW-to-GCPW transition. The extension of the GCPW enables the on-wafer probe to reach the antenna due to the blockage of the transmit-array. The four SIW slots with transmit-array were measured in an anechoic chamber as shown in Figure 9.32b. The measurement system consists of a customized Cascade Microtech probe station, Agilent E8361A PNA with OML millimeter-wave extender module (75–110 GHz) as well as other accessories [68]. A horn is mounted on a rotating arm which rotates from −90° to 90° enabling the measurements of half-space pattern. For the gain measurement, the calibration was performed with two standard gain horn antennas. The standard receiving horn was then replaced by the SIW slot antennas and transmit-array. All the losses,

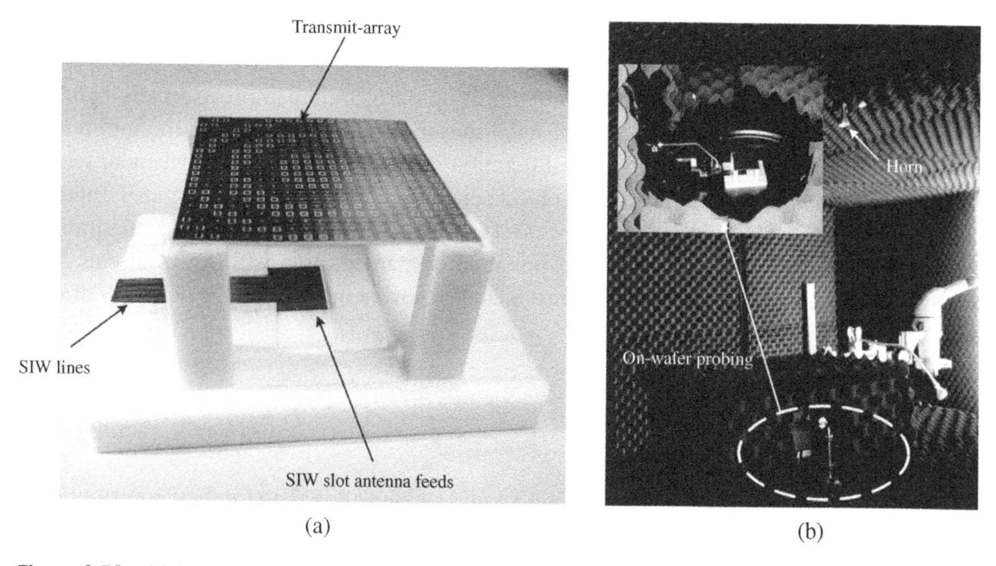

Figure 9.32 (a) Assembly of the co-planar transmit-array with four SIW slot primary feeds. (b) Set-up for radiation patterns and gain measurement.

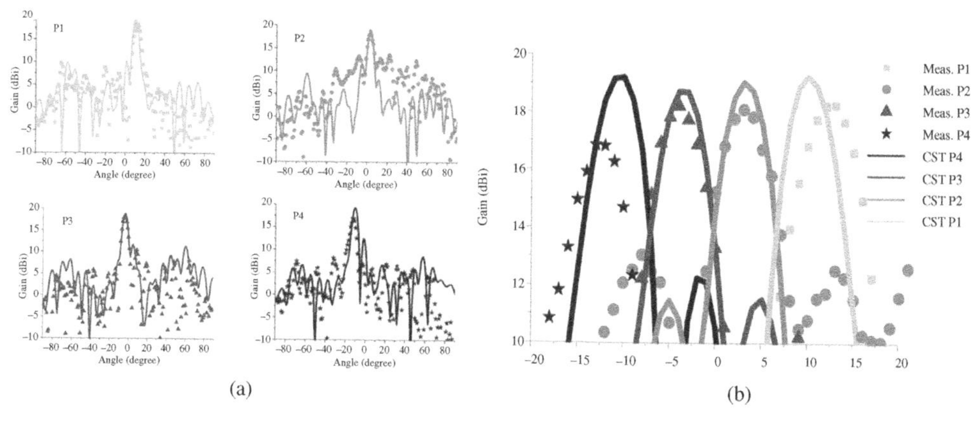

Figure 9.33 Simulated and measured radiation patterns of the co-planar transmit-array with four primary feeds at 76.5 GHz. (a) ±90° span of the four antennas, respectively, and (b) ±20° span.

such as waveguide adaptor loss and probe loss, were considered in the measurement. The gain measurements are de-embedded without including the losses incurred by the extended GCPW.

The *E*-plane patterns of the transmit-array with four SIW slot primary feeds were measured, which demonstrates the beam scanning characteristics. The *H*-plane patterns were not measured because of the limitation of the measurement set-up. Figure 9.33 shows the measured *E*-plane co-polar radiation patterns of the transmit-array with four SIW slot primary feeds. Minor discrepancies occur with one of the measured beams as well as slightly higher side lobe level, which may be attributed to several possibilities such as PCB fabrication tolerances of the SIW slot or the co-planar transmit-arrays, the alignment of the SIW slots to the transmit-array, as well as the transmit-array to the horn. Overall, the measured results agree quite well with the simulations. The gain of the co-planar transmit-array antenna prototype is 18.5 dBi for the ports at 76.5 GHz, each with 7° beamwidth and combined 3-dB beam span of about ±15° for the four beams is achieved.

9.5 Summary

Millimeter wave automotive radars are more and more popularly used to improve drivers' comfort and safety. High-performance and cost-effective radar sensors are always desired for the automotive industry. The antenna system is one of the key differentiating factors of an automotive radar sensor that defines the FOV and achievable angular object separation. The substrate integrated antenna is an excellent candidate for automotive radar sensors owing to the merits of planar configuration, light weight, low cost for material and fabrication, and easy integration with RF front-end. The antenna system technology of mmW radar sensors are fast advancing to meet the demands of vast automotive radar sensor market.

Acknowledgments

The authors would like to appreciate Dr. Junfeng Xu and Dr. Siew Bee Yeap for their contributions to the designs of single-layer 24 GHz SIW slot array and 77 GHz transmit-array, respectively.

References

1 Grimes, D.M. and Jones, T.O. (1974). Automotive radar: a brief review. *IEEE Proc.* 62 (6): 804–821.
2 Meinel, H.H. and Juergen, D. (2013). Automotive radar: from its origins to future directions. *Microw. J.* 56 (9): 24–407.
3 Meinel, H.H. (2014). Evolving automotive radar: from the very beginnings into the future. *Eur. Conf. Antennas Propag.*: 3107–3114.
4 Patole, S.M., Torlak, M., Wang, D., and Ali, M. (2017). Automotive radars: a review of signal processing techniques. *IEEE Signal Proces. Mag.* 34 (2): 22–35.
5 Rasshofer, R.H. and Gresser, K. (2005). Automotive radar and lidar systems for next generation driver assistance functions. *Adv. Radio Sci.* 3: 205–209.
6 Bloecher, H.L., Dickmann, J., and Andres, M. (2009). Automotive active safety & comfort functions using radar. *IEEE Int. Conf. Ultra-Wideband*: 490–4944.
7 Dudek, M., Nasr, I., Bozsik, G. et al. (2015). System analysis of a phased-array radar applying adaptive beam-control for future automotive safety applications. *IEEE Trans. Veh. Technol.* 64 (1): 34–47.
8 Gresham, I., Jain, N., Budka, T. et al. (2001). A compact manufacturable 76-77-GHz radar module for commercial ACC applications. *IEEE Trans. Microwave Theory Tech.* 49 (1): 44–58.
9 Kühnle, G., Mayer, H., Olbrich, H. et al. (2003). Low-cost long-range radar for future driver assistance systems. *AutoTechnol.* 4: 76–79.
10 Russell, M.E., Crain, A., Curran, A. et al. (1997). Millimeter-wave radar sensor for automotive intelligent cruise control (ICC). *IEEE Trans. Microwave Theory Tech.* 45 (12): 2444–2453.
11 Tokoro, S., Kuroda, K., and Kawakubo, A. (2003). Electronically scanned millimeter-wave radar for pre-crash safety and adaptive cruise control system. *IEEE Intell. Veh. Symp.*: 304–309.
12 Winkler, V., Feger, R., and Maurer, L. (2008). 79 GHz automotive short range radar sensor based on single-chip SiGe-transceivers. In: *2008 38th European Microwave Conference, Amsterdam*, 1616–1619.

13 Harsch, J., Topak, E., Schnabel, R. et al. (2012). Millimeter-wave technology for automotive radar sensors in the 77 GHz frequency band. *IEEE Trans. Microwave Theory Tech.* 60 (3): 845–860.

14 ETSI EN 300 400 V2.1.1 (2018)– *Radio Equipment to be Used in the 1 GHz to 40 GHz Frequency Range*. The European Telecommunications Standards Institute (ETSI), *France*. https://www.etsi.org/deliver/etsi_en/300400_300499/300440/02.02.01_60/en_300440v020201p.pdf.

15 FCC 47 CFR 15.245 (2011)– Operation within the bands 902–928 MHz, 2435–2465 MHz, 5785–5815 MHz, 10500–10550 MHz, and 24075–24175 MHz. https://www.govinfo.gov/app/details/CFR-2011-title47-vol1/CFR-2011-title47-vol1-sec15-245 (accessed 19 December 2020).

16 FCC report and order (2017)– radar services in the 76-81 GHz band, ET docket No. 15-26.

17 ETSI EN 301 091 V2.1.1 (2017)– *Radar Equipment Operating in the 76 GHz to 77 GHz Range*. The European Telecommunications Standards Institute (ETSI), *France*. https://www.etsi.org/deliver/etsi_en/301000_301099/30109101/02.01.01_60/en_30109101v020101p.pdf (accessed 19 December 2020).

18 ARIB *STD-T48 Version 2. 1 (2015), Millimeter-wave Radar Equipment for Specified Low Power Radio Station*. Association of Radio Industries and Businesses. *Japan*. https://www.arib.or.jp/english/std_tr/telecommunications/desc/std-t48.html (accessed 19 December 2020).

19 Ramasubramanian, K., Ramaiah, K., and Aginskiy, A. (2017). Moving from legacy 24 GHz to state-of-the-art 77 GHz radar. https://www.ti.com/lit/wp/spry312/spry312.pdf (accessed 19 December 2020).

20 Menzel, W. and Moebius, A. (2012). Antenna concepts for millimeter-wave automotive radar sensors. *IEEE Proc.* 100 (7): 2372–2395.

21 Binzer, T., Klar, M., and GroQ, V. (2007). Development of 77 GHz radar lens antennas for automotive applications based on given requirements. In: *2007 2nd International ITG Conference on Antennas, Munich*, 205–209.

22 Millitech Corporation (1994). Crash avoidance FLR sensors. *Microw. J.* 37 (12): 122–126.

23 Sarkas, I., Khanpour, M., Tomkins, A. et al. (2009). W-band 65-nm CMOS and SiGe BiCMOS transmitter and receiver with lumped I-Q phase shifters. In: *2009 IEEE Radio Frequency Integrated Circuits Symposium, Boston, MA*, 441–444.

24 Wagner, C., Hartmann, M., Stelzer, A., and Jaeger, H. (2008). A fully differential 77 GHz active IQ modulator in a silicon-germanium technology. *IEEE Microwave Wireless Compon. Lett.* 18: 362–364.

25 Steinhauer, M., Ruo, H.O., Irion, H., and Menzel, W. (2008). Millimeter-wave radar sensor based on a transceiver array for automotive applications. *IEEE Trans. Microwave Theory Tech.* 56 (2): 261–269.

26 Sakakibara, K., Sugawa, S., Kikuma, N., and Hirayama, H. (2010). Millimeter wave microstrip array antenna with matching-circuit-integrated radiating-elements for travelling-wave excitation. In: *Proceedings of the Fourth European Conference on Antennas and Propagation, Barcelona*, 1–5.

27 Han, L. and Wu, K. (2012). 24 GHz bandwidth-enhanced microstrip array printed on a single-layer electrically-thin substrate for automotive applications. *IEEE Trans. Antennas Propag.* 60 (5): 2555–2558.

28 Xu, J.F., Hong, W., Chen, P., and Wu, K. (2009). Design and implementation of low sidelobe substrate integrated waveguide longitudinal slot array antennas. *IET Microwave Antennas Propag.* 3 (5): 790–797.

29 Xu, J.F., Chen, Z.N., Qing, X., and Hong, W. (2011). Bandwidth enhancement for a 60 GHz substrate integrated waveguide fed cavity array antenna on LTCC. *IEEE Trans. Antennas Propag.* 59 (3): 826–832.

30 Chen, P., Hong, W., Kuai, Z. et al. (2009). A multibeam antenna based on substrate integrated waveguide technology for MIMO wireless communications. *IEEE Trans. Antennas Propag.* 57 (6): 1813–1821.
31 Shin, D.H., Park, S.J., Ahn, J.W. et al. (2015). Design of shorted parasitic rhombic array antenna for 24 GHz rear and side detection system. *IET Microwave Antennas Propag.* 9 (14): 1581–1586.
32 Hayashi, Y., Sakakibara, K., Nanjo, M. et al. (2011). Millimeter wave microstrip comb-line antenna using reflection-canceling slit structure. *IEEE Trans. Antennas Propag.* 59 (2): 398–406.
33 Shin, D.H., Kim, K.B., Kim, J.G., and Park, S.O. (2014). Design of null-filling antenna for automotive radar using genetic algorithm. *IEEE Antennas Wirel. Propag. Lett.* 14: 738–741.
34 Lee, W., Kim, J., and Yoon, Y.J. (2011). Compact two-layer Rotman lens-fed microstrip antenna array at 24 GHz. *IEEE Trans. Antennas Propag.* 59 (2): 460–466.
35 Deckmyn, T., Caytan, O., Bosman, D. et al. (2018). Single-fed 3 × 3 substrate-integrated waveguide parasitic antenna array for 24 GHz radar applications. *IEEE Trans. Antennas Propag.* 66 (11): 5955–5963.
36 Tekkouk, K., Ettorre, M., and Sauleau, R. (2018). Multibeam pillbox antenna integrating amplitude-comparison monopulse technique in the 24 GHz band for tracking applications. *IEEE Trans. Antennas Propag.* 66 (5): 2616–2321.
37 Yu, Y., Hong, W., Zhang, H. et al. (2018). Optimization and implementation of SIW slot array for both medium- and long-range 77 GHz automotive radar application. *IEEE Trans. Antennas Propag.* 66 (7): 3769–3774.
38 Xu, J., Hong, W., Zhang, H. et al. (2017). An array antenna for both long- and medium-range 77 GHz automotive radar applications. *IEEE Trans. Antennas Propag.* 65 (12): 7207–7216.
39 Rengarajan, S.R., Josefsson, L.G., and Elliott, R.S. (1999). Waveguide-fed slot antennas and arrays: a review. *Electromagnetics* 19 (1): 3–22.
40 Elliott, R. and O'Loughlin, W. (1986). The design of slot arrays including internal mutual coupling. *IEEE Trans. Antennas Propag.* 34 (9): 1149–1154.
41 Zhang, T., Zhang, Y., Cao, L. et al. (2015). Single-layer wideband circularly polarized patch antennas for Q-band applications. *IEEE Trans. Antennas Propag.* 63 (1): 409–414.
42 Ku, B.H., Schmalenberg, P., Inac, O. et al. (2014). A 77–81-GHz 16-element phased-array receiver with 50 beam scanning for advanced automotive radars. *IEEE Trans. Microwave Theory Tech.* 62 (11): 2823–2832.
43 Trotta, S., Li, H., Trivedi, V., and John, J. (2009). A tunable flipflop-based frequency divider up to 113 GHz and a fully differential 77 GHz push-push VCO in SiGe BiCMOS technology. In: *2009 IEEE Radio Frequency Integrated Circuits Symposium, Boston, MA*, 47–50.
44 Starzer, F., Fischer, A., Forstner, H. et al. (2010). A fully integrated 77-GHz radar transmitter based on a low phase-noise 19.25-GHz fundamental VCO. In: *2010 IEEE Bipolar/BiCMOS Circuits and Technology Meeting (BCTM), Austin, TX*, 65–68.
45 Knapp, H., Treml, M., Schinko, A. et al. (2012). Three-channel 77 GHz automotive radar transmitter in plastic package. In: *2012 IEEE Radio Frequency Integrated Circuits Symposium, Montreal, QC*, 119–122.
46 Wagner, C., Böck, J., Wojnowski, M. et al. (2012). A 77 GHz automotive radar receiver in a wafer level package. In: *2012 IEEE Radio Frequency Integrated Circuits Symposium, Montreal, QC*, 511–514.

47 Nicolson, S., Chevalier, P., Sautreuil, B., and Voinigescu, S. (2008). Single-chip W-band SiGe HBT transceivers and receivers for doppler radar and millimeter-wave imaging. *IEEE J. Solid-State Circuits* 43 (10): 2206–2217.

48 Wagner, C., Forstner, H.P., Haider, G. et al. (2008). A 79-GHz radar transceiver with switchable TX and LO feed through in a silicon-germanium technology. In: *2008 IEEE Bipolar/BiCMOS Circuits and Technology Meeting, Monteray, CA*, 105–108.

49 Fischer, A., Tong, Z., Hamidipour, A. et al. (2014). 77-GHz multi-channel radar transceiver with antenna in package. *IEEE Trans. Antennas Propag.* 62 (3): 1386–1394.

50 Trotta, S., Wintermantel, M., Dixon, J. et al. (2012). An RCP packaged transceiver chipset for automotive LRR and SRR systems in SiGe BiCMOS technology. *IEEE Trans. Microwave. Theory Tech.* 60 (3): 778–794.

51 Wojnowski, M., Lachner, R., Böck, J. et al. (2011). Embedded wafer level ball grid array(eWLB) technology for millimeter-wave applications. In: *2011 IEEE 13th Electronics Packaging Technology Conference, Singapore*, 423–429.

52 Wojnowski, M., Wagner, C., Lachner, R. et al. (2012). A 77-GHz SiGe single-chip four-channel transceiver module with integrated antennas in embedded wafer-level BGA package. In: *2012 IEEE 62nd Electronic Components and Technology Conference, San Diego, CA*, 1027–1032.

53 Xu, J., Chen, Z.N., and Qing, X. (2014). CPW center-fed single-layer SIW slot antenna array for automotive radars. *IEEE Trans. Antennas Propagat.* 62 (9): 4528–4536.

54 Hashimoto, K., Hirokawa, J., and Ando, M. (2010). A post-wall waveguide center-feed parallel plate slot array antenna in the millimeter-wave band. *IEEE Trans. Antennas Propagat.* 58 (11): 3532–3538.

55 Sehyun, P., Tsunemitsu, Y., Hirokawa, J., and Ando, M. (2006). Center feed single layer slotted waveguide array. *IEEE Trans. Antennas Propagat.* 54 (5): 1474–1480.

56 Tsunemitsu, Y., Matsumoto, S., Kazama, Y. et al. (2008). Reduction of aperture blockage in the center-feed alternating-phase fed single-layer slotted waveguide array antenna by E- to H-plane cross-junction power dividers. *IEEE Trans. Antennas Propagat.* 56 (6): 1787–1790.

57 Yeap, S.B., Qing, X., and Chen, Z.N. (2015). 77-GHz dual-layer transmit-array for automotive radar applications. *IEEE Trans. Antennas Propagat.* 63 (6): 2833–2837.

58 Rutledge, D. (1985). Substrate-lens coupled antennas for millimeter and sub-millimeter waves. *IEEE Antennas Propag. Soc. Newslett.* 27 (4): 4–8.

59 Porter, B.G., Rauth, L.L., Mura, J.R., and Gearhart, S.S. (1999). Dual-polarized slot-coupled patch antennas on Duroid with Teflon lenses for 76.5-GHz automotive radar system. *IEEE Trans. Antennas Propag.* 47 (12): 1832–1846.

60 McGrath, D.T. (1986). Planar three-dimensional constrained lenses. *IEEE Trans. Antennas Propag.* 34 (1): 46–50.

61 Pozar, D.M. (1996). Flat lens antenna concept using aperture coupled microstrip patches. *Electronics Lett.* 32 (23): 2109–2111.

62 Huder, B. and Menzel, W. (1988). Flat printed reflector antenna for mm-wave applications. *Electronic Lett.* 24 (6): 318–319.

63 Park, Y.J., Herschlein, A., and Wiesbeck, W. (2003). Offset cylindrical reflector antenna fed by a parallel-plate Luneburg lens for automotive radar applications in mmW. *IEEE Trans. Antennas Propag.* 51 (9): 2481–2483.

64 Padilla, P., Munoz-Acevedo, A., and Sierra-Castaner, M. (2010). Passive planar transmit-array microstrip lens for microwave purpose. *Microwave Opt. Tech. Lett.* 52 (4): 940–947.

65 Ryan, C.G.M. and Chaharmir, M.R. (2010). A wideband transmit-array using dual-resonant double square rings. *IEEE Trans. Antennas Propag.* 58 (5): 1486–1493.

66 Abbaspour-Tamijani, A., Sarabandi, K., and Rebeiz, G.M. (2007). A millimeter-wave bandpass filter-lens array. *IET Microwave Antennas Propag.* 1 (2): 388–395.

67 Kaouach, H., Dussopt, L., Lanteri, J. et al. (2001). Wideband low-loss linear and circular polarization transmit-arrays in V-band. *IEEE Trans. Antennas Propag.* 59 (7): 2531–2523.

68 Qing, X. and Chen, Z.N. (2014). Measurement setups for millimeter-wave antennas at 60/140/270 GHz bands. In: *2014 International Workshop on Antenna Technology: Small Antennas, Novel EM Structures and Materials, and Applications (iWAT), Sydney, NSW*, 281–284.

10

Sidelobe Reduction of Substrate Integrated Antenna Arrays at Ka-Band

Teng Li

State Key Laboratory of Millimeter Waves, School of Information Science and Engineering, Southeast University, Nanjing 210096, People's Republic of China.

10.1 Introduction

The sidelobe suppression technology plays an essential role in radar and communication systems that effectively reduces the noise and interference from undesired directions. There are two key issues to be addressed to lower the sidelobe levels (SLLs) of antenna arrays: how to determine the power aperture distribution of antenna array for low SLLs and how to realize the desired power distribution by feeding networks.

The first problem relates to the pattern synthesis with a low-sidelobe antenna array factor, which has been developed over a half century. Classical methods have been developed for linear and planar arrays, such as Schelkunov's form, Woodward synthesis, Fourier transform method, Dolph-Chebyshev synthesis, and Taylor line source/circular array synthesis. For the aforementioned methods, it is assumed that the patterns of all the antenna elements are identical. However, in a practical antenna array, the pattern of a central element might be quite different from that of the edge ones due to the mutual coupling between the elements. The differences of the element patterns cannot be ignored in a finite-size array and should be considered for an accurate pattern synthesis. Various methods, including space-mapping method, particle swarm optimization algorithm, genetic algorithms, and self-adaptive differential evolution algorithm, can be employed to address this issue. The methods have been proposed for the beam shaping applications [1–3].

The second issue is the architecture of antenna array for the desired power distribution where the feeding network is the key factor once the antenna element is selected. However, it is a challenge to obtain the wideband feeding network with accurate power ratio and phase balance at mmW bands.

In this chapter, the techniques of sidelobe reduction are introduced in the substrate integrated antenna array design. The standing wave antenna arrays with SIW based feeding network are discussed. In Section 10.2, the state-of-the-art techniques of feeding networks for low SLLs are reviewed and summarized at first. Then the design method and performance of the power dividers with balanced/unbalanced outputs are discussed. After that, two types of series feeding techniques for a small antenna array and a monopulse antenna array operating at Ka-band with sidelobe reduction are presented in Section 10.3.

Substrate-Integrated Millimeter-Wave Antennas for Next-Generation Communication and Radar Systems, First Edition.
Edited by Zhi Ning Chen and Xianming Qing.

10.2 Feeding Networks for Substrate Integrated Antenna Array

The feeding network is one of the key techniques to achieve a low SLL of an antenna array. As the operating frequency increases to mmW bands, the SIW-based feeding networks are attractive with advantages of high integration and low loss compared with the rectangular waveguide (RWG) and microstrip line. According to the operating mechanism, they can be categorized into series feed [4–8], parallel/corporate feed [9–15], and flat lens/reflector-based quasi-optics feed [16–18]. The features of each type of the feeding network are reviewed in detail. As an essential component of feeding network, the power dividers with arbitrary power ratio, especially the large ones, and phase balance are discussed.

10.2.1 Series Feeding Network

The series feeding architectures are quite classical and have been widely used in the SIW slot antenna arrays. According to the power flow, they can be mainly categorized as *H*- and *E*-plane series feedings. The longitudinal radiating slots are located along the SIW and inherently series-fed along the *H*-plane. Such a type of feeding structure is the most popular configuration for the SIW array design with high efficiency [4–11]. The slots are spaced with a half guided wavelength (λ_{gr}) of the radiating SIW or $\lambda_{gr}/4$ from the short end and alternatively offset for in-phase radiating, as shown in Figure 10.1. They can be equivalent to the shunt admittance that is the same as the waveguide ones [19]. The design considerations can be simply summarized as follows. The slot is resonant when the susceptance (b) is zero by modifying the slot length (l). Then normalized conductance (g) is proportional to the radiating power and the offset distance (x) from the centerline. Assuming the SIW is single-ended fed, the total normalized conductance should be one for impedance matching.

For the *E*-plane series feeding architecture, the inclined slots or SIW T-junctions are frequently applied, as shown in Figure 10.2a,b, respectively. The inclined coupling slots, inspired by the metal waveguide slot arrays, are etched on the bottom ground plane of each radiating SIW. A half-guided wavelength (λ_{gc}) of the coupling SIW is selected to be the spacing between slots. The in-phase excitation for each radiating SIW is achieved by alternately rotating the coupling slots. Similar to the radiating slots, the coupling slot is considered to be resonant if the reactance (x) is zero, and it can be realized by modifying the slot length. The coupled power is determined by the rotation angle and is proportional to the equivalent resistance. For the single-fed coupling SIW, the total normalized resistance is one for impedance matching. The coupling slots are usually located between the radiating slots for wideband operation, namely center feed. Unlike the series coupling slots, the series T-junctions operate at a quasi-traveling wave state and terminated by the SIW *H*-plane bend.

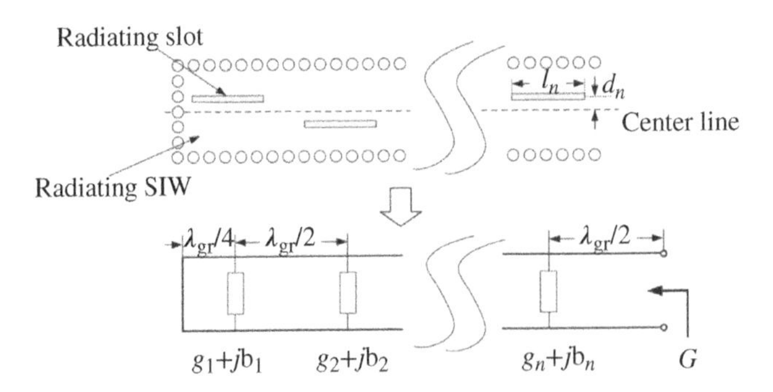

Figure 10.1 Series-fed SIW slot antenna array along the *H*-plane.

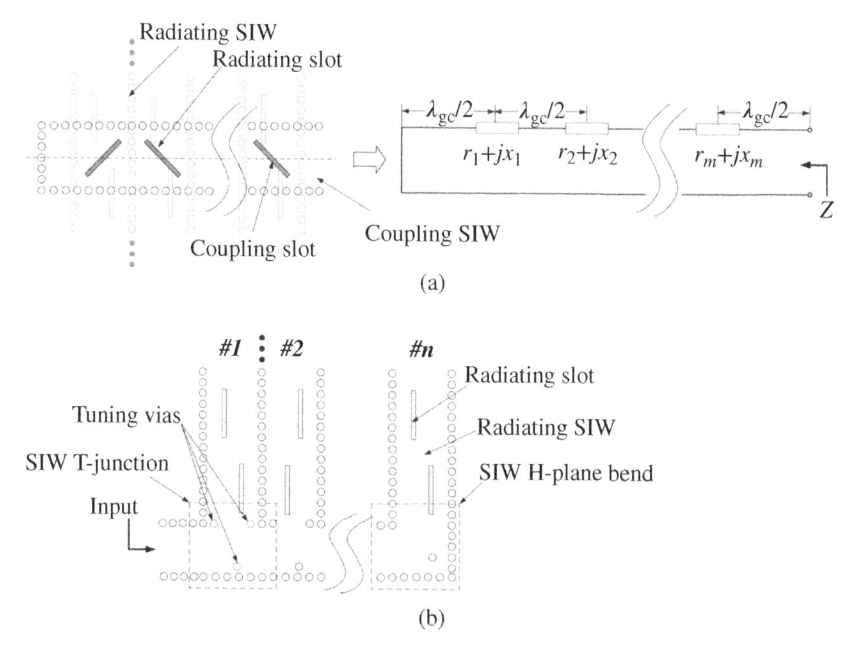

Figure 10.2 SIW based *E*-plane series feeding network. (a) Series coupling slots and (b) series T-junctions.

Therefore, the desired output ratios are calculated from the power of rest linear arrays and have no concern with the former ones. By modifying the positions of tuning vias nearby the window, the output ratio of the T-junction is adjustable. Furthermore, a half-guided wavelength (λ_{gT}) of T-junction is usually chosen as the period. Due to the out-of-phase output of the adjacent T-junctions, the radiating slots between the adjacent linear arrays are reversed offset for in-phase radiation.

Based on the aforementioned series feeding architectures, the desired power distribution of an SIW slot array can be achieved for the low SLL. Figure 10.3a shows a 3×6 SIW slot antenna array at X-band; based on the series coupling slots and a hybrid optimization method, the SLL in the *H*-plane is suppressed to −20 dB [4]. As operating frequency increases to mmW and the antenna array becomes larger for higher gain, the double-layer SIW feeding structure becomes more challenging due to the possible leakage between radiating and coupling SIWs. The hybrid feeding structure based on SIW and metal RWG is a good alternative that permits the low cost of SIW and RWG. In [5], a 16×56 slot antenna array with SLLs of −16 and − 20 dB in the *E*- and *H*-plane is designed, where the slot antenna elements grouped into eight subarrays are located on the SIW layer with the RGW feeding network beneath them. Compared with the multi-layer and hybrid feedings, the series T-junctions can be integrated into the same layer of the radiating SIW at the cost of enlarged aperture area. In Figure 10.3c, a 16×16 SIW slot array fed by series T-junctions is designed at X-band with SLLs less than −30 dB in both the *E*- and *H*-plane [7]. After that, a 12×12 SIW slot array based on the similar architecture, as depicted in Figure 10.3d, achieves the SLLs less than −15 and − 25 dB at 60 GHz in the *E*- and *H*-plane, respectively [8].

As a result, the series longitudinal slot array is a convenient and efficient approach to realize the series feeding in the *H*-plane. For the series coupling slots method, it is more suitable for the applications with a demand of high aperture efficiency. On the other hand, the series T-junctions method is a good candidate for single-layer applications featuring low profile and low cost. In spite of this, the series feeding network is limited by narrow bandwidth, typically around 3–5% due to the increased phase errors.

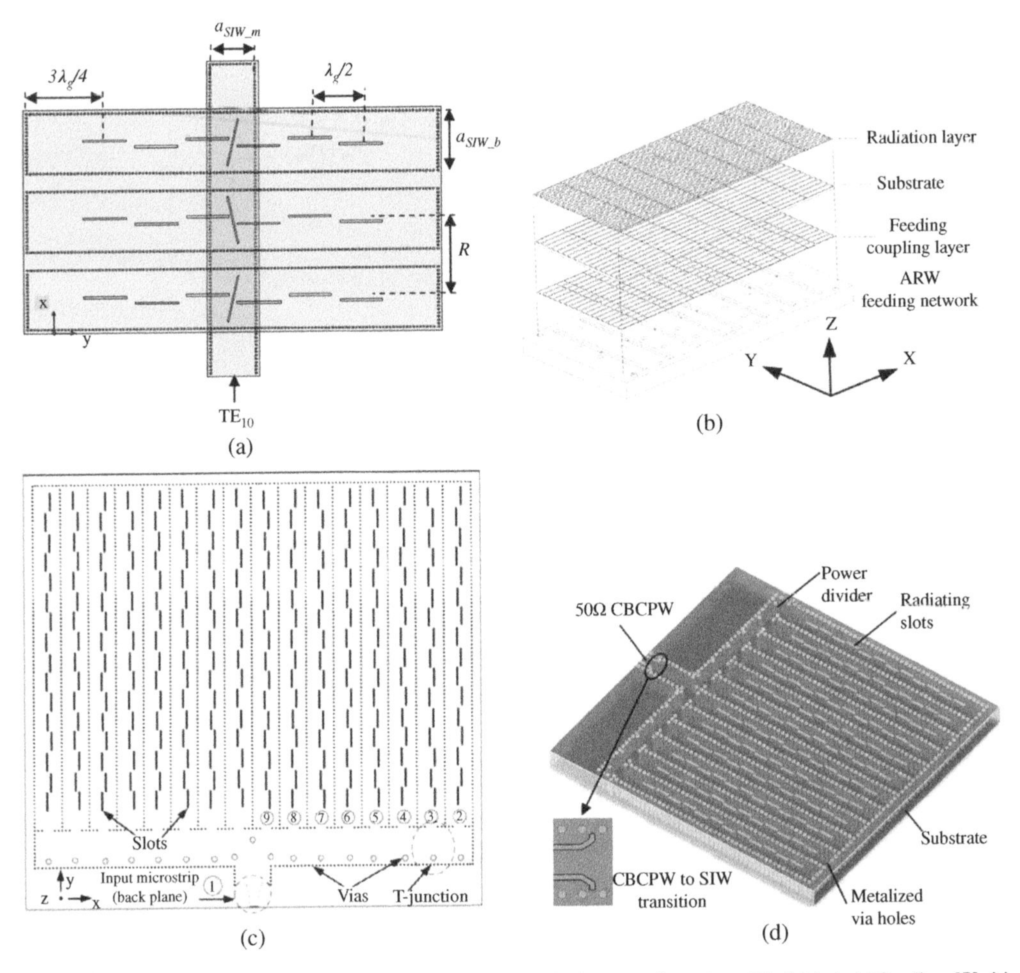

Figure 10.3 Examples of series-fed SIW slot arrays. (a) Series coupling slots [4], (b) hybrid feeding [5], (c) and (d) series T-junctions [7, 8].

10.2.2 Parallel/Corporate Feeding Network

To achieve a wide bandwidth of antenna array, the parallel-fed structure is a preferred selection and is categorized into two groups according to the feeding frames: the partially parallel feeding [9–13] and the fully corporate feeding [14, 15]. The partially parallel feeding indicates that the antenna is parallel-fed in one plane (along one direction) but series-fed in the other plane. Figure 10.4a shows a classical partially parallel-fed SIW-based feeding network that is realized by a four-stage 16-way equal output power divider [9]. At the output ports, the SIW series-fed longitudinal slot array is connected; therefore, the parallel feeding is achieved in the E-plane. Unlike the series-fed SIW T-junctions, the input port and one output port of a parallel-fed T-junction are exchanged. To achieve the non-uniform outputs for low-SLL application, the unequal power dividers can be employed, as shown in Figure 10.4b, and the SLL of −25 dB is realized in the E-plane [10]. The partially parallel feeding network can be integrated into the same layer of an SIW antenna array or folded into the next layer to save the space [11]. Based on this technique, the operating bandwidth can be certainly improved but is still limited by the series-fed subarray.

The fully corporate feed denotes that the SIW antenna or subarray is parallel-fed for each element and a wider bandwidth is anticipated. A typical corporate feeding architecture is depicted in

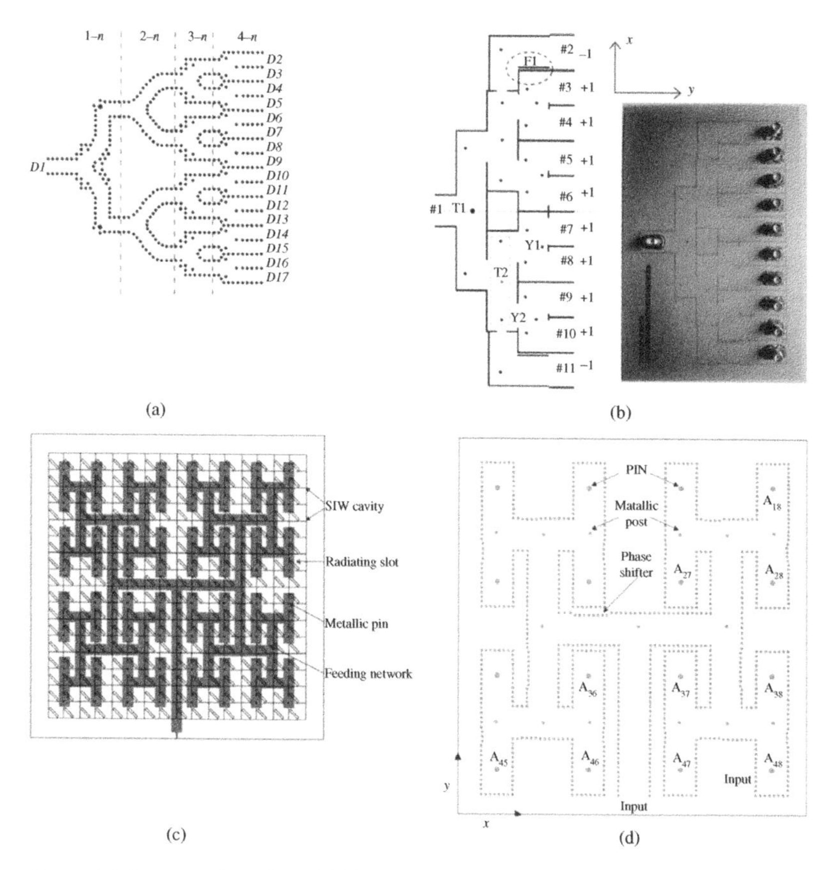

Figure 10.4 Examples of parallel/corporate SIW feeding network. (a) Uniform parallel feeding [9], (b) non-uniform parallel feeding [10], (c) uniform corporate feeding [14] and (d) non-uniform corporate feeding [15].

Figure 10.4c where the 2×2 subarray with 45° slots is uniformly fed for radiation [14]. Another non-uniform corporate feeding network is shown in Figure 10.4d where the unequal T-junctions are utilized to achieve the SLL of −17 dB over a bandwidth of 15% [15]. Compared with the partially parallel feeding network, the corporate feeding network is usually placed beneath the radiating layer with higher aperture usage and cost.

10.2.3 Flat Lens/Reflector-Based Quasi-Optics Feeding Network

Planar quasi-optics feeding is the two-dimensional lens/reflector based on SIW technology. Unlike the traditional circuit feeding based on TE_{10} mode, the quasi-optics feeding relies on the SIW-based parallel plates and supports a quasi-TEM mode. Tapered power distribution is naturally achieved by the illumination of SIW port or SIW horn, which is beneficial to the realization of low SLL. Furthermore, the multi-beam function can be easily achieved by the offset feedings due to the inherent feature of spatial feeding architecture.

Figure 10.5a shows a typical SIW-based Rotman lens, including seven input ports and nine output ports, for multi-beam applications where a series-fed 4×9 SIW slot antenna array is connected [16]. Each input port generates a directional beam, and seven beams are generated in the E-plane. Figure 10.5b shows the SIW-based pillbox antenna where the parabolic reflectors and feeding horns are proposed for power distribution and multi-beam [17]. Two subarrays with series-fed SIW slots

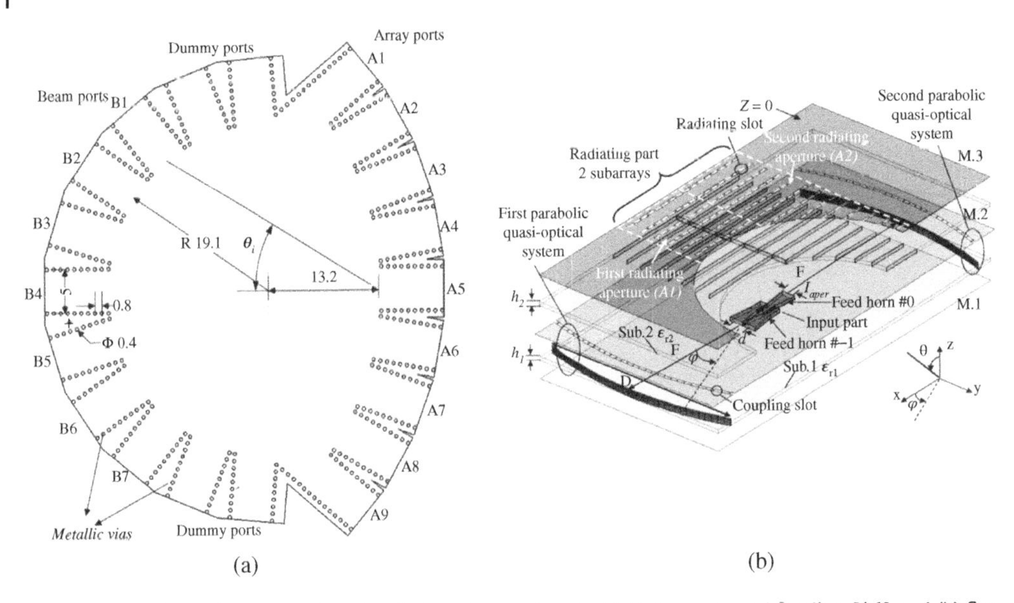

Figure 10.5 Examples of quasi-optics SIW feeding network. (a) Flat lens-based feeding [16] and (b) flat parabolic reflector-based feeding [17].

are designed in the top layer and excited by the coupling slots. The realized SLL of the center beam in the E-plane is lower than −24 dB.

As a result, the planar quasi-optics method is a preferred choice for multi-beam applications, and the inherent feature of tapered power distribution facilitates the realization of low SLL. Although the operating mechanism of quasi-optics feeding network is different from the circuit-based one, the series-fed radiating slot antenna array is usually connected indicating a similar characteristic to the partially parallel feeding network.

10.2.4 Power Dividers

The power dividers are the essential components of a feeding network to achieve the desired power distribution as depicted in Figure 10.4. There are mainly two types of SIW power dividers: T-junction and Y-junction. Take the T-junction as an example; the equal output one has been widely studied in the literature and the traditional configuration is as shown in Figure 10.6a. Port 1 is the input port and ports 2 and 3 are output ports. Three metallic inductive vias and split via are employed for impedance matching and power dividing. Due to the symmetric structure, the balanced outputs are naturally obtained without phase error and a wide bandwidth of more than 60% is anticipated by tuning the positions of these vias [20].

The classical unbalanced SIW T-junction can be evaluated from the equal one by shifting the split via toward one output port and slightly adjusting the inductive vias. However, the output ratio and bandwidth are limited for such a type of unequal SIW T-junction, for example, 12% bandwidth with an output ratio of 6 dB and balanced phase output [21]. Inspired by the SIW coupler and H-plane right-angle bend structure, a developed unequal T-junction is proposed for large output ratio and bandwidth enhancement, as shown in Figure 10.6b. Three vias are also employed as the inductive vias and arranged almost in a straight line. An SIW step section is also introduced for impedance matching. By optimizing the dimensions and positions of step section and inductive vias, more than 60% bandwidth with an output ratio of 9 dB and balanced phase output is anticipated which

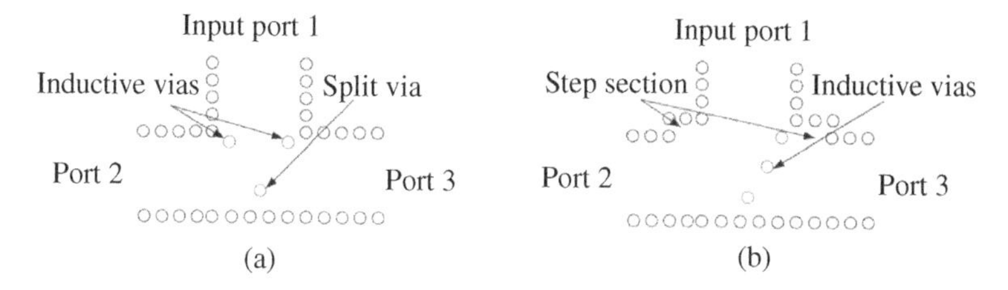

Figure 10.6 Configuration of SIW T-junctions. (a) Conventional equal output T-junction and (b) unequal output T-junction with large output ratio and bandwidth enhancement [20].

is obviously beyond the classical one [20]. As a result, a wideband feeding network with arbitrary power distribution and balanced phase output is achievable for low SLL array.

10.3 SIW Antenna Arrays with Sidelobe Reduction at Ka-Band

To introduce the design process of SIW antenna array with sidelobe reduction, as an example, two double-layer series-fed SIW slot antenna arrays at Ka-band are presented in detail. The antenna array is designed at 35 GHz and the operating band falls into an atmospheric window for radar systems, such as synthetic aperture radar and monopulse radar. As mentioned in the Introduction, the Taylor distribution is a good candidate for sidelobe reduction and is selected as the initial power distribution. The PCB technology is used for low-cost applications. An 8×8 SIW slot array is first proposed with the coupling slots for high aperture usage. Then a 16×16 SIW slot array with the series T-junctions feeding in the same layer and the sum-difference network in the second layer is presented.

10.3.1 Double-Layer 8×8 SIW Slot Array

Figure 10.7a shows the overall explosion view of the double-layer 8×8 SIW slot antenna array. To ensure the assembly reliability, a top square metal frame and a bottom metal plate with screw and aligned holes are employed to hold the radiating layer and feeding layer. Furthermore, a wideband SIW-to-RWG transition is integrated into the bottom plate to facilitate external connection [22]. Figure 10.7b shows the top and bottom views of the radiating layer where all the 8×8 radiating slots are etched onto the top copper layer along the radiating SIW. On the other hand, the top coupling slots are placed at the bottom center of radiating SIW and alternatively rotated along the x-axis. Due to the symmetry of the antenna array, only half part is taken into account, and the radiating waveguide is numbered from 1 to 4. The feeding layer is depicted in Figure 10.7c where the bottom coupling slots are broadened for assembly tolerance, and a wideband balanced T-junction is employed for excitation. To simplify the design complexity, the width of radiating and feeding SIW is in the same value but still satisfies the in-phase excitation condition, as mentioned in Section 10.2.1. The substrates used here are Rogers RT5880 with $\varepsilon_r = 2.2$ and $\tan\delta = 0.009$ at 10 GHz. The Taylor distribution of -30 dB with $n = 5$ is employed for low SLL. We will focus on the design process; all related values of parameters can be found in [23].

10.3.1.1 Parameter Extraction of Radiating Slots

In order to obtain the desired power distribution for low SLL, an accurate parameter extraction method of radiating slots is required. Various methods have been proposed for parameter extraction, such as the experiments and Elliott's formulas. Here, a simple efficient method based on

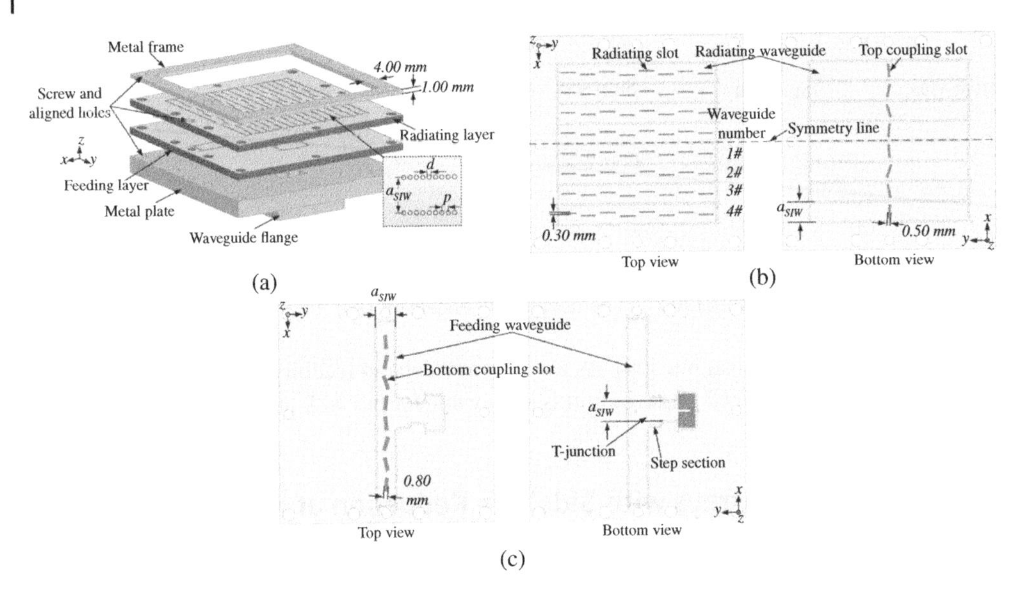

Figure 10.7 Configuration of the double-layer 8 × 8 SIW slot antenna array. (a) 3-D view, (b) top and bottom views of the radiating layer, and (c) feeding layer.

equivalent circuit and S-parameters is introduced, which can be easily obtained by the full-wave electromagnetic simulation.

The schematic diagram and equivalent circuit of the end-fed SIW slot array are shown in Figure 10.1. Assuming each slot is with the same size, the total normalized admittance for a radiating waveguide i is

$$G_i = ng_i + jnb_i = \frac{1 - S_{ii}}{1 + S_{ii}}, \tag{10.1}$$

where n is the number of slots in the radiating waveguide i, S_{ii} is the reflection coefficient of port i, g_i, and b_i are the normalized conductance and susceptance of a slot on waveguide i, respectively. According to the equivalent circuit of the radiating slot, a slot is considered to be resonant with $b_i = 0$. Therefore, the conductance of a resonant slot can be obtained with $\text{Imag}(S_{ii}) = 0$. For a certain slot offset, there would be a proper slot length for the resonance.

In order to improve the accuracy, a quarter of slot array with symmetric boundary and end-fed ports is employed. The feeding ports are assigned with the real power ratios, and the mutual coupling between adjacent slots is considered accordingly. Finally, the parameter extracting curves, namely relationships between slot offset, slot length, and conductance, are obtained accordingly, as shown in Figure 10.8. It can be found that the curves of resonant slot length are almost equal to each other, but the normalized conductance curves become diverging with the increase of slot offset This phenomenon indicates the mutual coupling between the radiating waveguides is significant for a large slot offset.

According to the equivalent circuit of the radiating slots, the total normalized conductance of the radiating waveguide is one, and the conductance (g_{ij}) of slot ij is determined by the radiating power

$$g_{ij} = \frac{p_{ij}}{q_i}, \tag{10.2}$$

where p_{ij} is the calculated radiating power of slot from a Taylor distribution and q_i is the assigned power for radiating waveguide i, namely $\sum_{j=1}^{4} p_{ij}$. As a result, the initial size of radiating slots can be calculated from these curves by the interpolation method.

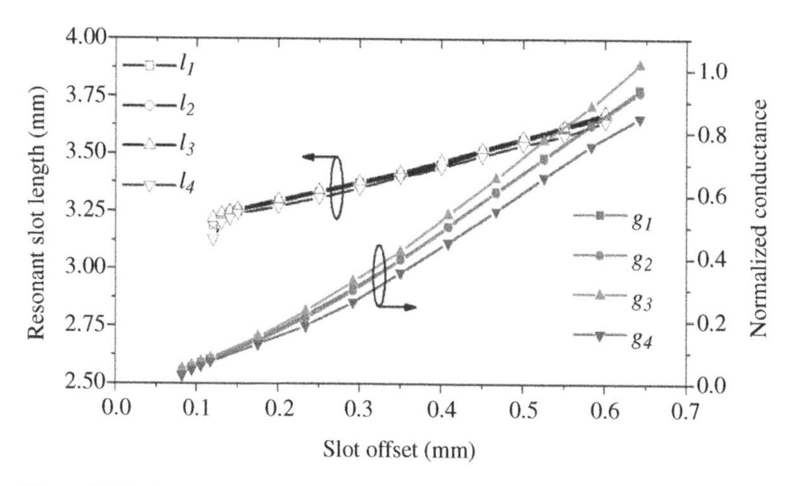

Figure 10.8 Parameter extracting curves.

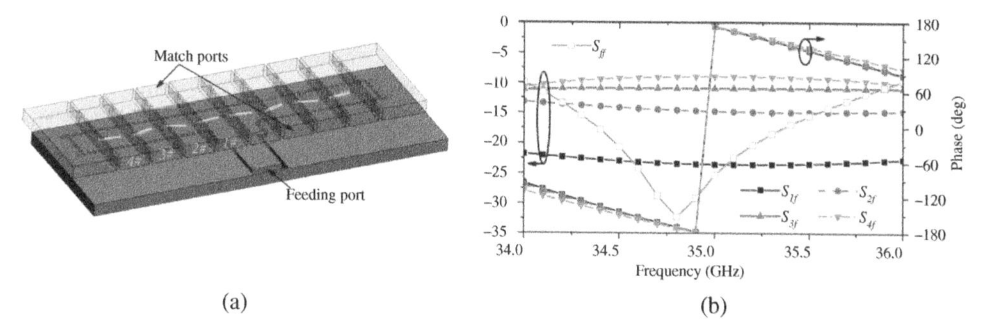

Figure 10.9 Feeding network of 8 × 8 SIW slot array. (a) Simplified model and (b) simulated S-paramters of the feeding network. The subscript f denotes feeding port.

10.3.1.2 Feeding Network

To simplify the design procedure of the feeding network, the coupling slots, feeding waveguide, and radiating waveguides without radiating slots are modeled, as shown in Figure 10.9a. Furthermore, all the radiating waveguides are terminated with matching ports, and only the marked waveguides are considered due to the symmetry. As mentioned in Section 10.2, the coupling slots are equivalent to series impedance. Similar to the radiating slots, the coupling slot is considered to be resonant if the reactance equals zero, and the coupling power is proportional to the resistance. Similarly, it is possible to obtain the parameter curves of coupling slots, namely the relationship between the tilt angle, slot length, and series resistance.

After selecting the initial size of coupling slots, a fine adjustment is carried out for optimization and a general rule is summarized as the larger the tilt angle, the larger the coupling power. In addition, the slot length refers to the phase response. Figure 10.9b shows the acceptable S-parameters of feeding network after several rounds of optimization. The desired power ratios are achieved over 34–36 GHz with a balanced phase response and −10 dB input reflection coefficient.

10.3.1.3 Simulations and Experiments

Although good results of the radiating slots and feeding network are individually and easily realized, the internal coupling between the coupling slots and the nearby radiating slots is not taken into account in the design procedure. Furthermore, the coupling between radiating waveguides is not considered for feeding network design. Therefore, the joint optimization of the entire array is carried out for low SLL and impedance matching. Considering that the interference mainly affects the conjunction parts, the center coupling slots and radiating slots are selected for optimization, which refers to the SLL in the E- and H plane, respectively. The impedance matching can be adjusted by modifying the dimensions of all slots simultaneously and the power distribution is assumed to be hardly changed. Based on these effective optimizations, the acceptable performance of an antenna array can be achieved.

Figure 10.10 shows the fabricated prototype of the optimized 8×8 SIW slot array and the aperture size is around $35\,\text{mm} \times 35\,\text{mm}$. The simulated and measured S-parameters are compared in Figure 10.12. Suppose the antenna array is designed for high power transmitter, the $-20\,\text{dB}$ input reflection coefficient is preferred. It can be found that the measured $-20\,\text{dB}$ impedance bandwidth is 1.32 GHz (33.97–35.29 GHz), which shifts 300 MHz to the lower frequency band.

The measured and simulated normalized radiation patterns at 35 GHz are summarized in Figure 10.11. It can be observed that the simulated SLLs are −27.8 and −26.9 dB in the E- and H-plane that are close to the desired Taylor distribution. However, the measured SLLs are −19.8 and −17.7 dB in the E- and H-plane, respectively. The distortions are mainly attributed to the fabrication errors of PCB. The metallic vias and slots are processed in different technologies that might introduce alignment errors. Here, the position error between slots and vias along the x-axis in the radiating layer, namely err_x, is analyzed, as shown in Figure 10.13. It can be found that the SLL in the H-plane becomes worse as err_x increases, especially around 60°, which is similar to the measurements. The measured gain is 19.9 dBi, that is, 1.2 dB lower than the simulation. As a result, the 8×8 SIW slot array with low SLL is realized.

Figure 10.10 Fabricated 8×8 SIW slot array.

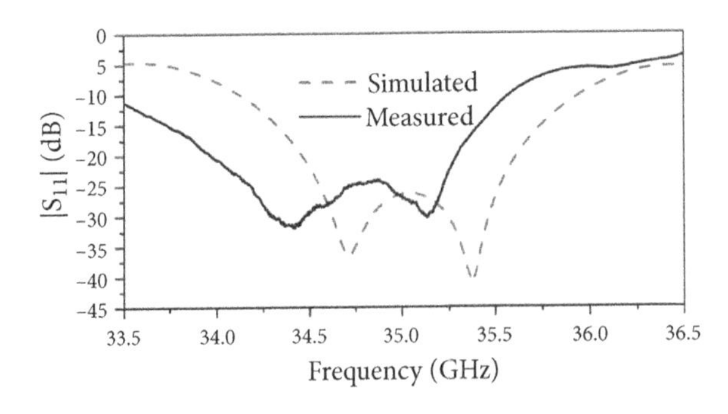

Figure 10.11 Simulated and measured reflection coefficients of the SIW slot array.

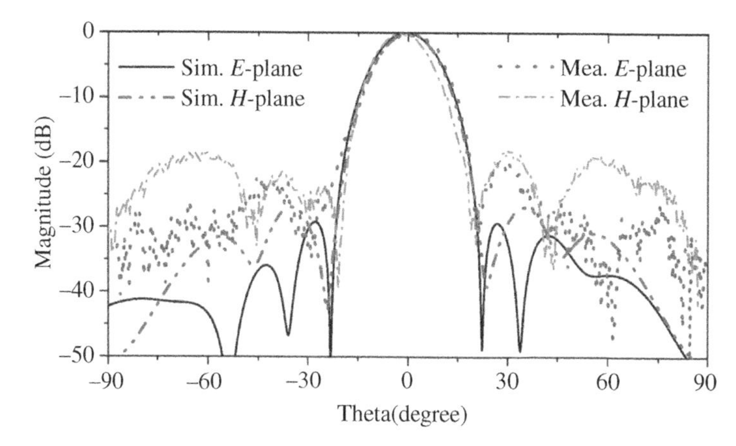

Figure 10.12 Simulated and measured radiation pattern of the SIW slot array.

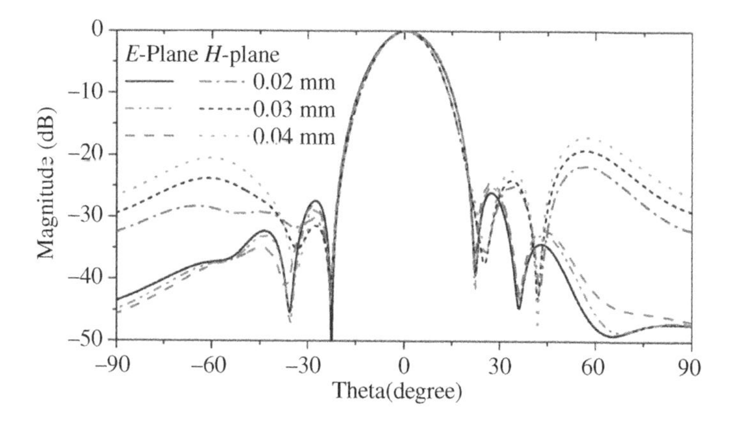

Figure 10.13 Parametric study of fabrication error err_x for the SIW slot array.

10.3.2 16 × 16 Monopulse SIW Slot Array

As an alternative series feeding technique, the series T-junction feeding network is combined with SIW slot array in the same layer featuring low profile, low cost, and high integration. Figure 10.14 shows the configuration of a16 × 16 monopulse SIW slot array where two layers of substrate are employed [24]. The radiating slot array, consisting of four sub-arrays, and a series T-junction feeding network with unbalanced output for low SLL are designed in the first layer, as depicted in Figure 10.14a. The sum-difference network realized by phase shifters and 3-dB directional couplers is designed on the second layer, as shown in Figure 10.14b. Four coupling slots are used for the connection between the SIW slot array and sum-difference network. Compared with the 8 × 8 array, the fewer coupling slots improves the reliability. The design procedure of the SIW slot antenna array is similar to Section 10.3.1.1.

Here, the Taylor distribution with −28 dB SLL is applied and the simulated radiation patterns of the array are shown in Figure 10.15 where a simplified model of a subarray with symmetric electric and magnetic boundaries is employed. The series feeding network is out of consideration, and the SIW radiating waveguides of 1–8 are, respectively excited by waveguide ports with power ratios of −4.905, −5.896, −7.003, −9.344, −11.593, −17.975, −18.675, and −23.018 dB. It is found that the

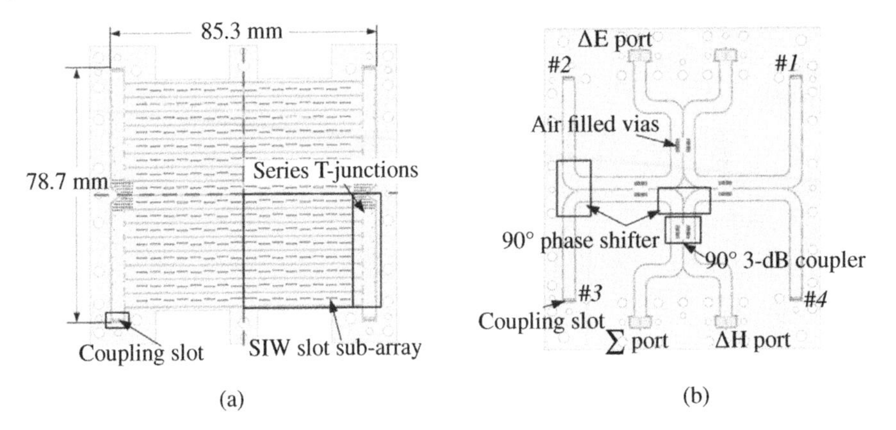

Figure 10.14 Configuration of a 16 × 16 SIW monopulse slot array. (a) First layer of radiating slots and feeureding network and (b) sum-difference network.

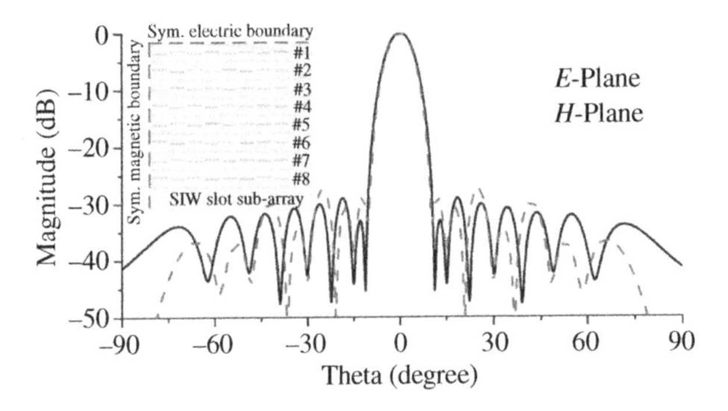

Figure 10.15 Simulated radiation pattern of the SIW radiating slots based on the simplified model at 35 GHz.

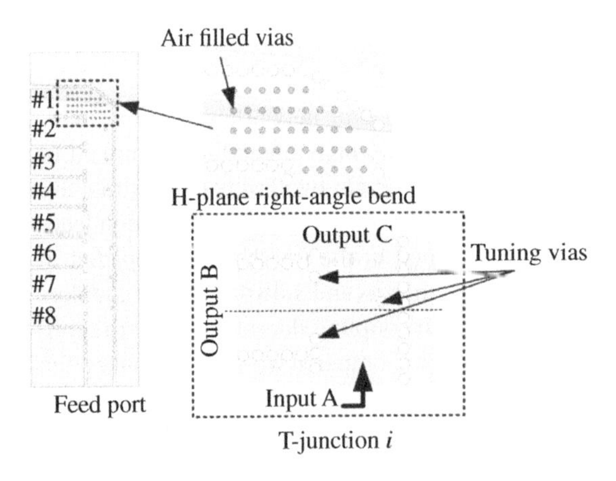

Figure 10.16 Configuration of the series T-junction feeding network for 16 × 16 SIW monopulse slot array.

SLLs are suppressed to −27.45 dB, close to that by the Taylor distribution. The next step is to design a series T-junction feeding network with desired power ratios and phase responses.

10.3.2.1 Series T-Junction Feeding Network

Figure 10.16 shows the detailed configuration of the series T-junction feeding network including seven series T-junction and an *H*-plane SIW bend structure. As one kind of *E*-plane series feeding architecture, it is discussed in Section 10.2.1 with a brief explanation of operating mechanism. The

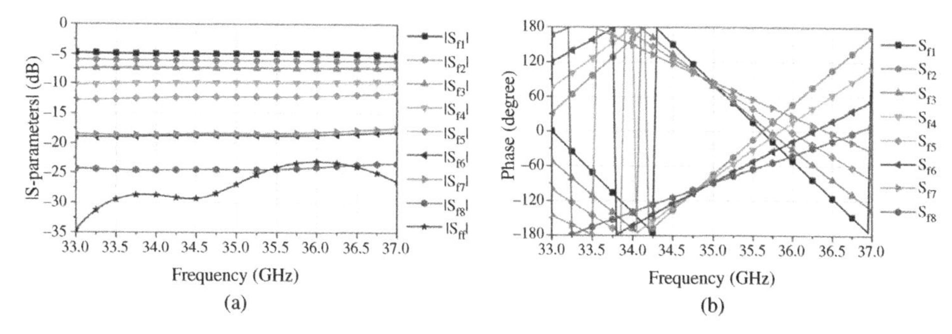

Figure 10.17 Simulated results of series T-junction based feeding network. (a) Amplitude response and (b) phase response.

design procedure is summarized as follows. First, to determine the power dividing ratios of each T-junction. It can be derived from the power distribution

$$\Delta T_i = \frac{P_B}{P_C} = \frac{ri}{\sum_{n=1}^{i-1} r_n}, i = 2, 3, \ldots, \tag{10.3}$$

where i is the number of T-junction, P_B and P_C are the output power of output ports B and C for the T-junction. r_i is the specific power for radiating waveguide i. The corresponding phase responses are Phase(Si_{BA}) = Constant and Phase(Si_{CA}) = $\pm\pi$. Then the T-junctions are separately designed with the certain specification by modifying the position and size of tuning vias. An SIW H-plane bend structure is proposed for end-load where the through phase response is $\pm\pi$ as well. Here, a metamaterial-based H-plane right-angle bend is proposed and the air-filled vias are introduced to change the effective permittivity where the desired phase response and impedance matching are both achieved. Finally, all T-junctions and H-plane bend structures are cascaded together as the feeding network.

Figure 10.17 shows the simulated results of series T-junction based feeding network. The curves of amplitude response are quite flat, as shown in Figure 10.17a, which indicates a wideband stable power dividing. The reflection coefficient of the input port is less than −23 dB over the observation band. Figure 10.17b depicts the phase responses of output pots which are alternatively changed with out-of-phase around 35 GHz. As operating frequency shifts, the phase errors increased rapidly due to the series architecture.

10.3.2.2 Sum-Difference Network

It has been studied in the former work that a planar sum-difference network is achieved by the 90° 3-dB directional coupler and 90° phase shifter [6, 9]. The outputs from the sum or difference port to the four subarrays should be of equal amplitude with certain phase differences. A sum beam for emission will be generated by exciting the sum port and difference beams in the E- and H-plane for receiving will be excited by the difference ports. By comparing the simultaneously received signals, the angular location of a target is determined, and this is the operating mechanism of monopulse radar.

Figure 10.18a shows the configuration of a 3-dB SIW directional coupler where the air-filled vias are employed, and the detailed dimensions can be found in [25]. The 90° phase shifter is realized by changing the radius (R_1-R_4) of H-plane bend structures, as shown in Figure 10.18b. By optimizing the size of bend radius, a wide bandwidth can be achieved. The complete sum-difference network is formed by four 90° 3-dB directional couplers and four 90° phase shifters, as shown in Figure 10.14b.

Figure 10.19 shows the simulated S-parameters of the sum-difference network where the subscripts 1–4 indicate the subarray number, and the symbols $\sum$ and Δ denote sum and difference ports, respectively, which are labeled in Figure 10.14b as well. For the sum port, the reflection coefficient is below −22.5 dB and the stably balanced outputs to subarrays are realized over 33–37 GHz

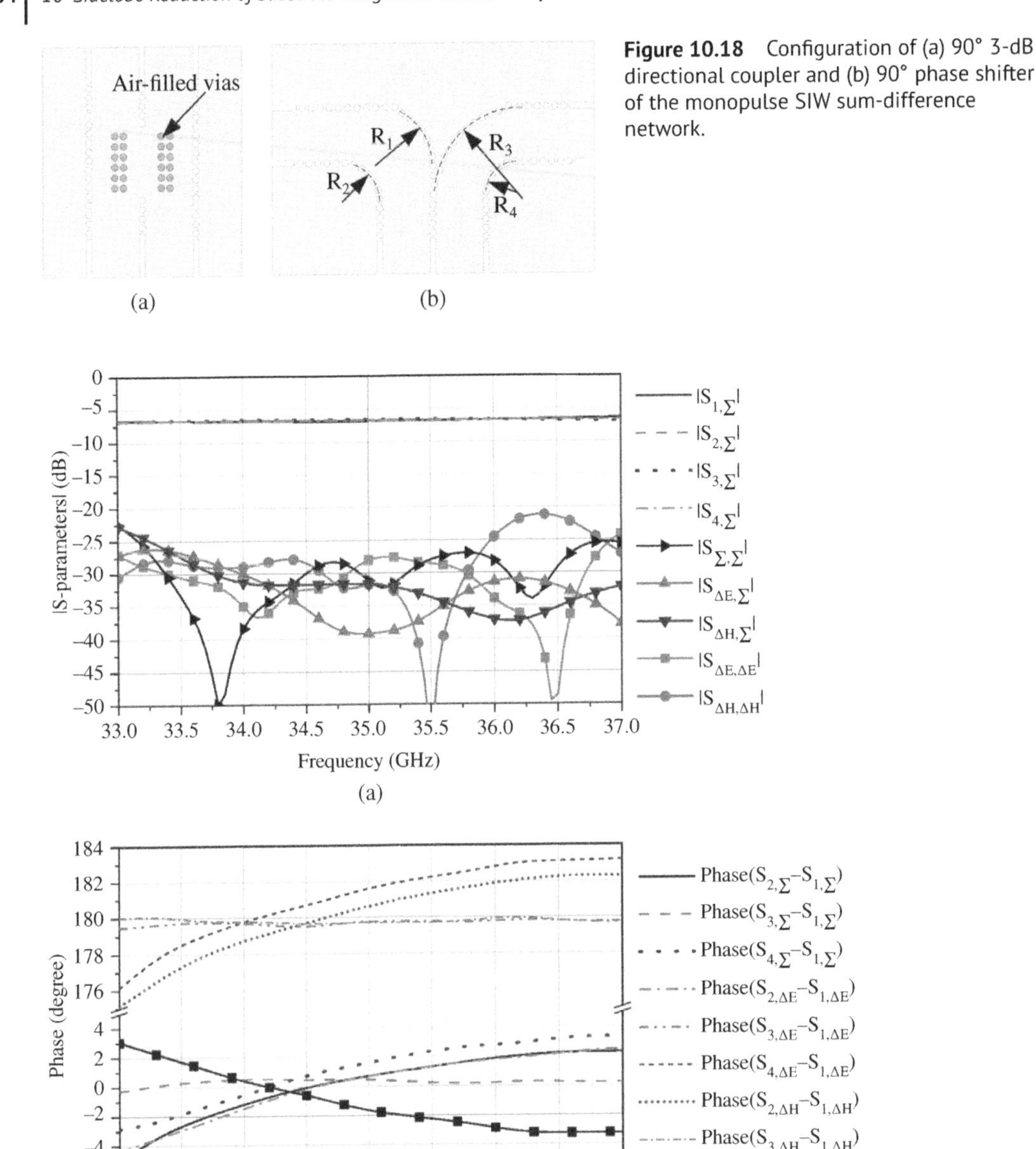

Figure 10.18 Configuration of (a) 90° 3-dB directional coupler and (b) 90° phase shifter of the monopulse SIW sum-difference network.

Figure 10.19 Simulated (a) amplitude response and (b) phase response of monopulse SIW sum-difference network.

with phase differences less than ±5°. The isolation between the sum and difference ports are all higher than 22 dB, especially 32 dB at 35 GHz. In addition, the phase errors between subarrays and sum and difference ports are all less than ±5° over the bandwidth of interest. As a result, the sum-difference network with wideband balanced outputs based on SIW is realized.

10.3.2.3 Simulations and Experiments

To validate the proposed design, a prototype of 16 × 16 monopulse SIW slot array is fabricated on the double-layer substrate based on PCB and laminated together by screws, as shown in Figure 10.20.

Figure 10.20 Photography of the fabricated 16 × 16 monopulse SIW slot array.

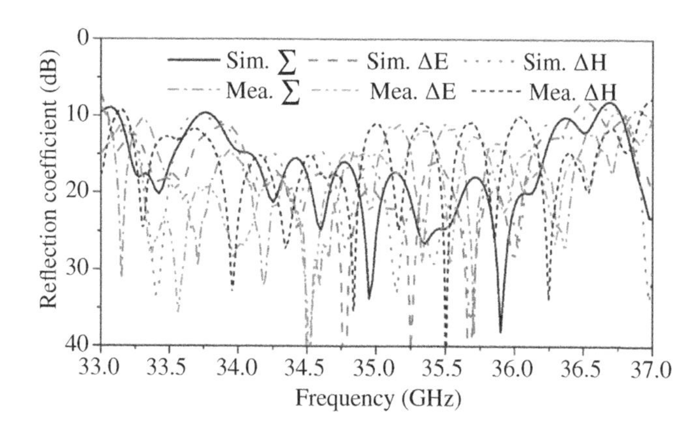

Figure 10.21 Simulated and measured reflection coefficients.

The substrate used here is F4BMX220 with dielectric constant of 2.2, loss tangent of 0.001 at 10 GHz, and thickness of 1.5 mm. Several alignment pinholes are placed around the array for accurate assembly. Figure 10.21 shows the simulated and measured reflection coefficients of sum and difference ports where good agreements are observed and all curves are below −10 dB over 34–36 GHz. The corresponding isolations between sum and difference ports are all higher than 25 dB, which are not shown for brevity.

The simulated and measured radiation patterns at 35 GHz are shown in Figure 10.22. The simulated SLLs of the sum beam are −26 and − 28.4 dB in the *E*- and *H*-plane, respectively, which are close to the performance of array without feeding network in Figure 10.15. However, the measured SLLs are −20 and − 22.5 dB in the *E*- and *H*-plane, respectively, which are seriously deteriorated due to the fabrication and assembly errors. The position tolerance of metallic vias is ±0.05 mm, which might affect the power dividing ratios of series T-junction feeding network where the different sizes of tuning vias might increase the tolerance. Furthermore, the alignment errors would affect the power ratios for subarray and introduce the asymmetric radiation patterns. The measured

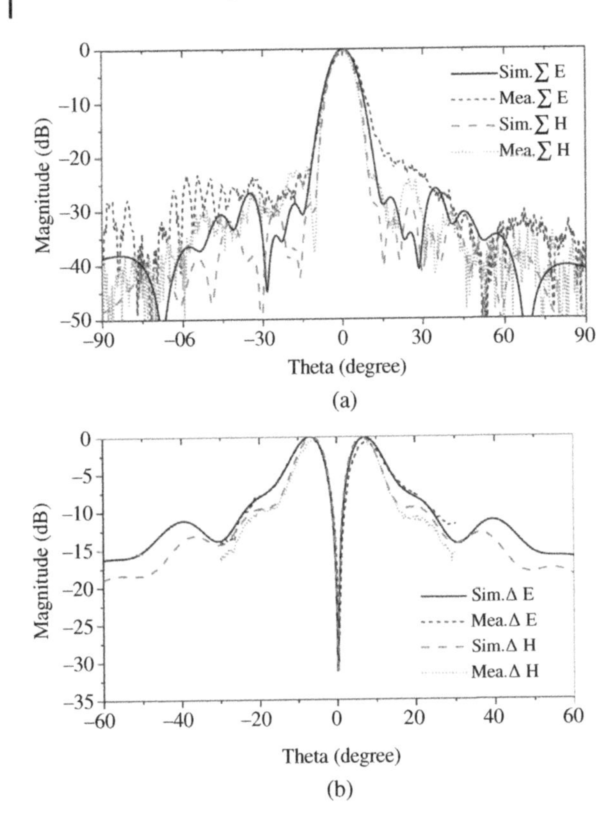

Figure 10.22 Simulated and measured radiation pattern of (a) sum beams and (b) difference beams at 35 GHz.

gain is 24 dBi at 35 GHz, which is 2.3 dB less than the simulations. The loss is mainly caused by the increased insertion loss of SIW, transition and coupling slots between the double layer. The obtained gain of difference beams is 4 dB less than the sum one. The measured null depths of difference beams are 31 and 28 dB while the simulated ones are around 35 dB. As a result, the monopulse SIW slot array with low SLL is realized based on PCB technology and is a good candidate for radar systems due to low cost, low profile, and light weight. The performance of antenna array is expected to be improved using the same size of metallic vias and conductive bonding layer between the double-layer substrate.

10.4 Summary

This chapter has reviewed the pattern synthesis technology and feeding technology based on SIW for sidelobe reduction. The feeding networks have been categorized into series feed, parallel/corporate feed, and flat lens/reflector-based quasi-optics feed. The features of these feeding networks have been discussed for practical applications. A design of 8 × 8 SIW slot antenna array based on the series coupling slot feeding network has been presented for low sidelobe level consideration. Then a 16 × 16 SIW slot array with low sidelobe level has been demonstrated for a Ka-band monopulse radar application. The detailed design procedures of SIW slot array, two kinds of series feeding networks, and sum-difference network have been discussed.

References

1 Hao, Z.C., He, M., and Hong, W. (2016). Design of a millimeter-wave high angle selectivity shaped-beam conformal array antenna using hybrid genetic/space mapping method. *IEEE Antennas Wirel. Propag. Lett.* 15: 1208–1212.

2 Echeveste, J.I., González de Aza, M.A., and Zapata, J. (2016). Shaped beam synthesis of real antenna arrays via finite-element method floquet modal analysis and convex programming. *IEEE Trans. Antennas Propag.* 64 (4): 1279–1286.

3 Zhang, Z.Y., Liu, N.W., Zuo, S. et al. (2015). Wideband circularly polarised array antenna with flat-top beam pattern. *IET Microwave Antennas Propag.* 9 (8): 755–761.

4 Hosseininejad, S.E., Komjani, N., and Mohammadi, A. (2015). Accurate design of planar slotted SIW array antennas. *IEEE Antennas Wirel. Propag. Lett.* 14: 261–264.

5 Ding, Z., Xiao, S., Tang, M.C., and Liu, C. (2018). A compact highly efficient hybrid antenna array for W-band applications. *IEEE Antennas Wirel. Propag. Lett.* 17 (8): 1547–1551.

6 Liu, B., Hong, W., Kuai, Z.Q. et al. (2009). Substrate integrated waveguide (SIW) monopulse slot antenna array. *IEEE Trans. Antennas Propag.* 57 (1): 275–279.

7 Xu, J.F., Hong, W., Chen, P., and Wu, K. (2009). Design and implementation of low sidelobe substrate integrated waveguide longitudinal slot array antennas. *IET Microw. Antennas Propag.* 3 (5): 790–797.

8 Chen, X.-P., Wu, K., Han, L., and He, F. (2010). Low-cost high gain planar antenna array for 60-GHz band applications. *IEEE Trans. Antennas Propag.* 58 (6): 2126–2129.

9 Cheng, Y.J., Hong, W., and Wu, K. (2012). 94 GHz substrate integrated monopulse antenna array. *IEEE Trans. Antennas Propag.* 60 (1): 121–128.

10 Yang, H., Montisci, G., Jin, Z.S. et al. (2015). Improved design of low sidelobe substrate integrated waveguide longitudinal slot array. *IEEE Antennas Wirel. Propag. Lett.* 14: 237–240.

11 Navarro-Mendez, D.V., Carrera-Suarez, L.F., Baquero-Escudero, M., and Rodrigo-Penarrocha, V.M. (2010). Two layer slot-antenna array in SIW technology. *Eur. Microw. Conf.*: 1492–1495.

12 Park, S.J., Shin, D.H., and Park, S.O. (2016). Low side-lobe substrate-integrated-waveguide antenna array using broadband unequal feeding network for millimeter-wave handset device. *IEEE Trans. Antennas Propag.* 64 (3): 923–932.

13 Chang, L., Li, Y., Zhang, Z. et al. (2017). Low-sidelobe air-filled slot array fabricated using silicon micromachining technology for millimeter-wave application. *IEEE Trans. Antennas Propag.* 65 (8): 4067–4074.

14 Guan, D.F., Qian, Z.P., Zhang, Y.S., and Jin, J. (2015). High-gain SIW cavity-backed array antenna with wideband and low sidelobe characteristics. *IEEE Antennas Wirel. Propag. Lett.* 14: 1774–1777.

15 Guan, D.F., Ding, C., Qian, Z.P. et al. (2015). An SIW-based large-scale corporate-feed array antenna. *IEEE Trans. Antennas Propag.* 63 (7): 2969–2976.

16 Cheng, Y.J., Hong, W., Wu, K. et al. (2008). Substrate integrated waveguide (SIW) Rotman lens and its Ka-band multibeam array antenna applications. *IEEE Trans. Antennas Propag.* 56 (8): 2504–2513.

17 Tekkouk, K., Ettorre, M., Gandini, E., and Sauleau, R. (2015). Multibeam pillbox antenna with low sidelobe level and high-beam crossover in SIW technology using the split aperture decoupling method. *IEEE Trans. Antennas Propag.* 63 (11): 5209–5215.

18 Tekkouk, K., Ettorre, M., Le Coq, L., and Sauleau, R. (2015). SIW pillbox antenna for monopulse radar applications. *IEEE Trans. Antennas Propag.* 63 (9): 3918–3927.

19 Li, T., Meng, H., and Dou, W. (2014). Design and implementation of dual-frequency dual-polarization slotted waveguide antenna array for Ka-band application. *IEEE Antennas Wirel. Propag. Lett.* 13: 1317–1320.

20 Li, T. and Dou, W. (2015). Broadband substrate-integrated waveguide T-junction with arbitrary power-dividing ratio. *Electron. Lett.* 51 (3): 259–260.

21 Contreras, S. and Peden, A. (2013). Graphical design method for unequal power dividers based on phase-balanced SIW tee-junctions. *Int. J. Microwave Wireless Technol.* 5: 603–610.

22 Li, T. and Dou, W. (2014). Broadband right-angle transition from substrate-integrated waveguide to rectangular waveguide. *Electron. Lett.* 50 (19): 1355–1356.

23 Li, T. and Dou, W.B. (2015). Millimetre-wave slotted array antenna based on double-layer substrate integrated waveguide. *IET Microwave Antennas Propag.* 9 (9): 882–888.

24 Li, T., Dou, W., and Meng, H. (2016). A monopulse slot array antenna based on dual-layer substrate integrated waveguide (SIW). In: *Proc. IEEE 5th Asia–Pacific Conf. Antennas Propag. (APCAP)*, 373–374.

25 Li, T. and Dou, W. (2017). Substrate integrated waveguide 3 dB directional coupler based on air-filled vias. *Electron. Lett.* 53 (9): 611–613.

11

Substrate Edge Antennas

Lei Wang[1] and Xiaoxing Yin[2]

[1] *Institute of Sensors, Signals and Systems, Heriot-Watt University, Edinburgh EH14 4AS, United Kingdom*
[2] *State Key Lab of Millimeter Waves, Southeast University, Nanjing 210096, People's Republic of China*

11.1 Introduction

Usually dielectric substrate edges of PCBs are treated as an ideal open circuit at frequencies below 1 GHz because the EM radiation from substrate edge is limited. However, when the operating frequency increases, for instance to 28 GHz, the radiation from the substrate edge as shown in Figure 11.1 cannot be ignored [1] and even can be intentionally used to design antennas.

With the rapid development of the 5G network, SIW horn antennas and arrays become promising because of their compact size, high gain, low transmission loss, and easy integration [2]. As shown in Figure 11.2, the substrate edge antennas (SEAs) like H-plane horn antennas [3] can be implemented and integrated with other passive and active components such as filters, couplers, oscillators, and amplifiers in substrate. Comparing with patch antennas radiating at boresight, SEAs provide another degree of freedom for antenna placement in systems. The application of SEAs also suppresses the electromagnetic interference (EMI) from antenna into other components. Up to date, the SEAs have been proposed as monopulse arrays [4] and dual-polarization phased arrays in 5G handsets [5].

Acting as an antenna, an open-end SIW, however, suffers from low radiation efficiency because of the severe impedance mismatching between the substrate and the free space. In order to improve the matching, technologies of using dielectric loadings [4, 6] and printed strips [7, 8] have been presented. With regard to the gain of the SIW horns, apart from the impedance matching, aperture efficiency is another key parameter. The radiating aperture of the SIW horn is limited since the substrate is usually electrically thin. Lens and other loading slabs [9, 10] have been used to increase the gain of SIW horns with the phase correction and extension of radiating horn apertures with increased entire antenna size.

In addition, the radiation from the substrate edges can also be in forms of leaky-waves [11, 12]. Most leaky-wave antennas (LWAs) steer the radiation beams against operating frequencies. Such beam-steering performance can be useful for frequency-scanning antennas in radar systems rather than the point-to-point communications over a wide frequency band. To overcome this limit, a prism lens implemented by metasurfaces is loaded on the substrate edge. The dispersion of LWAs is compensated by using the dispersive prism lens. A more than 20% frequency bandwidth is achieved at a specific direction.

Substrate-Integrated Millimeter-Wave Antennas for Next-Generation Communication and Radar Systems, First Edition. Edited by Zhi Ning Chen and Xianming Qing.

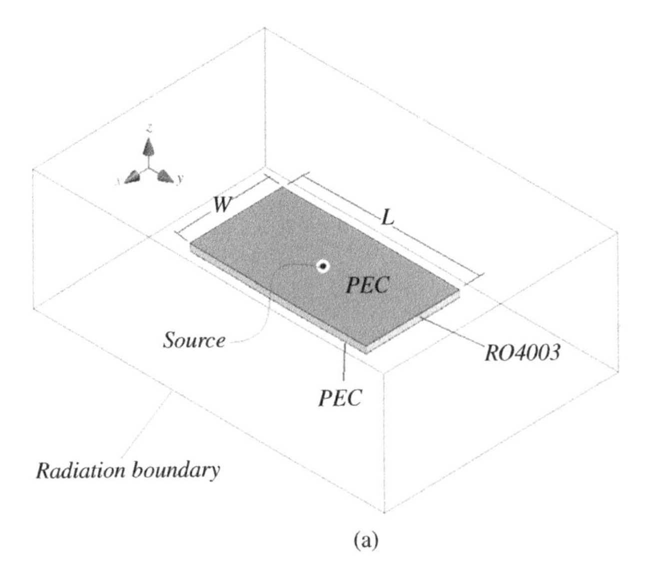

(a)

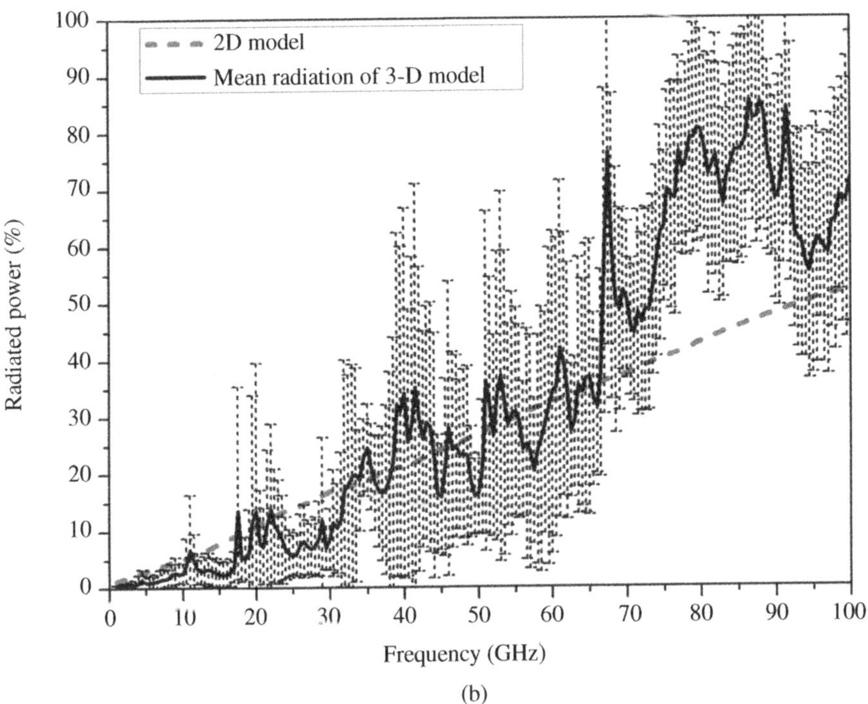

(b)

Figure 11.1 Radiation from substrate edges [1]. (a) Three-dimensional (3-D) model in full-wave simulation and (b) radiated power versus operating frequencies from a piece of RO4003C substrate sheet of size of 2 mm × 15 mm × 0.508 mm with 3.55 dielectric constant.

Large-scale of SIW horn arrays might cause difficulty from a computation point of view due to the large amount of meshes for grounded vias. Some efficient numerical methods for SIW components and antennas have been proposed. For instance, a hybrid method based on a contour integral method and Huygens' principle was proposed [13], which demonstrates to be numerically more efficient than three-dimensional full-wave simulations by two to three orders and applicable with reasonable accuracy.

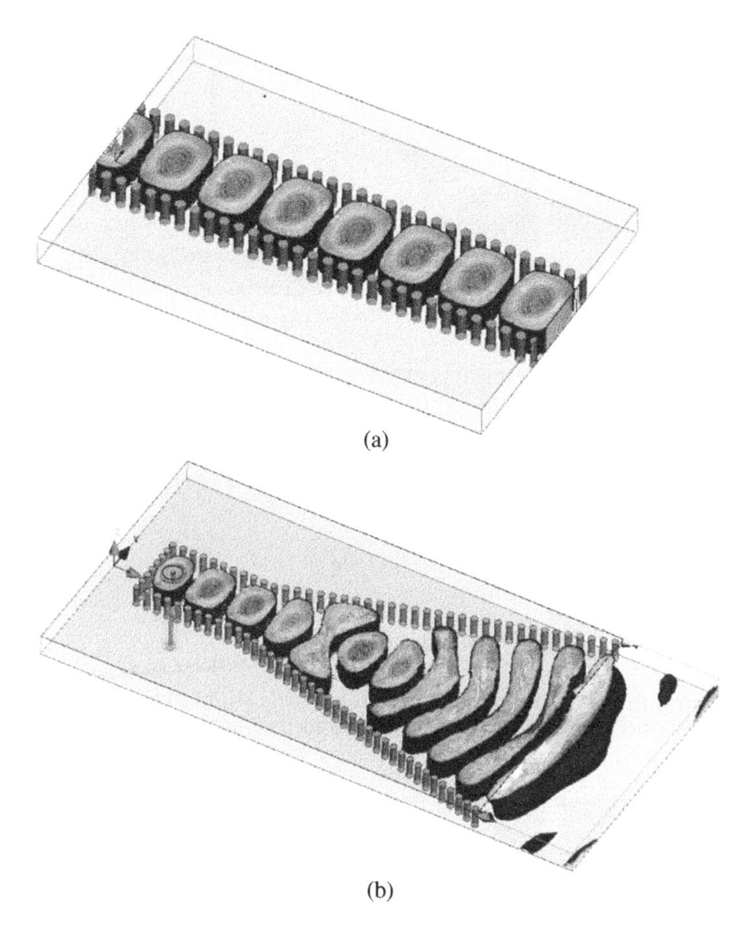

Figure 11.2 Propagating electric field in (a) SIW and (b) SIW *H*-plane horn antenna.

11.2 State-of-the-Art

11.2.1 End-Fire SEAs

As implied in Figure 11.1b, EM waves can be radiated from the substrate edges, but the total radiated power is not sufficient, for instance less than 20% at 30 GHz. This is caused by the severe mismatching between the edges of electrically thin substrate and the free space [14].

To improve the impedance matching, the horn aperture is loaded with rectangular or elliptical dielectric slabs [4, 9, 15] by naturally extending the substrate as shown in Figure 11.3a. An add-on Polycarbonate dielectric in Figure 11.3b was employed for both impedance matching and enhancement of front-to-back ratio (FTBR) [6]. A dielectric constant modified loading by drilling air holes in Figure 11.3c was proposed for an orthomode SIW horn [16]. Similarly, another loading dielectric slab with air holes modified in graded dielectric constants was used to enhance the bandwidth as shown in Figure 11.3d [17].

Moreover, several transitions of rectangular strips shown in Figure 11.4a were printed in front of the SIW horn aperture to improve the impedance matching at Ku-band [7, 8]. In Figure 11.4b, the two arrays of triangular strips were further developed and printed on the horn aperture, which extend the bandwidth and suppress the back radiation [18]. Furthermore, Figure 11.4c shows that the Ku-band offset double-sided parallel patches are printed in front of the horn [19]. Periodic

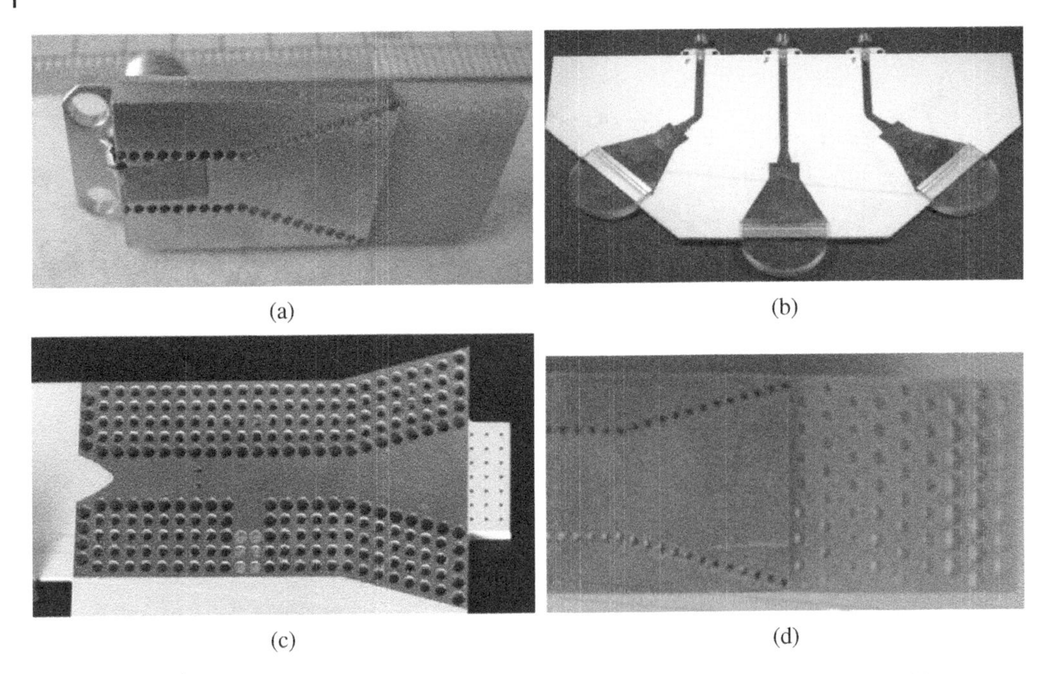

Figure 11.3 Dielectric loaded SIW horn antennas. (a) Rectangular dielectric loading [4], (b) add-on Polycarbonate dielectric loading [6], (c) air-hole dielectric lens [16], and (d) graded air-hole dielectric loading [17].

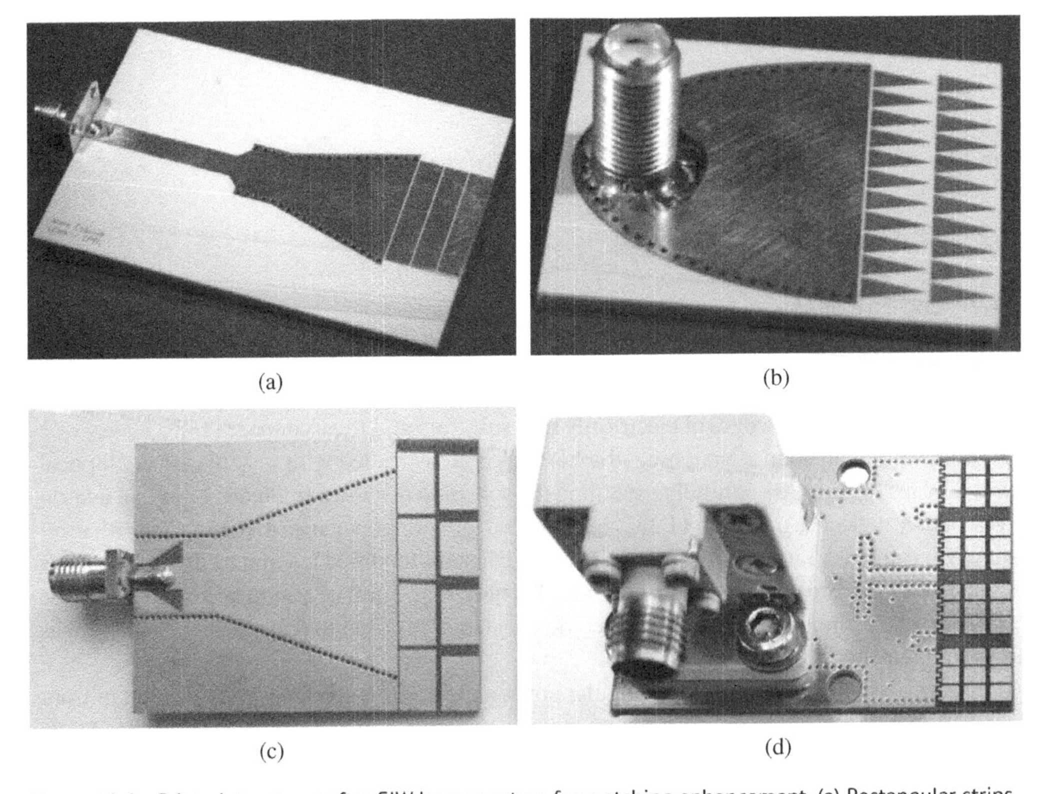

Figure 11.4 Printed structures after SIW horn aperture for matching enhancement. (a) Rectangular strips [8], (b) triangular strips [18], (c) offset double-sided parallel-strip lines [19], and (d) rectangular patches using a characteristic mode method [20].

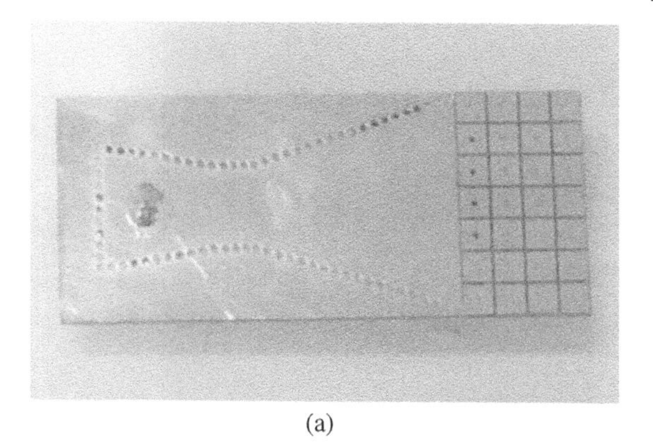

(a)

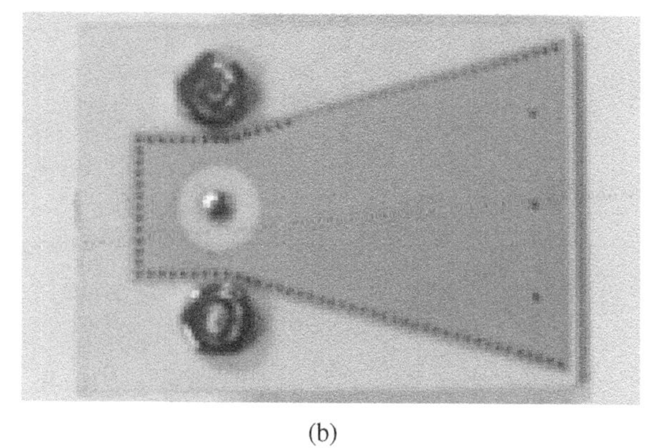

(b)

Figure 11.5 Metallic via structures for impedance matching. (a) Mushroom metasurface [10] and (b) embedded metallic vias [21].

patches as shown in Figure 11.4d were analyzed using a characteristic mode method and employed in the SIW horn apertures and arrays [20].

Furthermore, a mushroom metasurface with metallic vias as shown in Figure 11.5a was used to improve the impedance matching with backward dispersion, leading to backward end-fire radiation patterns [10]. Three metallic vias were inserted inside the SIW horn and close to the aperture as shown in Figure 11.5b, aiming to tune the impedance matching too [21].

11.2.2 Leaky-Wave SEAs

SIW LWAs offer interesting features for mmW wireless applications because of their high directivity and low losses using a single-feeding planar SIW guide [22]. As opposed to resonant antennas, their impedance bandwidth is wide due to their traveling-wave radiation mechanism, which is another general advantage of LWAs. Due to their dispersive nature, the directive radiated beam is scanning with frequencies in all previous directive SIW LWA designs. Substrate edge radiating LWAs can operate in different modes such as TE_{10} mode [23], half TE_{10} mode [11], TE_{20} mode [12], and half TE_{20} mode [24]. Figure 11.6a shows a leaky-wave SEA by reducing the via number of one SIW wall, and the dispersion of the radiation angles is shown in Figure 11.6b. In Figure 11.6c,d, by changing the width of the leaky-wave SEA, the radiation pattern was synthesized with a bandpass function [25].

However, the frequency-dependent beam steering is a drawback for highly directive point-to-point broadband wireless links, since the pattern bandwidth (PBW) is narrow due to the beam squinting. Normally when the SIW LWA is more directive, the half-power beamwidth is

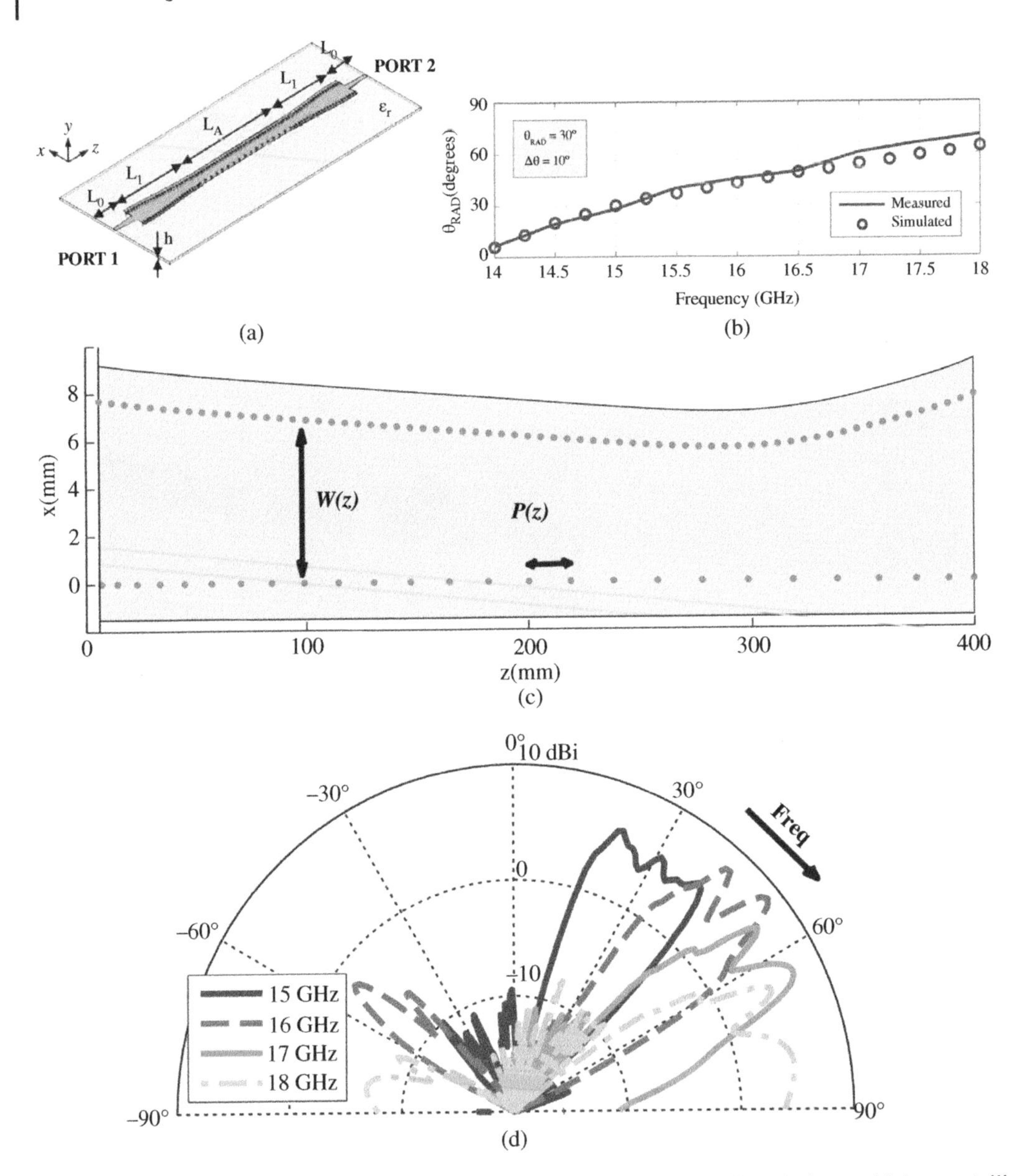

Figure 11.6 Leaky-wave antennas radiating from substrate edges. (a, b) Uniform leakage with less metallic vias [23] and (c, d) modified width LWA for angular-filtering pattern shaping [25].

narrower. As a result, SIW LWAs are not suitable for applications in point-to-point high-throughput communications [26], which require broadband radiation at a fixed direction. Thus, the beam squint reduces the effective bandwidth of directive electrically long leaky-wave SEAs for wideband fixed-direction applications.

11.3 Tapered Strips for Wideband Impedance Matching

In this section, two types of tapered strips are introduced to improve the impedance matching of SIW horn antennas. First, slotted triangular strips are used to enhance the impedance matching

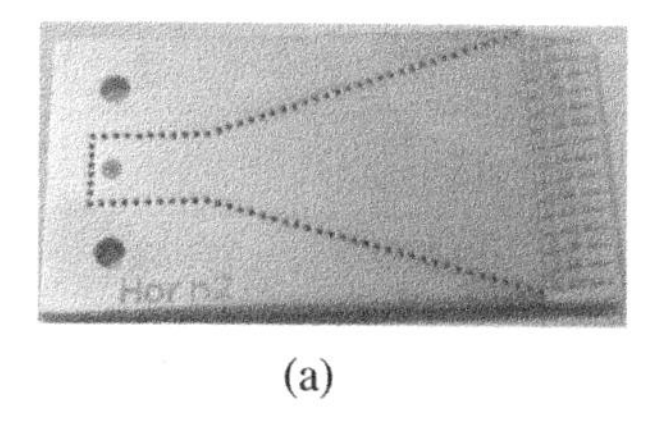

(a)

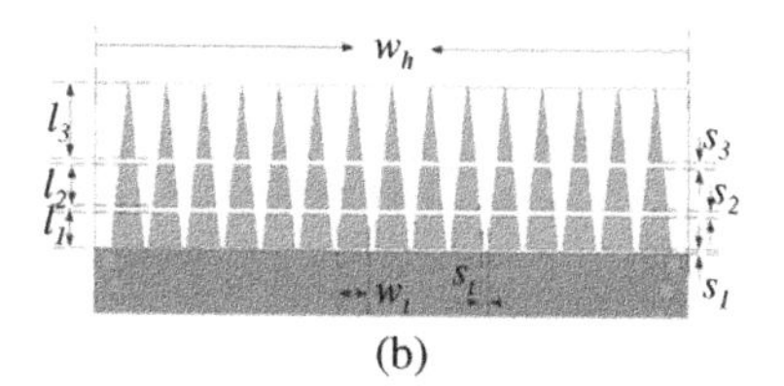

(b)

Figure 11.7 Tapered triangular strips with slots for wideband matching [28]. (a) SIW horn with triangular strips and (b) structure of the triangular strips.

and FTBRs. Then, tapered periodic rectangular strips are studied for significantly broadening the antenna bandwidth.

11.3.1 Tapered Triangular Strips

Two arrays of triangular strips have been used to improve impedance matching and FTBRs [18, 27]. An array of triangular strips is further explored to control the radiation and matching, as illustrated in Figure 11.7. The SIW horn is designed based on a piece of RO4003C ($\varepsilon_r = 3.55$). The width of the horn aperture is 20.5 mm ($w_h = 2.3\lambda_0$), and the substrate thickness is 1.524 mm ($0.17\lambda_0$), where λ_0 is the operating wavelength in the free space. The overall length of the antenna is 39.3 mm ($4.4\lambda_0$).

A tapered-ladder-like transition is printed in front of the horn aperture as shown in Figure 11.7. To cope with the substrate thickness of $0.17\lambda_0$ and provide an even better matching to free space, thin slots are incorporated to the original triangles introduced in [18]. A strong electric field appears in the first two rows of slots with the widths of s_1 and s_2 and then dilutes progressively in the last row of slots with the width of s_3. As a result, good matching to the free space and the effectiveness of the slotted triangles are observed.

Figure 11.8 shows that $|S_{11}|$ is improved from −3 dB to less than −15 dB over a wide bandwidth. The bandwidth for $|S_{11}|$ less than −10 dB reaches up to 35% as the parallel slots are with $l_1 = 1.5$ mm, $l_2 = 1.3$ mm, and $l_3 = 1.8$ mm.

The slotted triangles are also capable of controlling the FTBRs. If the overall length of the triangular strip is fixed, the positions of the slots affect the FTBRs, as indicated in Table 11.1. The FTBRs vary from 9.2 to 25.3 dB at 30 GHz for three different slot positions, when $w_h = 20.5$ mm, $w_t = 1.1$ mm, $s_t = 0.1$ mm, $s_1 = 0.1$ mm, $s_2 = 0.2$ mm, and $s_3 = 0.2$ mm.

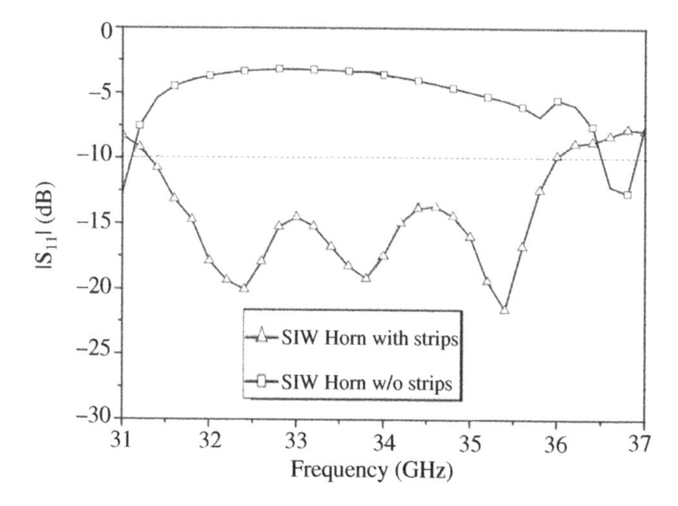

Figure 11.8 Reflection coefficients with the slotted triangular strips.

Table 11.1 Slots' positions and the antenna performance.

l_1 (mm)	l_2 (mm)	l_3 (mm)	Bandwidth (GHz)	FTBRs (dB)		
				30 GHz	34 GHz	38 GHz
1.0	1.5	1.9	4.62 (13%)	9.2	6.2	10.0
1.3	1.6	1.7	11.1 (29%)	21.3	10.1	12.9
1.5	1.3	1.8	10.5 (35%)	25.3	12.1	16.3

11.3.2 Tapered Rectangular Strips

In this section, a method using two-strip pair with each width of $\lambda_0/\left(4\sqrt{\varepsilon_r}\right)$ as shown in Figure 11.9 is discussed for improving the impedance matching. The method is that the wave reflected by the first strip is canceled or nearly so by the wave reflected by the second strip when the phase difference of the two waves is 180°. The total reflected wave is

$$E_r = E_{r1} + E_{r2} = a_1 e^{-j\phi_0} + a_2 e^{-j\phi_0} e^{-j4\pi L/\lambda_g}, \tag{11.1}$$

where E_r is the total reflected wave, E_{r1} and E_{r2} are the reflected waves from the first and second slots, a_1 and a_2 are the corresponding amplitude, ϕ_0 is the phase at the first slot, λ_g is the guided wavelength in the dielectric, L is the width of the strip, and ε_r is the relative dielectric constant. If $L = \lambda_0/\left(4\sqrt{\varepsilon_r}\right) = \lambda_g/4$, then

$$E_r = (a_1 - a_2)\, e^{-j\phi_0}. \tag{11.2}$$

If the widths of the two slots are the same (as proposed in [8]), then $a_2 < a_1$ and $E_r \neq 0$. However, by increasing the width of the second slot, the reflection E_{r2} increases. Hence a_2 increases. Thus, an approximation of $a_1 = a_2$ is achievable, which leads to $E_r \approx 0$.

As shown in Figure 11.9, an SIW in a piece of RO4003C substrate slab with a height of 1.524 mm and a width of 5 mm is constructed by two rows of metallic vias with a radius of 0.25 mm and spacing of 0.8 mm. Three types of printed strips are used to improve the matching. The widths of one wide strip, two uniform strips, and two tapered strips are $0.44\lambda_g$, $0.25\lambda_g$, and $0.25\lambda_g$, respectively. The bandwidths and fraction bandwidths are compared in Table 11.2. It is seen that the design with the two tapered strips achieves the largest bandwidth of 13%.

Furthermore, more tapered strips are employed and printed in front of the open-SIW aperture, as shown in Figure 11.10. To simplify the description, pairs of tapered strip-slot are defined with

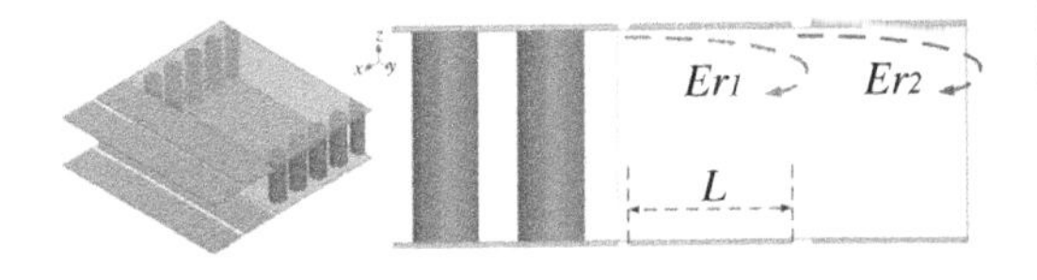

Figure 11.9 Matching structure of two rectangular strips [29]

Table 11.2 Achieved bandwidths of three impedance matching structures.

	One strips	Two uniform strips	Two tapered strips
Bandwidth (GHz)/%	1.22/4.36	2.79/8.72	4.17/13.03

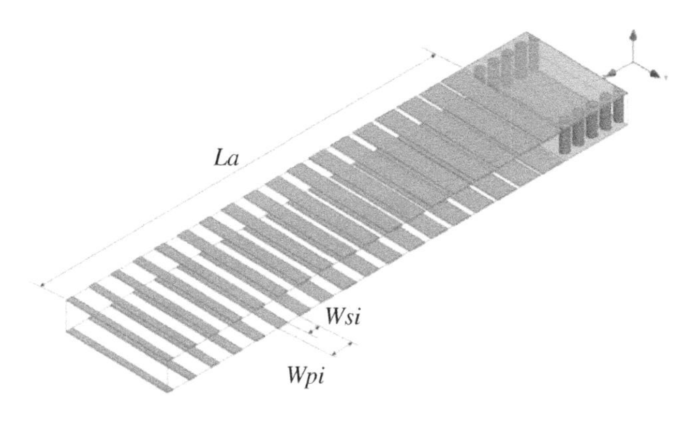

Figure 11.10 Impedance matching structure of a group of tapered strips [29].

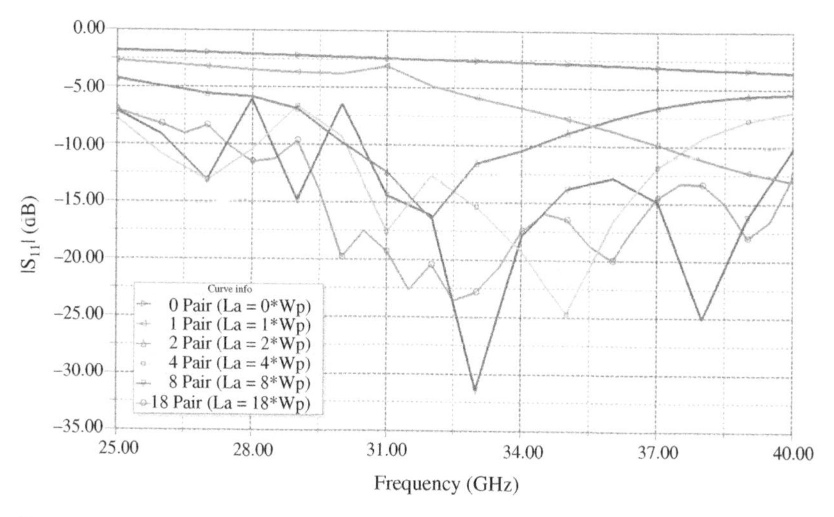

Figure 11.11 Reflection coefficients of different numbers of the strip-slot pairs, with different overall lengths L_a and a constant w_p of 1.3 mm [29].

the parameters of the slot width w_{si} and the total pair width w_{pi}. The total length of the strip-slot pairs is L_a. A group of tapered strips with a step of 0.05 mm and a total pair width of 1.3 mm are chosen. The dimensions of different strip-slot pairs are designed as

$$w_{si} = 0.1 + 0.05\,(i - 1) \ \ \text{mm}, \qquad i = 1, 2, \ldots, 18. \tag{11.3}$$

The influence of the different numbers of the strip-slot pairs on the impedance matching is shown in Figure 11.11, indicating that the impedance bandwidth is enhanced with the increasing number of the strip-slot pairs. It is found that $|S_{11}|$ of the original SIW without tapered structures is larger than −5 dB from 25 GHz to 40 GHz. However, with the 18 tapered strips the impedance bandwidth is greater than 35%.

Based on the above analysis, an H-plane horn antenna with 18 tapered strips is presented as illustrated in Figure 11.12a. The flaring angle of the horn is 22°, leading to a narrower H-plane beamwidth comparing to the one of the uniform-width strips in Figure 11.10. The electric fields in Figure 11.12b viewed from the top and the side both demonstrate the strips have smoothly guided the EM waves out from the SIW into the free space. The top view of electric field also proves that an H-plane horn radiation is realized by flaring the length of strips.

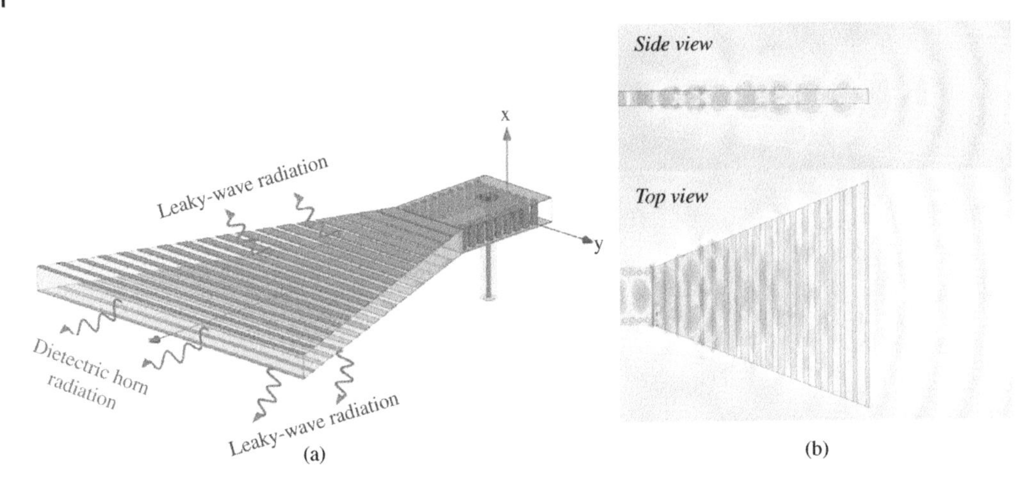

Figure 11.12 Illustration of the wideband *H*-plane horn antenna with a group of tapered strips [29]. (a) Antenna structure and (b) electric fields around the horn.

Figure 11.13 Prototype of the wideband *H*-plane horn antenna with a group of tapered strips [29].

A horn antenna using a piece of RO4003C substrate is shown in Figure 11.13. The thickness of the substrate is 1.524 mm and the width of the SIW is 5 mm. A connector with a metallic pin of a 0.15 mm radius is inserted into the SIW to excite EM waves in TE_{10} mode.

The measured reflection coefficient agrees very well with the simulated result as shown in Figure 11.14. The frequency ranges from 29.3 to 43.0 GHz for $|S_{11}|$ below −10 dB. Stable *H*-plane radiation patterns over wide frequency band are plotted in Figure 11.15, with low side lobe levels (less than −20 dB). The *E*-plane beamwidths are reduced due to the leaky-wave radiation.

11.4 Embedded Planar Lens for Gain Enhancement

With the flaring of horn arms, the amplitude and phase of the electric fields on the horn aperture become non-uniform, which reduces the aperture efficiency and antenna gain. One common way to overcome this problem is to use lens to correct the phase distributions on aperture. The SIW horns loaded with printed lenses [4, 6, 9] increases in size at a price. In addition, the horn antennas lose

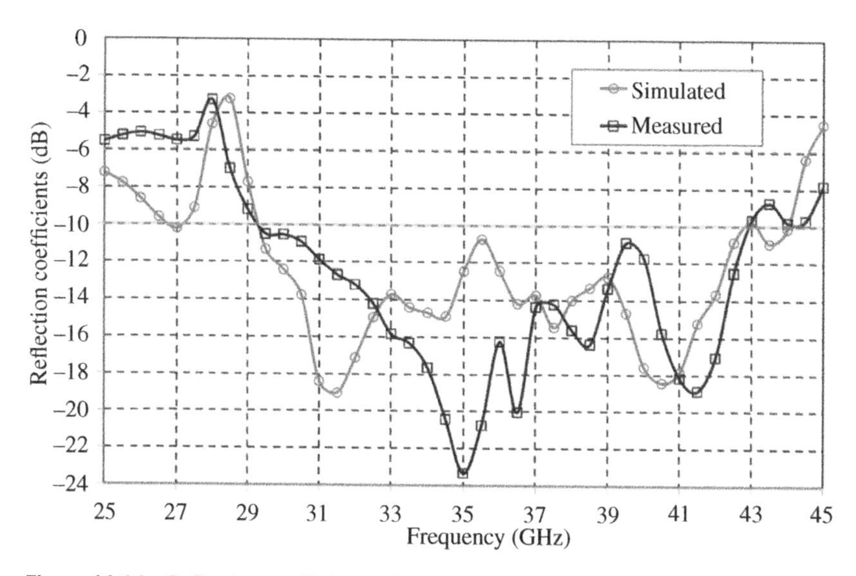

Figure 11.14 Reflection coefficients of the wideband horn antenna with tapered strips [29].

a degree of freedom to print other structures, e.g. printed strips for FTBRs improvement compared with the previous technique.

11.4.1 Embedded Metallic Lens

In 1946, a metallic lens was proposed to correct the phase distribution on horn aperture [30]. Figure 11.16a depicts that a group of short waveguides with different widths are located on the horn aperture, to correct the phase distribution of the electric field on the horn aperture. The concept of the phase correction is illustrated in Figure 11.16b. It is known that the phase velocity of the dominant mode TE_{10} in the rectangular dielectric filled waveguide could be controlled by the space d, as shown in the following:

$$v = \frac{v_0}{\sqrt{1 - [\lambda/(2d)]^2}} \tag{11.4}$$

where v_0 is the velocity of the light in the dielectric, v is the phase velocity in the dielectric filled waveguide, d is the width of the waveguide, and λ is the operating wavelength in the dielectric. As a result, the phase velocity can be tuned by changing the waveguide width d. So, metal lenses can be designed by tuning the width and length of the waveguides to correct the aperture phase.

In Figure 11.16, the phase velocity between the paralleled metal plates with a uniform distance is constant but faster than that in the air. The phase velocity in the narrow waveguide is faster than that in the wide waveguide. Consequently, all the waves can arrive at the horn aperture with the same phase. A combination solution is proposed by embedding the metallic lens inside the horn. The metallic waveguide lens is implemented by the rows of metallic vias, as shown in Figure 11.17a,b.

To demonstrate the phase correction effect of the metallic lens, a conventional SIW horn in the same geometry is manufactured together with the SIW horn antenna with embedded metallic lens. Table 11.3 compares the measured gain of the SIW horn antennas. The one using the embedded metallic lens shows the gain enhancement of more than 2 dB.

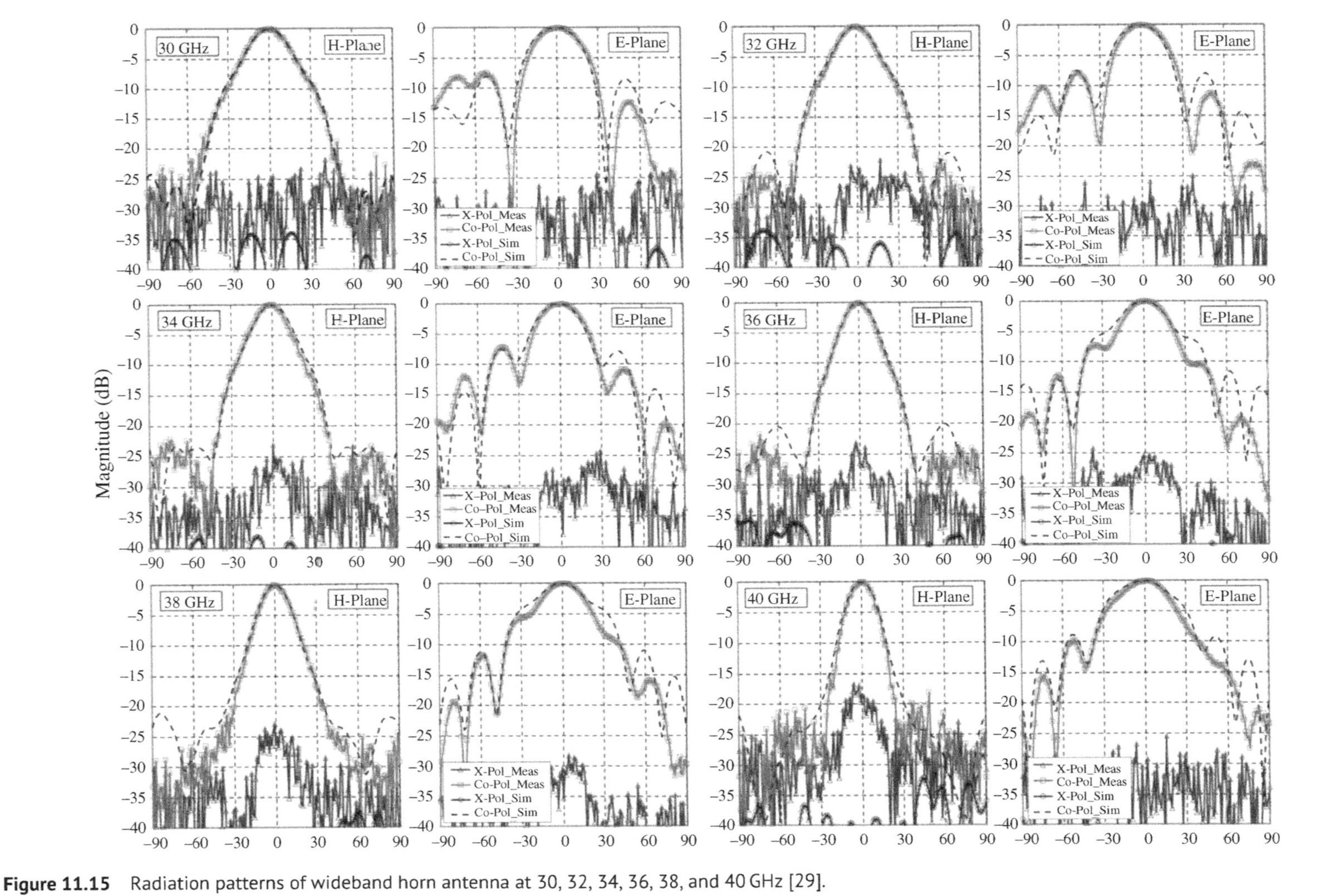

Figure 11.15 Radiation patterns of wideband horn antenna at 30, 32, 34, 36, 38, and 40 GHz [29].

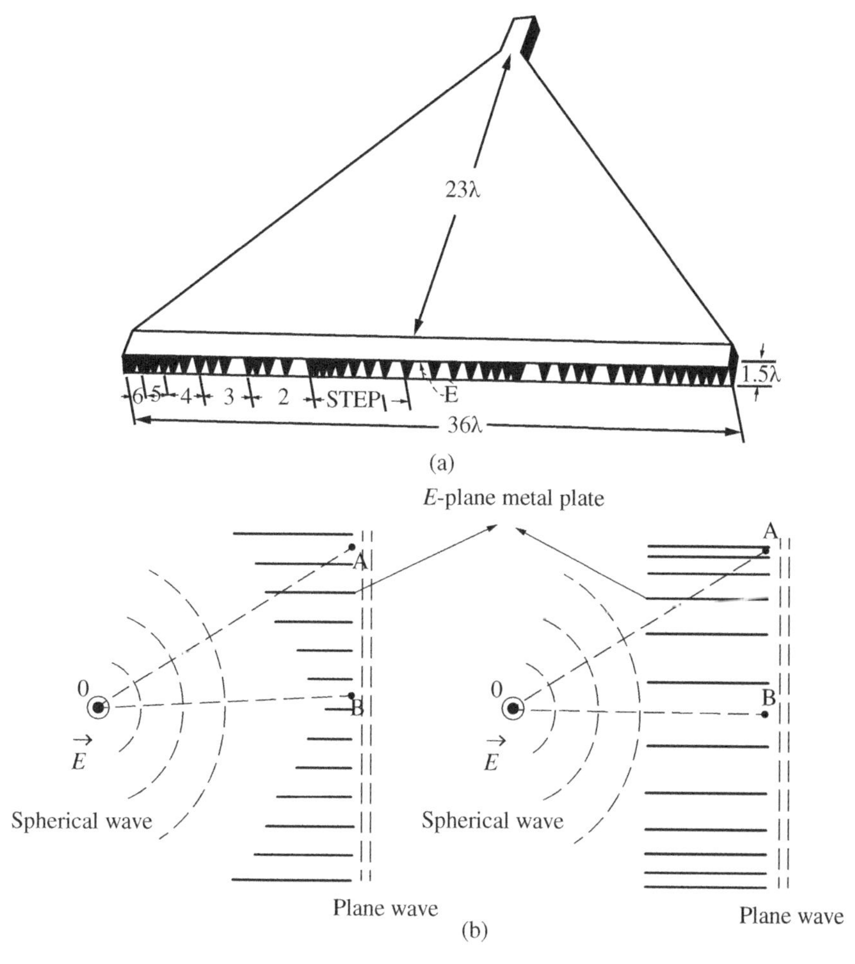

Figure 11.16 (a) Sector horn loaded with a waveguide lens [30] and (b) analysis of the lens [31].

11.4.2 Embedded Gap Lens

Similarly, an embedded gap lens is implemented by etching the gaps inside SIW horn antennas. As shown in Figure 11.18, the gap lens is embedded inside the horn that additional triangular strips are available to be load for better matching and FTBRs suppression. The triangular trips are designed as discussed in Section 11.3.1. The following starts with the basic idea of the SIWs with symmetrical and asymmetrical with gaps as illustrated in Figure 11.19.

Figure 11.19 compares the difference between the symmetrical and asymmetrical gap SIWs. The symmetrical gap SIW behaves the same as classical SIW that without gaps. However, the phase velocity in different parts (wide and narrow half-mode SIWs) of the asymmetrical gap SIW is different, which can be used to tune the phase distributions.

Following the same design procedure as the embedded metallic lens, the gap lens is embedded into the SIW horn as shown in Figure 11.18. Figure 11.20 depicts the electric field and phase distributions after the phase correction by using the gap lens. It is obvious that the phase on the aperture becomes linear and the amplitude of electric fields turn to be uniform, contributing to high aperture efficiency of the SIW horn antenna.

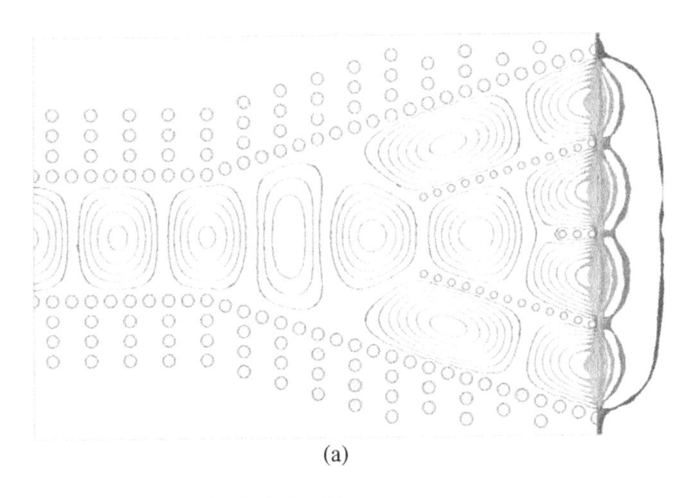

(a)

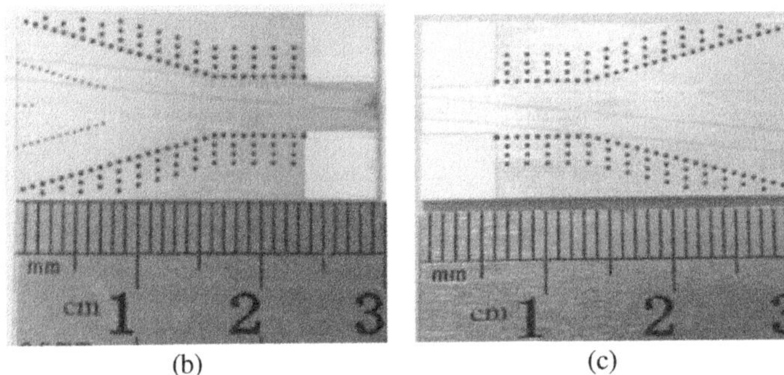

(b) (c)

Figure 11.17 (a) Electric fields inside the SIW horn with metallic via lens, (b) prototype of the SIW horn with embedding metallic lens, and (c) traditional SIW horn. [31].

Table 11.3 Comparison of the measured antenna gain.

Frequency (GHz)	**28**	**30**	**34**	**35**	**36**
Gain of SIW horn with metallic lens (dBi)	8.16	10.83	8.29	9.20	8.75
Gain of conventional SIW horn (dBi)	5.26	6.94	7.17	7.09	6.91

Four SIW horn prototypes with different gap lens and tapering strips are shown in Figure 11.21. A coaxial connector with a feeding pin is used to test the four horns. Figure 11.22 compares the simulated and measured reflection coefficients as well as the measured gain of the antennas.

Both Horns 1 and 2 exhibit good impedance matching as shown Figure 11.22a, and the effect of the phase correction is demonstrated in Figure 11.22b. It is found that around 2-dB gain enhancement is achieved over the operating frequency band. Stable radiation patterns from 30 to 36 GHz are shown in Figure 11.23.

11.5 Prism Lens for Broadband Fixed-Beam Leaky-Wave SEAs

In order to overcome the beam squint effect and increase the bandwidth of leaky-wave SEAs for directional applications, Figure 11.24 illustrates the mechanisms of a leaky-wave SEA and a prism lens, in which the leaky-wave SEA and the prism lens have the compensated dispersion [33–36].

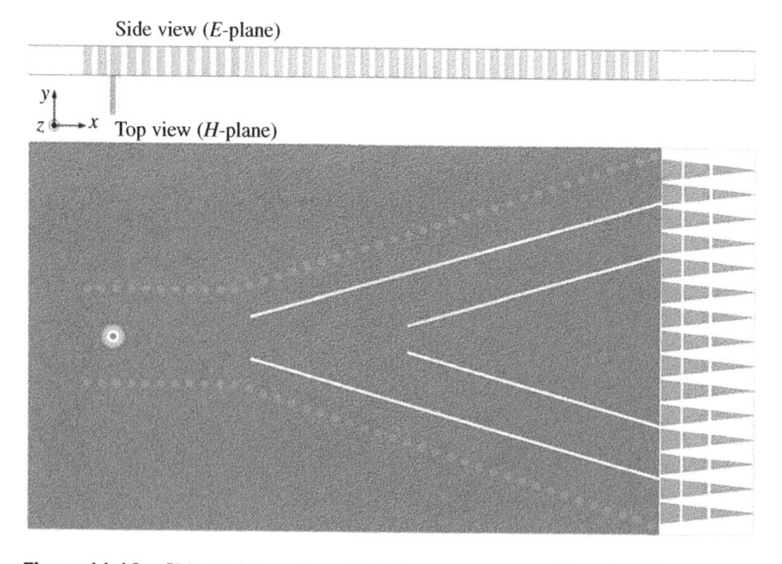

Figure 11.18 Side and top view of SIW horn antenna with embedding gap lens [28, 32].

Figure 11.19 Electric-field distributions inside and on the cross section of (a) the central symmetrical and (b) the asymmetrical gap SIW [28].

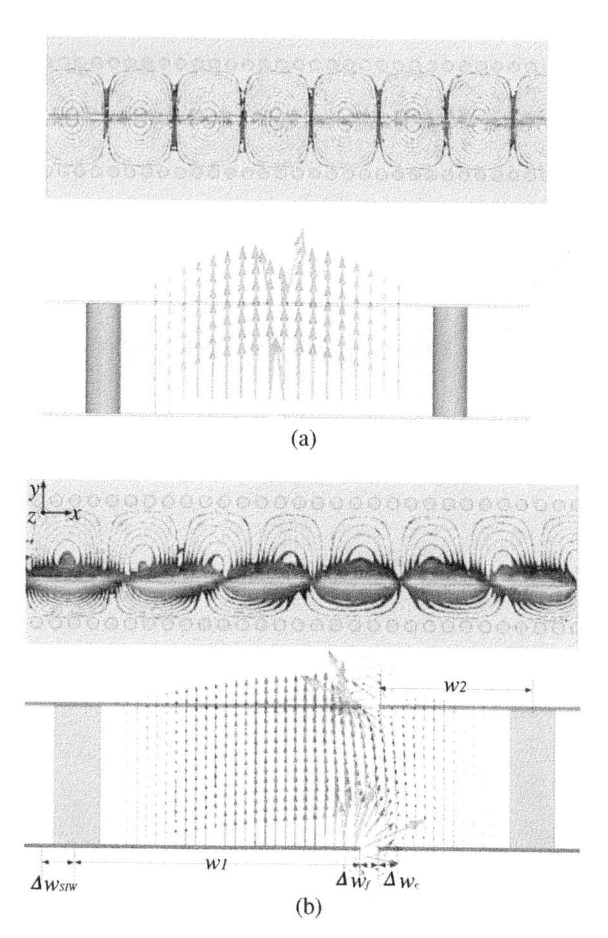

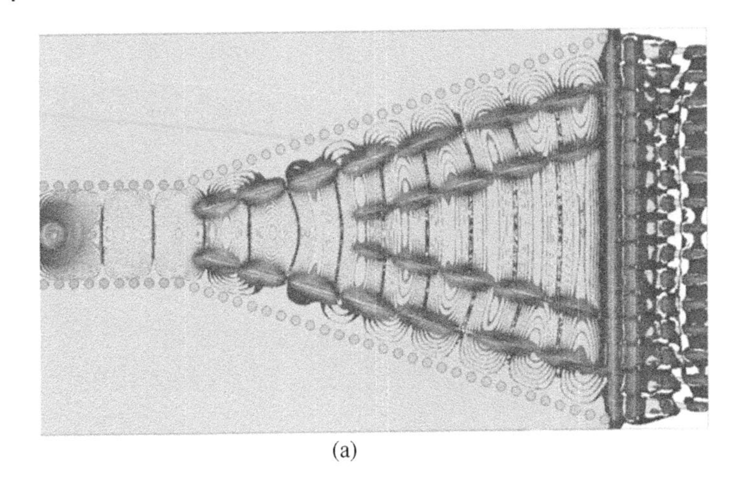

(a)

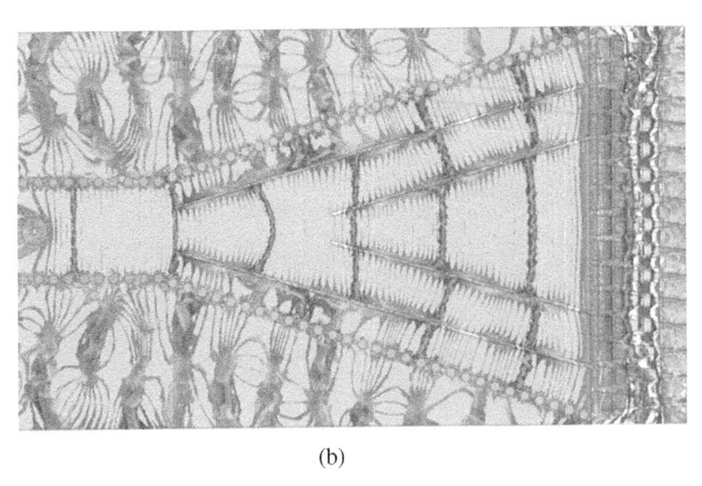

(b)

Figure 11.20 Electric field (a) and phase (b) distributions of the proposed SIW horn embedded with a gap lens [28].

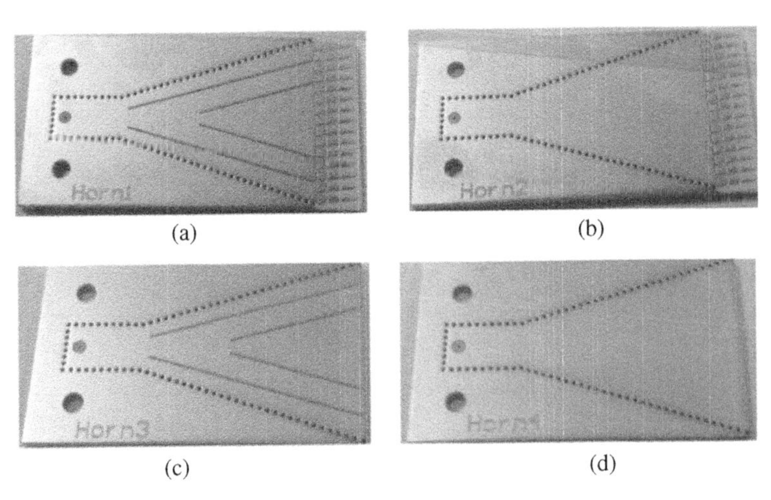

(a) (b)

(c) (d)

Figure 11.21 Prototypes of the four SIW horn antennas with/without gap lens and with/without tapering strips [28]. (a) The gap SIW horn with tapered strips, (b) the traditional SIW horn with tapered strips, (c) the gap SIW horn, and (d) the traditional SIW horn.

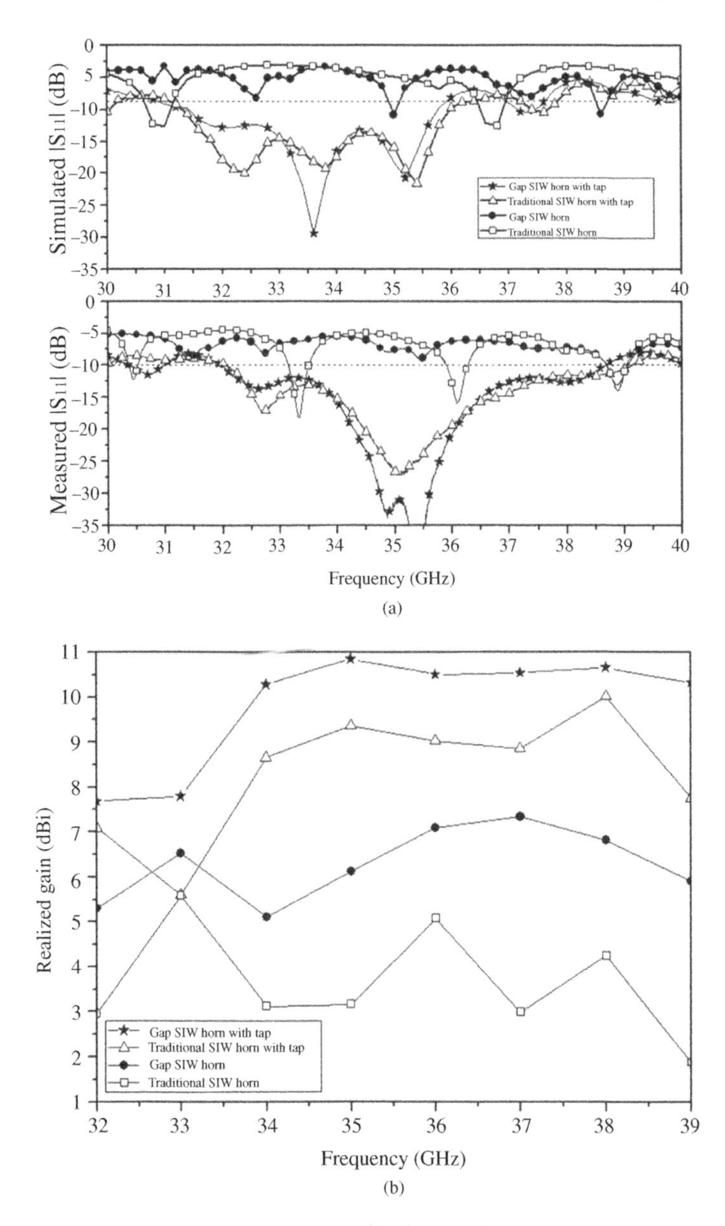

Figure 11.22 (a) Comparison of reflection coefficients of four prototypes both in simulation and measurement and (b) comparison of realized antenna gains in measurement [28].

By combining the leaky-wave SEA and the prism lens together, a fixed beam leaky-wave SEA in a wide frequency band is achieved, which is demonstrated below.

The frequency-scanning properties of the leaky-wave SEAs are illustrated in Figure 11.24a. The EM waves inside an SIW travel in the y-axial direction and the leaky waves radiate at a direction, $\phi_L(f)$ against frequencies. This is due to the inherent dispersive nature of the propagating mode in the SIW. The propagation and the radiation angles are defined as

$$k_0^2 \times \varepsilon_{r,L} = \beta_x^2 + \beta_y^2, \tag{11.5}$$

$$\sin \phi_L(f) = \frac{\beta_y}{k_0 \times \sqrt{\varepsilon_{r,L}}}, \tag{11.6}$$

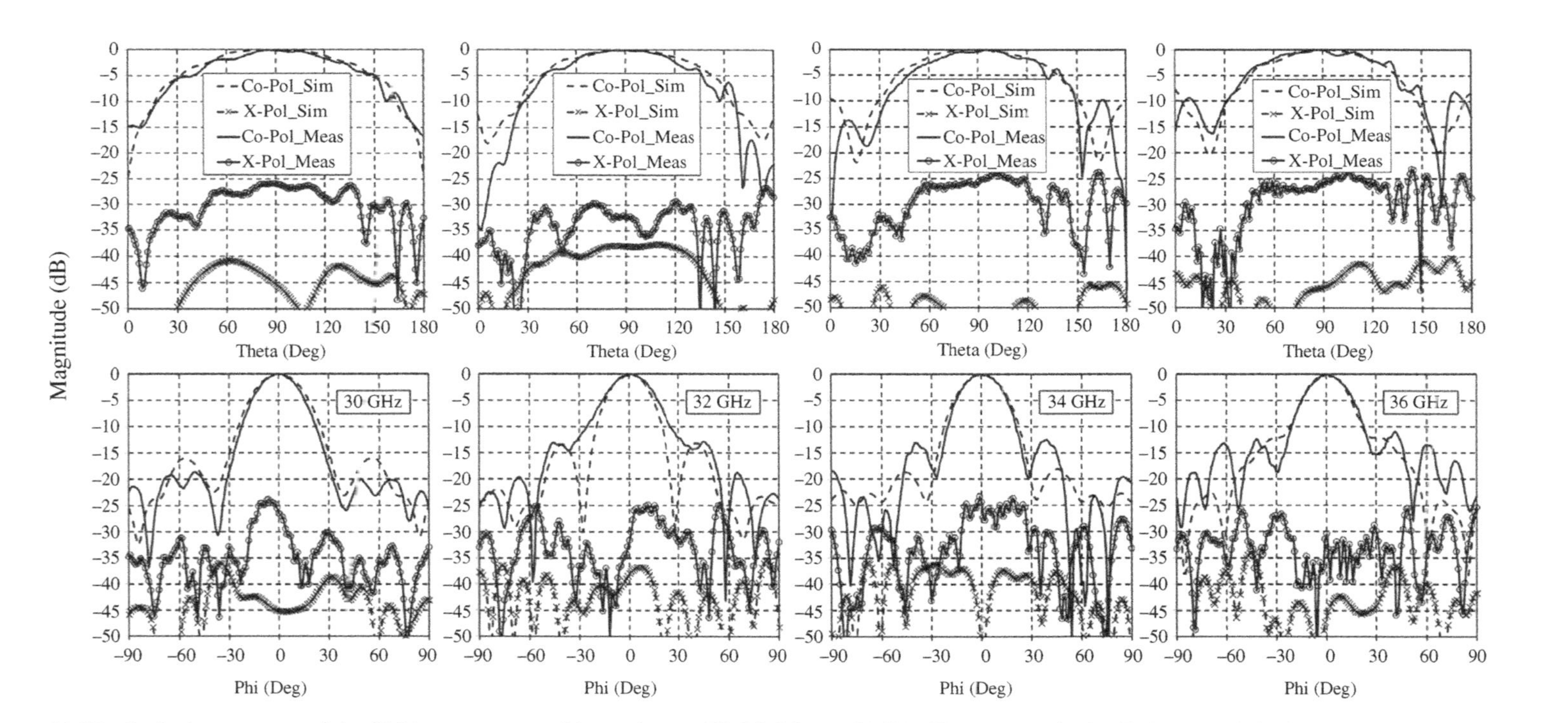

Figure 11.23 Radiation patterns of the SIW horn antenna with gap lens at 30, 32, 34, and 36 GHz. The top row: in the E-planes and the bottom row: in the H-planes [28].

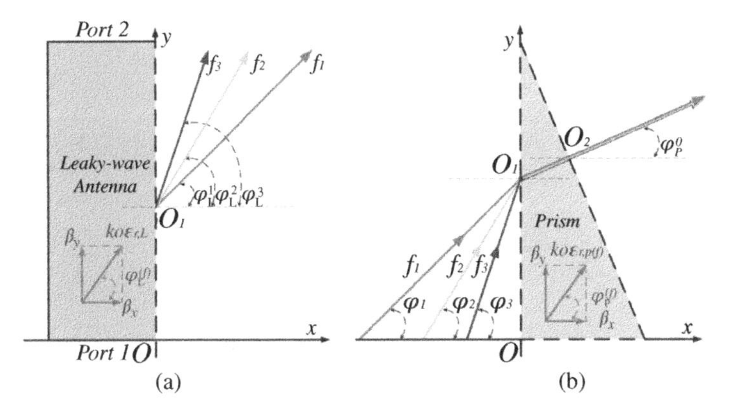

Figure 11.24 Illustration of the principle of operation. (a) Radiation dispersion in a leaky-wave SEA and (b) combination of rays with different angular directions depending on the frequency using a dispersive prism lens [33].

where k_0 is the wave number constant in the free space, $\varepsilon_{r,L}$ is the relative dielectric constant in the leaky-wave SEA, and β_x and β_y are the propagation constants in the x-axis and y-axis direction, respectively.

Optical prisms made of dispersive materials can break up the light into different directions depending on the color. Each color represents a different frequency. Equivalently, it is expected that prisms can be employed to focus the rays of different frequencies, arriving from different directions, to the same propagating direction, as depicted in Figure 11.24b. The equivalent permittivity $\varepsilon_{r,P}(f)$ of the prism must be frequency-dependent and follows

$$k_0^2 \times \varepsilon_{r,P}(f) = \beta_x^2 + \beta_y^2, \tag{11.7}$$

$$\sin \phi_P(f) = \frac{\beta_y}{k_0 \times \sqrt{\varepsilon_{r,P}(f)}}. \tag{11.8}$$

If a leaky-wave SEA and a dispersive prism lens are combined, the β_y is linked in both structures according to the boundary condition between two media

$$\beta_y = k_0 \times \sqrt{\varepsilon_{r,L}} \times \sin \phi_L(f) = k_0 \times \sqrt{\varepsilon_{r,P}(f)} \times \sin \phi_P(f), \tag{11.9}$$

and therefore

$$n_L \times \sin \phi_L(f) = n_P(f) \times \sin \phi_P(f), \tag{11.10}$$

which is Snell's Law, where the equivalent refractive indexes are $n_L = \sqrt{\varepsilon_{r,L}}$ and $n_P(f) = \sqrt{\varepsilon_{r,P}(f)}$. The refractive index n_L in the leaky-wave SIW is constant, whereas both the leaky-wave radiation direction $\phi_L(f)$ and the refractive index in the prism $n_P(f)$ are frequency-dependent. In order to obtain a constant radiation $\phi_P(f)$ for a given leaky-wave characteristic $\phi_L(f)$, the refractive index $n_P(f)$ of the prism should be

$$n_P(f) = \frac{n_L \times \sin \phi_L(f)}{\sin \phi_P(f)}. \tag{11.11}$$

Based on the above analysis, a metasurface prism is designed by using metallic vias in a RO4003C substrate. Figure 11.25a illustrates the dispersion diagram of the unit-cell of a leaky-wave SEA and unit-cell of a metallic via based metasurface. These two kind dispersion curves are located

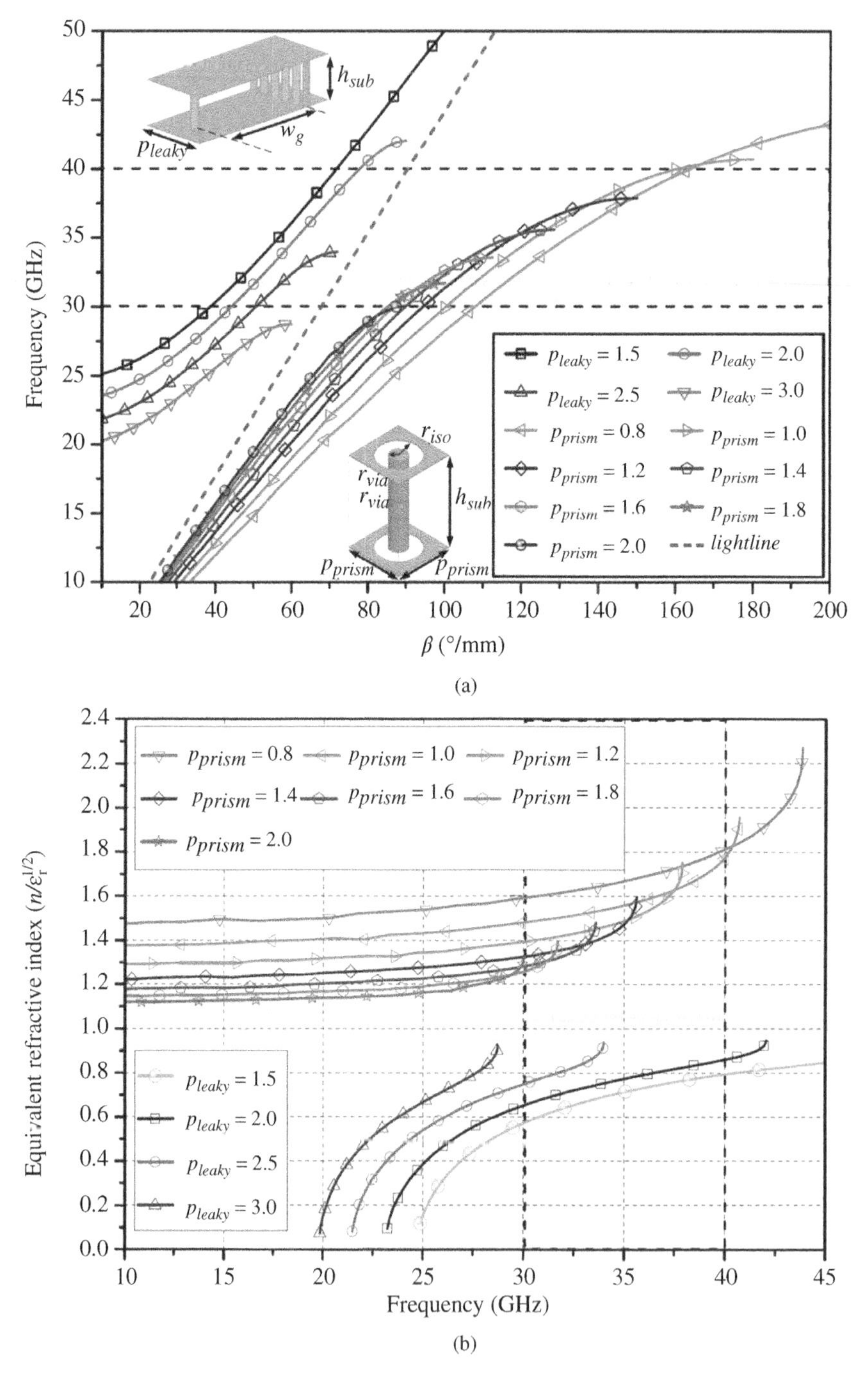

Figure 11.25 (a) Dispersion diagram and (b) equivalent refractive index of a number of leaky-wave SEAs with different periodicities p_{leaky} from 1.5 to 3.0 mm. Dispersion of pins with periodicities p_{prism} from 0.8 to 2.0 mm [33].

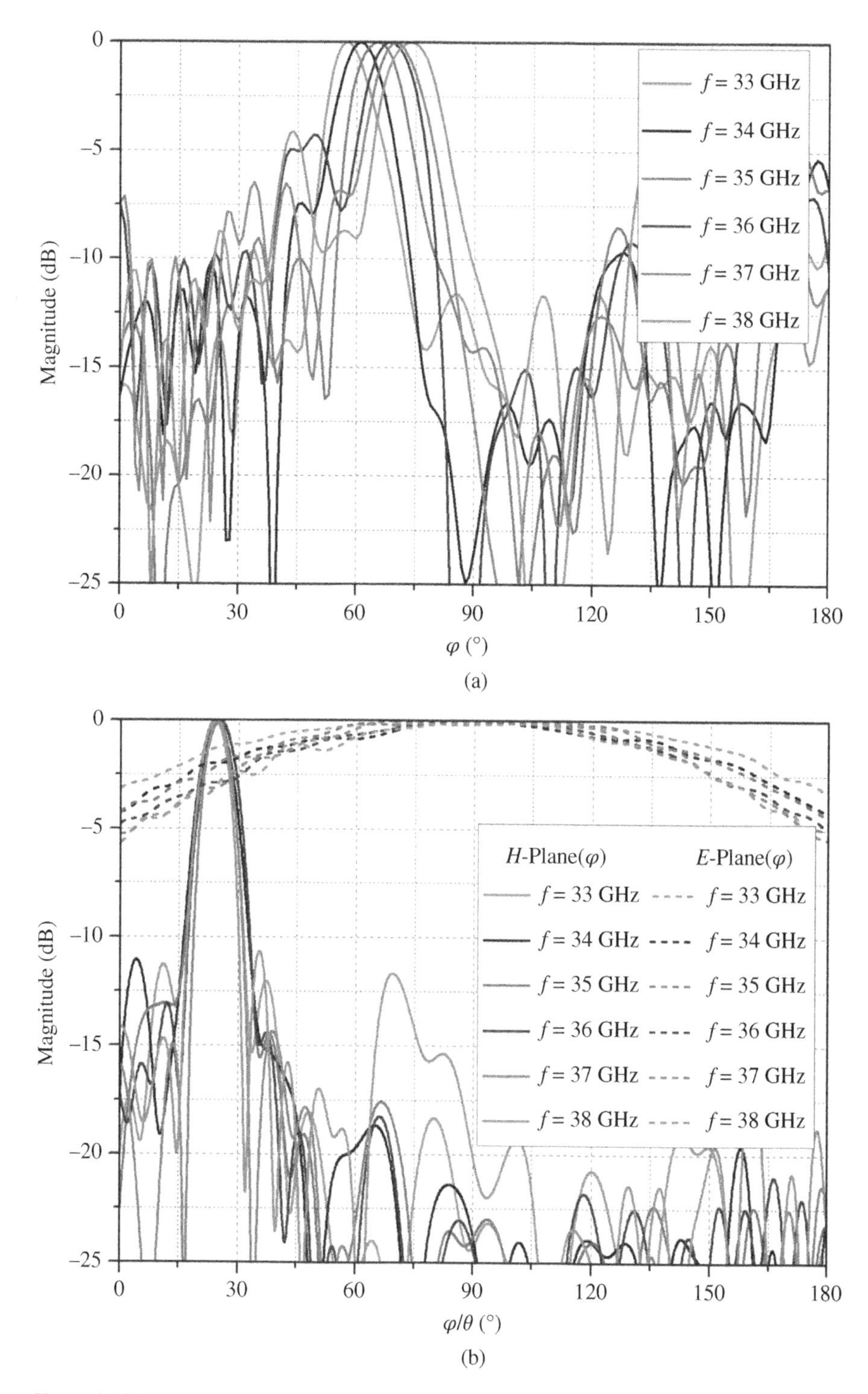

Figure 11.26 Radiation patterns between 33 and 38 GHz for the leaky-wave SEAs. (a) Without and (b) with the dispersive prism in full-wave simulation [33].

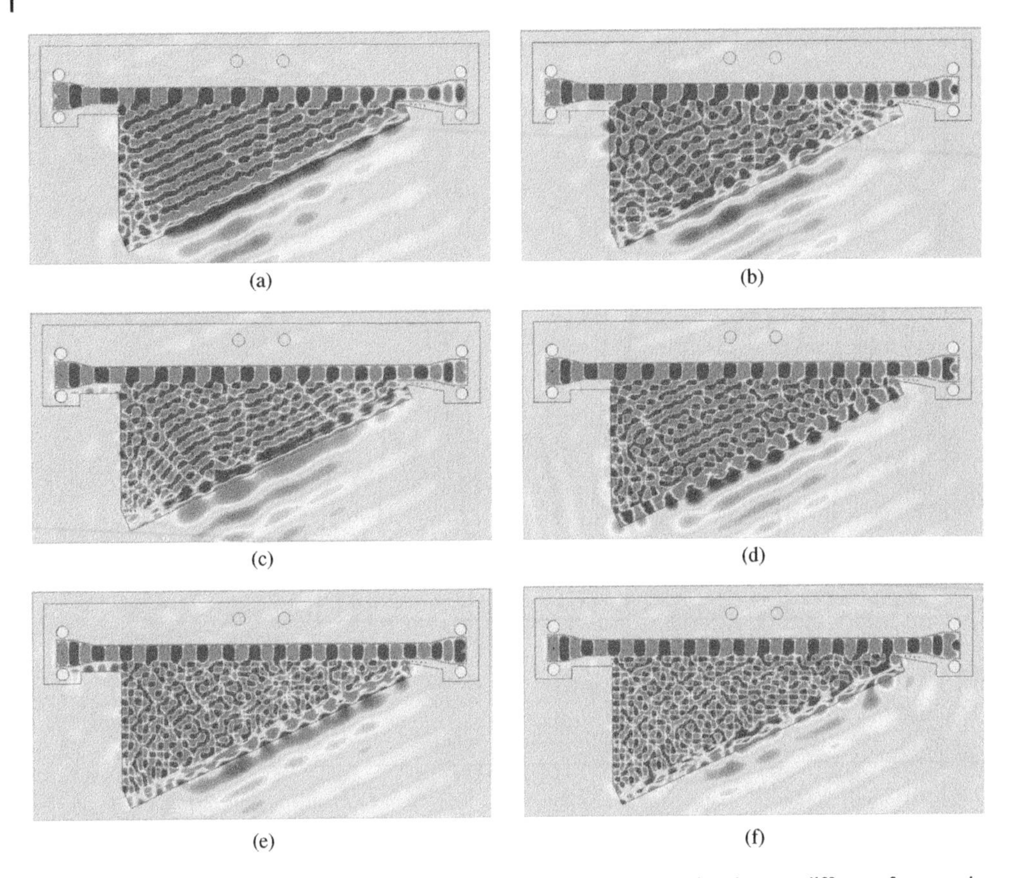

Figure 11.27 Electric-field distribution of the leaky-wave SEAs with prism lens at different frequencies. (a) 33 GHz, (b) 34 GHz, (c) 35 GHz, (d) 36 GHz, (e) 37 GHz, and (f) 38 GHz [33].

at two different sides of the lightline, and they have an opposite curve trend versus the frequency. It is found that the black-square curve and the pink right-triangle curve are two good candidates for the leaky-wave SEA and prism lens in the goal frequency band from 30 to 40 GHz. Computed from the dispersion diagram, the equivalent refractive indexes are shown in Figure 11.25b, while $p_{prism} = 1.0$ mm and $p_{leaky} = 2.0$ mm are chosen for the design.

The simulated radiation patterns are plotted in Figure 11.26. It shows that the original leaky-wave SEA without prism has a steering beam range of 16° from 33 to 38 GHz, whereas the leaky-wave SEA with prism has less than 1° beam steering in the H-plane. The side-lobe levels are below −10 dB.

With the loading prism, the electric field distributions at different frequencies over a wide bandwidth (33–38 GHz) in Figure 11.27 imply that the wave-front are all in parallel to the radiating aperture. It demonstrates again that the constant radiation performance in a wide bandwidth is achieved by loading the dispersive lens.

A prototype of the fixed-beam leaky-wave SEA loaded with a dispersive prism lens using RO4003 substrate of 1.524 mm is shown in Figure 11.28a. Two rectangular strips are printed on the radiating aperture to improve the matching. Fed by two coaxial connectors, the scattering parameters in Figure 11.28b imply that there is good impedance matching from 30 to 40 GHz in measurement. And the $|S_{21}|$ is also very low, maintaining good antenna efficiency.

(a)

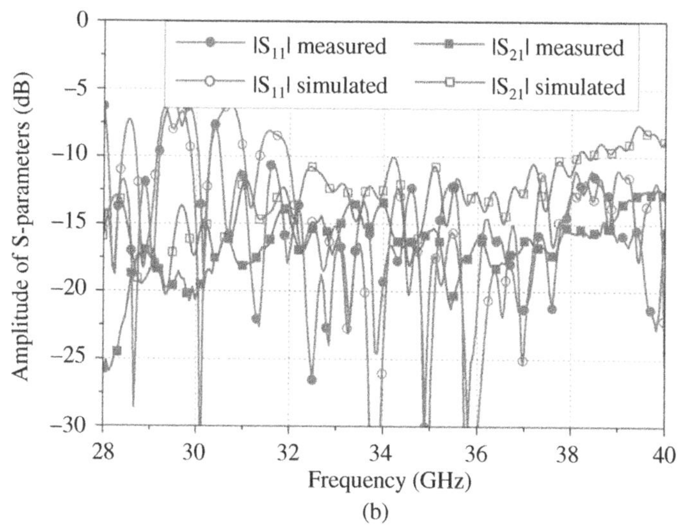

(b)

Figure 11.28 (a) Prototype and measurement setup of the leaky-wave SEA with the prism and (b) its scattering parameters [33].

Figure 11.29a shows the measured radiation patterns of the leaky-wave SEA. It is found that the patterns from 35 to 40 GHz are all radiating at $\phi = 31°$ with a $\pm 0.5°$ variation. The frequency bandwidth of 3-dB radiation drop-off at the direction of $\phi = 28°$, 29°, 30°, and 31° is plotted in Figure 11.29b. It implies that the radiation at $\phi = 31°$ has less than 1-dB difference from 35 to 40 GHz and 3-dB difference from 33.2 to 40 GHz. For example, if the radiation angle at $\phi = 30°$ is

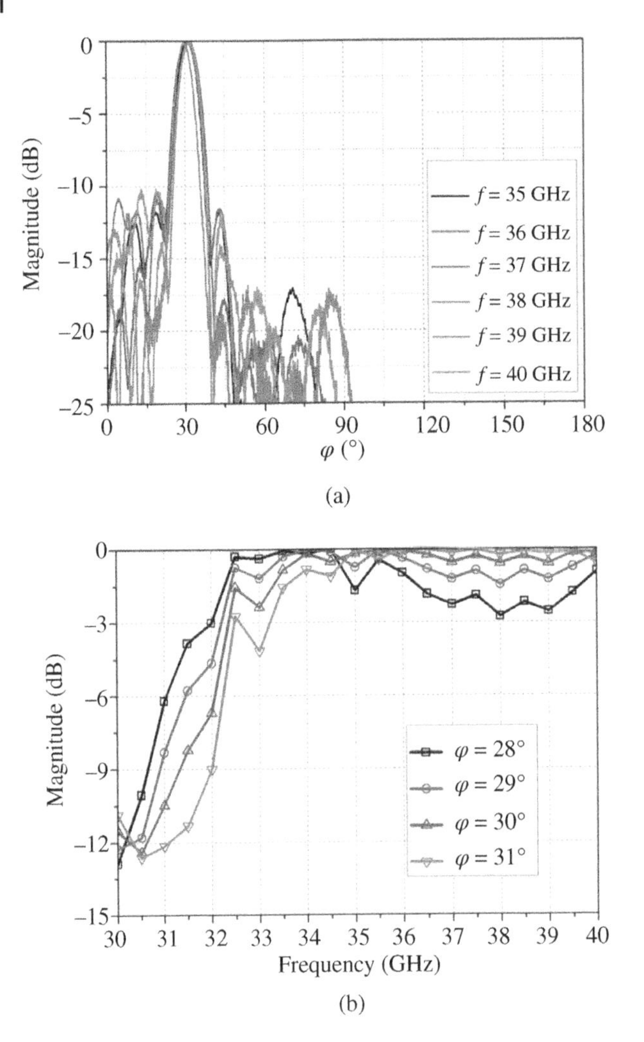

Figure 11.29 (a) Measured radiation patterns and (b) radiation levels at a specific direction as a function of the frequency [33].

used, it has a broad band of 34 ~ 40 GHz for the 1-dB radiation difference and 32.3 ~ 40 GHz (>21%) for 3-dB pattern variation.

11.6 Summary

Substrate integrated antennas with radiation from the substrate edges have been reviewed and discussed. Moving to mmW applications such as 5G and the next-generation radars, SEAs are very promising candidates because of the compact geometry, wide band, low EMI, and easy design for arrays [5].

Apart from the radiators, SEAs also can be used as substrate integrated feeding structures for other radiators, for instance for linearly tapered slot antennas [37–39], Yagi-Uda arrays [40], log-periodic dipole arrays [41], and broadside radiating slot arrays [42].

In addition to the mmW arrays and feeding applications, SEAs are also very suitable for antennas in packaging antennas on systems for future wireless communications operating above 100 GHz.

References

1 Wang, L. and Schuster, C. (2018). Investigation of radiated EMI from printed circuit board edges up to 100 GHz by using an effective two-dimensional approach. In: *International Symposium on Electromagnetic Compatibility (EMC EUROPE)*, 473–476.

2 Li, Z., Wu, K., and Denidni, T.A. (2004). A new approach to integrated horn antenna. In: *10th International Symposium on Antenna Technology and Applied Electromagnetics and URSI Conference*, 1–3.

3 Cao, Y., Cai, Y., Wang, L. et al. (2018). A review of substrate integrated waveguide end-fire antennas. IEEE Access 6: 66243–66253.

4 Wang, H., Fang, D., Zhang, B., and Che, W. (2010). Dielectric loaded substrate integrated waveguide (SIW) H-plane horn antennas. *IEEE Trans. Antennas Propag.* 58 (3): 640–647.

5 Zhang, J., Zhao, K., Wang, L. et al. (2020). Dual-polarized phased array with endfire radiation for 5G handset applications. *IEEE Trans. Antennas Propag.* 68, 4, 3277–3282.

6 Yousefbeiki, M., Domenech, A.A., Mosig, J.R., and Fernandes, C.A. (2012). Ku-band dielectric-loaded SIW horn for vertically-polarized multi-sector antennas. In: *6th European Conference on Antennas and Propagation (EuCAP)*, 2367–2371.

7 Morote, M.E., Fuchs, B., and Mosig, J.R. (2012). Printed Transition for SIW Horn Antennas — Analytical Model. In: *6th European Conference on Antennas and Propagation (EuCAP)*, 1–4.

8 Esquius-Morote, M., Fuchs, B., Zürcher, J., and Mosig, J.R. (2013). A printed transition for matching improvement of SIW horn antennas. *IEEE Trans. Antennas Propag.* 61 (4): 1923–1930.

9 Che, W., Fu, B., Yao, P., and Chow, Y.L. (2007). Substrate integrated waveguide horn antenna with dielectric lens. *Microwave Opt. Technol. Lett.* 49 (1): 168–170.

10 Cai, Y., Zhang, Y., Yang, L. et al. (2017). Design of low-profile metamaterial-loaded substrate integrated waveguide horn antenna and its array applications. *IEEE Trans. Antennas Propag.* 65 (7): 3732–3737.

11 Xu, J., Hong, W., Tang, H. et al. (2008). Half-mode substrate integrated waveguide (HMSIW) leaky-wave antenna for millimeter-wave applications. *IEEE Antennas Wirel. Propag. Lett.* 7: 85–88.

12 Xu, F., Wu, K., and Zhang, X. (2010). Periodic leaky-wave antenna for millimeter wave applications based on substrate integrated waveguide. *IEEE Trans. Antennas Propag.* 58 (2): 340–347.

13 Dahl, D., Brüns, H.D., Wang, L. et al. (2019). Efficient simulation of substrate-integrated waveguide antennas using a hybrid boundary element method. *IEEE J. Multiscale Multiphys. Comput. Tech.* 4: 180–189.

14 Yeh, C.I., Yang, D.H., Liu, T.H. et al. (2010). MMIC compatibility study of SIW H-plane horn antenna. In: *International Conference on Microwave and Millimeter Wave Technology*, 933–936.

15 Yeap, S.B., Qing, X., Sun, M., and Chen, Z.N. (2012). 140-GHz 2×2 SIW horn array on LTCC. In: *IEEE Asia-Pacific Conference on Antennas and Propagation*, 279–280.

16 Esquius-Morote, M., Mattes, M., and Mosig, J.R. (2014). Orthomode transducer and dual-polarized horn antenna in substrate integrated technology. *IEEE Trans. Antennas Propag.* 62 (10): 4935–4944.

17 Cai, Y., Qian, Z., Zhang, Y. et al. (2014). Bandwidth enhancement of SIW horn antenna loaded with air-via perforated dielectric slab. *IEEE Antennas Wirel. Propag. Lett.* 13: 571–574.

18 Esquius-Morote, M., Fuchs, B., Zürcher, J., and Mosig, J.R. (2013). Novel thin and compact H-plane SIW horn antenna. *IEEE Trans. Antennas Propag.* 61 (6): 2911–2920.

19 Cao, Y., Cai, Y., Jin, C. et al. (2018). Broadband SIW horn antenna loaded with offset double-sided parallel-strip lines. *IEEE Antennas Wirel. Propag. Lett.* 17 (9): 1740–1744.

20 Li, T. and Chen, Z.N. (2018). Wideband substrate-integrated waveguide-fed endfire metasurface antenna array. *IEEE Trans. Antennas Propag.* 66 (12): 7032–7040.

21 Sun, D., Xu, J., and Jiang, S. (2015). SIW horn antenna built on thin substrate with improved impedance matching. *Electron. Lett* 51 (16): 1233–1235.

22 Jackson, D.R., Arthur, A.O., and Balanis, C. (2008). Leaky-wave antennas. In: *Modern Antenna Handbook*, Wiley. 325–367.

23 Martinez-Ros, A.J., Gomez-Tornero, J.L., and Goussetis, G. (2012). Planar leaky-wave antenna with flexible control of the complex propagation constant. *IEEE Trans. Antennas Propag.* 60 (3): 1625–1630.

24 Liao, Q. and Wang, L. Switchable bidirectional/unidirectional LWA array based on half-mode substrate integrated waveguide. *IEEE Antennas Wirel. Propag. Lett.* 19 (7). 1261–1265.

25 Martinez-Ros, A.J., Gómez-Tornero, J.L., and Goussetis, G. (2017). Multifunctional angular bandpass filter SIW leaky-wave antenna. *IEEE Antennas Wirel. Propag. Lett.* 16: 936–939.

26 Zetterstrom, O., Pucci, E., Padilla, P. et al. (2020). Low-dispersive leaky-wave antennas for mmWave point-to-point high-throughput communications. *IEEE Trans. Antennas Propag.* 68 (3): 1322–1331.

27 Li, Y., Yin, X., Zhao, H. et al. (2014). Radiation enhanced broadband planar TEM horn antenna. In: *Asia-Pacific Microwave Conference*, 720–722.

28 Wang, L., Esquius-Morote, M., Qi, H. et al. (2017). Phase corrected H-plane horn antenna in gap SIW technology. *IEEE Trans. Antennas Propag.* 65 (1): 347–353.

29 Wang, L., Garcia-Vigueras, M., Alvarez-Folgueiras, M., and Mosig, J.R. (2017). Wideband H-plane dielectric horn antenna. *IET Microwaves Antennas Propag.* 11 (12): 1695–1701.

30 Kock, W.E. (1946). Metal-lens antennas. *Proc. IRE* 34 (11): 828–836.

31 Wang, L., Yin, X., Li, S. et al. (2014). Phase corrected substrate integrated waveguide H-plane horn antenna with embedded metal-via arrays. *IEEE Trans. Antennas Propag.* 62 (4): 1854–1861.

32 Wang, L., Esquius-Morote, M., Yin, X., and Mosig, J.R. (2015). Gain enhanced H-plane gap SIW horn antenna with phase correction. In: *9th European Conference on Antennas and Propagation (EuCAP)*, 1–5.

33 Wang, L., Gómez-Tornero, J.L., and Quevedo-Teruel, O. (2018). Substrate integrated waveguide leaky-wave antenna with wide bandwidth via prism coupling. *IEEE Trans. Microwave Theory Tech.* 66 (6): 3110–3118.

34 Wang, L., Gómez-Tornero, J.L., and Quevedo-Teruel, O. (2018). Dispersion reduced SIW leaky-wave antenna by loading metasurface prism. In: *International Workshop on Antenna Technology (iWAT)*, 1–3.

35 Wang, L., Gómez-Tornero, J.L., Rajo-Iglesias, E., and Quevedo-Teruel, O. (2018). On the use of a metasurface prism in gap-waveguide technology to reduce the dispersion of leaky-wave antennas. In: *12th European Conference on Antennas and Propagation (EuCAP 2018)*, 1–3.

36 Wang, L., Gómez-Tornero, J.L., Rajo-Iglesias, E., and Quevedo-Teruel, O. (2018). Low-dispersive leaky-wave antenna integrated in groove gap waveguide technology. *IEEE Trans. Antennas Propag.* 66 (11): 5727–5736.

37 Iigusa, K., Li, K., Sato, K., and Harada, H. (2012). Gain enhancement of H-plane sectoral post-wall horn antenna by connecting tapered slots for millimeter-wave communication. *IEEE Trans. Antennas Propag.* 60 (12): 5548–5556.

38 Wang, L., Yin, X., and Zhao, H. (2015). A planar feeding technology using phase-and-amplitude-corrected SIW horn and its application. *IEEE Antennas Wirel. Propag. Lett.* 14: 147–150.

39 Wang, L., Yin, X., Esquius-Morote, M. et al. (2017). Circularly polarized compact LTSA array in SIW technology. *IEEE Trans. Antennas Propag.* 65 (6): 3247–3252.

40 Wang, L., Yin, X., and Zhao, H. (2014). Quasi-Yagi array loaded thin SIW horn antenna with metal-via-array lens. In: *IEEE Antennas and Propagation Society International Symposium (APS)*, 1290–1291.

41 Chen, Q., Yin, X., and Wang, L. (2018). Compact printed log-periodic dipole arrays fed by SIW horn. In: *12th European Conference on Antennas and Propagation (EuCAP 2018)*, 1–3.

42 Ettorre, M., Sauleau, R., and Le Coq, L. (2011). Multi-beam multi-layer leaky-wave SIW pillbox antenna for millimeter-wave applications. *IEEE Trans. Antennas Propag.* 59 (4): 1093–1100.

Index

Substrate-Integrated Millimeter-Wave Antennas for Next-Generation Communication and Radar Systems, First Edition.
Edited by Zhi Ning Chen and Xianming Qing.

y

IEEE Press Series on Electromagnetic Wave Theory

Field Theory of Guided Waves, Second Edition
Robert E. Collin

Field Computation by Moment Methods
Roger F. Harrington

Radiation and Scattering of Waves
Leopold B. Felsen, Nathan Marcuvitz

Methods in Electromagnetic Wave Propagation, Second Edition
D. S. J. Jones

Mathematical Foundations for Electromagnetic Theory
Donald G. Dudley

The Transmission-Line Modeling Method: TLM
Christos Christopoulos

The Plane Wave Spectrum Representation of Electromagnetic Fields, Revised Edition
P. C. Clemmow

General Vector and Dyadic Analysis: Applied Mathematics in Field Theory
Chen-To Tai

Computational Methods for Electromagnetics
Andrew F. Peterson

Plane-Wave Theory of Time-Domain Fields: Near-Field Scanning Applications
Thorkild B. Hansen

Foundations for Microwave Engineering, Second Edition
Robert Collin

Time-Harmonic Electromagnetic Fields
Robert F. Harrington

Antenna Theory & Design, Revised Edition
Robert S. Elliott

Substrate-Integrated Millimeter-Wave Antennas for Next-Generation Communication and Radar Systems, First Edition.
Edited by Zhi Ning Chen and Xianming Qing.

Differential Forms in Electromagnetics
Ismo V. Lindell

Conformal Array Antenna Theory and Design
Lars Josefsson, Patrik Persson

Multigrid Finite Element Methods for Electromagnetic Field Modeling
Yu Zhu, Andreas C. Cangellaris

Electromagnetic Theory
Julius Adams Stratton

Electromagnetic Fields, Second Edition
Jean G. Van Bladel

Electromagnetic Fields in Cavities: Deterministic and Statistical Theories
David A. Hill

Discontinuities in the Electromagnetic Field
M. Mithat Idemen

Understanding Geometric Algebra for Electromagnetic Theory
John W. Arthur

The Power and Beauty of Electromagnetic Theory
Frederic R. Morgenthaler

Modern Lens Antennas for Communications Engineering
John Thornton, Kao-Cheng

Electromagnetic Modeling and Simulation
Levent Sevgi

Multiforms, Dyadics, and Electromagnetic Media
Ismo V. Lindell

Low-Profile Natural and Metamaterial Antennas: Analysis Methods and Applications
Hisamatsu Nakano

From ER to E.T.: How Electromagnetic Technologies Are Changing Our Lives
Rajeev Bansal

Electromagnetic Wave Propagation, Radiation, and Scattering: From Fundamentals to Applications, Second Edition
Akira Ishimaru

Time-Domain Electromagnetic Reciprocity in Antenna Modeling
Martin Štumpf

Boundary Conditions in Electromagnetics
Ismo V. Lindell, Ari Sihvola

Substrate-Integrated Millimeter-Wave Antennas for Next-Generation Communication and Radar Systems,
Edited by, Zhi Ning Chen, Xianming Qing

Printed and bound by CPI Group (UK) Ltd, Croydon, CR0 4YY
16/04/2025
14658604-0005